Large Scale Computations in Air Pollution Modelling

T0214398

NATO Science Series

*A Series presenting the results of activities sponsored by the NATO Science Committee.
The Series is published by IOS Press and Kluwer Academic Publishers, in conjunction
with the NATO Scientific Affairs Division.*

A. **Life Sciences**   IOS Press
B. **Physics**   Kluwer Academic Publishers
C. **Mathematical and Physical Sciences**   Kluwer Academic Publishers
D. **Behavioural and Social Sciences**   Kluwer Academic Publishers
E. **Applied Sciences**   Kluwer Academic Publishers
F. **Computer and Systems Sciences**   IOS Press

1. **Disarmament Technologies**   Kluwer Academic Publishers
2. **Environmental Security**   Kluwer Academic Publishers
3. **High Technology**   Kluwer Academic Publishers
4. **Science and Technology Policy**   IOS Press
5. **Computer Networking**   IOS Press

NATO-PCO-DATA BASE

The NATO Science Series continues the series of books published formerly in the NATO ASI
Series. An electronic index to the NATO ASI Series provides full bibliographical references (with
keywords and/or abstracts) to more than 50000 contributions from international scientists published
in all sections of the NATO ASI Series.
Access to the NATO-PCO-DATA BASE is possible via CD-ROM "NATO-PCO-DATA BASE" with
user-friendly retrieval software in English, French and German (© WTV GmbH and DATAWARE
Technologies Inc. 1989).

The CD-ROM of the NATO ASI Series can be ordered from: PCO, Overijse, Belgium.

Series 2: Environmental Security – Vol. 57

Large Scale Computations in Air Pollution Modelling

edited by

Z. Zlatev, J. Brandt
National Environmental Research Institute, Roskilde, Denmark

P.J.H. Builtjes
TNO Institute for Environmental Sciences, Apeldoorn, The Netherlands

G. Carmichael
University of Iowa, Iowa City, Iowa, U.S.A.

I. Dimov
Bulgarian Academy of Sciences, Sofia, Bulgaria

J. Dongarra
University of Tennessee, Knoxville, Tennessee, U.S.A.

H. van Dop
Utrecht University, Utrecht, The Netherlands

K. Georgiev
Bulgarian Academy of Sciences, Sofia, Bulgaria

H. Hass
Ford Forschungzentrum, Aachen, Germany

R. San Jose
Technical University of Madrid, Madrid, Spain

Kluwer Academic Publishers

Dordrecht / Boston / London

Published in cooperation with NATO Scientific Affairs Division

Proceedings of the NATO Advanced Research Workshop on
Large Scale Computations in Air Pollution Modelling
Sofia, Bulgaria
6-10 July 1998

A C.I.P. Catalogue record for this book is available from the Library of Congress

ISBN 0-7923-5677-2 (HB)
ISBN 0-7923-5678-0 (PB)

Published by Kluwer Academic Publishers,
P.O. Box 17, 3300 AA Dordrecht, The Netherlands.

Sold and distributed in North, Central and South America
by Kluwer Academic Publishers,
101 Philip Drive, Norwell, MA 02061, U.S.A.

In all other countries, sold and distributed
by Kluwer Academic Publishers,
P.O. Box 322, 3300 AH Dordrecht, The Netherlands.

Printed on acid-free paper

All Rights Reserved
© 1999 Kluwer Academic Publishers
No part of the material protected by this copyright notice may be reproduced or
utilized in any form or by any means, electronic or mechanical, including photo-
copying, recording or by any information storage and retrieval system, without written
permission from the copyright owner.

Printed in the Netherlands

TABLE OF CONTENTS

PREFACE

1. Contents of these proceedings. These proceedings contain most of the papers which were presented at the NATO ARW (Advanced Research Workshop) on *"Large Scale Computations in Air Pollution Modelling"*. The workshop was held, from June 6 to June 10, 1998, in Residence Bistritza, a beautiful site near Sofia, the capital of Bulgaria, and at the foot of the mountain Vitosha.

2. Participants in the NATO ARW. Scientists from 23 countries in Europe, North America and Asia attended the meeting and participated actively in the discussions. The total number of participants was 57. The main topic of the discussions was the role of the large mathematical models in resolving difficult problems connected with the protection of our environment.

3. Major topics discussed at the workshop. The protection of our environment is one of the most important problems facing modern society. The importance of this problem has steadily increased during the last two-three decades, and environment protection will become even more important in the next century. Reliable and robust control strategies for keeping the pollution caused by harmful chemical compounds under certain safe levels have to be developed and used in a routine way. Large mathematical models, in which **all** important physical and chemical processes are adequately described, can successfully be used to solve this task. However, the use of large mathematical models in which all important physical and chemical processes are adequately described leads - after the application of appropriate discretization and splitting procedures - to the treatment of huge computational tasks. In a typical simulation one has to perform several hundred runs, in each of these runs one has to carry out several thousand time-steps and at each time-step one has to solve numerically systems of coupled ordinary differential equations containing up to several million equations. Therefore, it is difficult to treat such large mathematical models numerically even when fast modern computers are available. Combined research from specialists from the fields of environmental modelling, numerical analysis and scientific computing must be initiated. The specialists from these fields must unite their efforts in attempts to solve these problems. Distinguished specialists from environmental modelling, numerical

ix

analysis and scientific computing attended this workshop and discussed the above problems.

In some areas of Europe, as well as in some other parts of the world, the pollution levels are so high that preventive actions are urgently needed. Therefore robust and reliable control strategies must be developed in order to find out where and by how much the emissions of harmful pollutants should be reduced. The solutions found must be optimal (or, at least, close to optimal), because the reduction of the emissions is as a rule an expensive process. Therefore, great economical problems may appear when the task of optimal reduction of the emissions is not correctly solved. Optimal (or nearly optimal) solutions can be successfully found only by carrying out long simulation experiments consisting of many hundreds runs performed by using comprehensive mathematical models. Then the results obtained by different models must be carefully compared in order to answer the following questions: (i) are there any discrepancies and (ii) what are the reasons for discrepancies. In the latter case, the needed corrections of some of the models have to be made, and the simulation experiments must be repeated for the corrected models. This shows that the process of finding an optimal solution will in general be very long and, thus, efficiency of the codes is highly desirable. Such an efficiency can only be achieved by co-ordinated efforts of specialists in environmental modelling, numerical analysis and computer science. One of the main topics of this workshop was the discussion of different ways for improving the co-operative work of scientists from these three fields.

Running models in real time can be of essential and even live saving importance in the case of accidental, hazardous releases (as, for example, the accident in Chernobyl in 1986). Real time calculations will also be needed in connection with the forthcoming EU Ozone Directive. High ozone concentrations occur in large parts in Europe during short summer periods and can cause damages. A special abatement strategy in such periods can be based on temporal reductions of emissions from specific sources; e.g. traffic regulations, the energy sector, large industries (such as the petrochemical industry). Reliable mathematical models have to be coupled with weather forecasting models and the coupled models have to be run operationally in order (i) to predict the appearance of high concentrations and (ii) to make the right decisions in such situations. Running the models in real time is another very difficult and very challenging problem that must urgently be solved.

The computer architectures are becoming faster and faster. We shall have in the near future supercomputers that have top performance of several Tflops. New and efficient numerical methods, by which the great potential power of the parallel computers can be better exploited, are gradually becoming available. However, there remain many unresolved problems. The most important problem is the great gap between the top performance of a modern supercomputer and the actually achieved speeds when large application codes are run. The task of achieving high speeds of computation, which are close to the top performance of the computer available, is very difficult and, at the same time, very challenging for the comprehensive environmental models. This is also true for the heterogeneous computations, where it is potentially possible to achieve very high computational speed. The time is now coming when specialists from the areas of supercomputing and numerical analysis should unite their efforts with environmental modelling specialists in order to develop fast and robust codes that can be used in the solution of many urgent and important for the modern society environmental problems.

Non-realistic simplifying assumptions are always made in all existing large-scale air pollution models in order to be able to treat them numerically on the computers available. Many such assumptions are no longer absolutely necessary (because both the computers and the numerical methods are much faster now and will become even faster in the near future). Therefore, it is important now to describe all important physical and chemical processes in an adequate way (according to the present knowledge) and to try to solve the problems by exploiting extensively the modern computational and numerical tools. This is a very difficult interdisciplinary task which must be solved with combined efforts from specialists from several different fields. The problems arising during the interaction of specialists from the appropriate fields and ways for improving this interaction were discussed at the meeting. High-qualified specialists from all related fields (large-scale air pollution modelling, numerical analysis, scientific computing, optimization and computer science) attended this meeting and contributed for its success. Every session was finished by a common discussion organized by a panel consisting of specialists from all related fields. Every panel prepared a report containing recommendations based on the results of the discussion. A summary of these reports is given in the end of these proceedings.

4. Overall conclusions. Overall, the workshop conclusion was that much closer connections and more extensive exchange of information among the specialists working in this field are required in the efforts to improve the performance of the existing large scale environmental models as well as in the efforts to develop new models which are both more efficient and more powerful. The discussions during the workshop as well as the conclusions drawn during these discussions are described in a special chapter at the end of these proceedings.

5. Contacts with the Panel of the Environmental Security Area of the NATO Scientific Programme. The fact that the Panel of the Environmental

Security Area within the NATO Scientific Programme held one of the their annual meetings at the same place, Residence Bistritza, during the workshop was an extra benefit for the participants. Some contacts with the members of the Panel were established. There were numerous informal discussions between groups of participants on the one side and members of the Panel on the other side. The Chairman of the Panel gave a talk at the workshop, in which he explained in detail the possibilities for funding joint scientific research by the NATO Scientific Programme.

6. Interest to the NATO ARW. It should be mentioned here that the NATO ARW on *"Large Scale Computations in Air Pollution Modelling"* was an object of interest for the broad public in Bulgaria. Announcements about this event were published in several central newspapers and in some journals of the Bulgarian Academy of Sciences. There was a report about the NATO ARW in the Channel 1 of the Bulgarian National Television. Deputy ministers from several ministries were present at the opening session of the workshop. In their talks they acknowledged the relevance of the topics and wished a successful workshop. The Vice-President of the Bulgarian Parliament was one of the participants and speakers at the workshop. It should also be mentioned that the Local Organizing Committee did an excellent job for dissemination of information about the workshop.

7. Ordering the papers in this volume. The original idea was to order the papers by topics. However, we discovered very quickly that this is an impossible task, because this is an interdisciplanary area and, thus, most of the papers deal with two or more topics. Indeed, the comprehensive paper of Gregory Carmichael and his co-workers deals with all topics discussed at this NATO ARW. Therefore the papers were ordered in an alphabetic order, but a subject index is provided at the end of the volume. An author index and a list of the participants are also included.

8. Who will be interested in these proceedings? This volume contains papers which describe many of the difficult problems which appear during the development and the treatment of large scale air pollution models as well as different ways for solving some of these problems. Moreover, the authors were asked, and most of them fulfilled this requirement, to make presentations which will be understandable from specialists both in the field of air pollution modelling and in all related fields. Therefore we believe that this volume will be useful both for specialists directly involved in air pollution modelling and for specialists from all related areas.

The editors

ACKNOWLEDGEMENTS

This meeting was fully supported by a grant ENVIR.ARW971731 from the NATO Scientific Programme. We acknowledge this generous support which allowed us to organize this interdisciplanary meeting.

The Local Organizing Committee, chaired by Prof. Krassimir Georgiev, selected a private company, the Company for International Meetings, which took care for all problems connected with the accommodation of the participants. The members of this committee carried out a lot of work investigating many sites (proposed by the Company for International Meetings), where the meeting could be held and to choose the best one, the Residence Bistritza. We should like to thank all the members of the Local Organizing Committee for their work to find the place and also for their work, both before the workshop and during the workshop.

The staff of the Company for International Meetings worked diligently: ensuring transfers of all participants from Sofia Airport to Residence Bistritza and from Residence Bistritza to Sofia Airport, keeping an open desk during the whole meeting, where all questions of the participants were kindly answered and making all needed arrangements in Residence Bistritza. Thus, the involved members of the Company for International Meetings contributed greatly to the success of this workshop. We should like to thank them very much for their efforts during the workshop.

The members the Panel of the Environmental Area within the NATO Scientific Programme were very kind in carefully answering numerous questions asked by the participants during the meals and the breaks. Many of the participants, especially the participants from the Partner Countries, learned a lot about the possibilities for obtaining NATO grants for joint scientific research. We should like to thank the members of the Panel both for their decision to hold a meeting at the same place and for their willingness to give us as much information about the NATO Scientific Programme as possible.

The editors

PARALLEL ALGORITHMS FOR SOME OPTIMIZATION PROBLEMS ARISING IN AIR POLLUTION MODELING

J. L. ALONSO and V. N. ALEXANDROV
Department of Computer Science · University of Liverpool · U.K.

Abstract

We will consider some optimization problems arising in air pollution modeling. An analogy with similar optimization problems, which can be formulated as Quadratic Optimization and Mixed Integer Linear Programming problems and arising in placement of components on a Printed Circuit Board (PCB) under different geometrical and technological constraints, will be made. We will also describe an efficient parallel Branch and Bound (B&B) algorithms to solve these MILP problems in a MIMD distributed memory environment. The algorithms are scalable and run on a cluster of workstations under PVM. The efficiency of the algorithms has been investigated and the test results on some placement tasks and from MIPLIB library are presented.

1. Introduction

In this paper we consider some optimization problems arising in air pollution modeling and we make an analogy with similar optimization problems, which can be formulated as Quadratic Optimization and Mixed Integer Linear Programming problems and arising in placement of components on a Printed Circuit Board (PCB) under different geometrical and technological constraints.

We have developed efficient parallel Branch and Bound (B&B) algorithms for Integer Linear Programming (ILP) and Mixed ILP (MILP) problems in a MIMD distributed memory environment.

An efficient solution (and efficient parallel algorithms respectively) of ILP and MILP problems is required since many practical optimization problems and especially many Combinatorial Optimization Problems can be formulated as ILP and MILP problems [10, 6]. On the other hand it is know that ILP and MILP problems are NP-hard and thus difficult to solve [10]. Practical needs also require solution of ILP

1

Z. Zlatev et al. (eds.), Large-Scale Computations in Air Pollution Modelling, 1–14.
© *1999 Kluwer Academic Publishers. Printed in the Netherlands.*

and MILP of larger and larger size. We study ILP and MILP problems with *sparse* constraint matrices.

Many publications appeared in the past ten years about the parallelisation of B&B for different Combinatorial Optimization Problems [7, 6]. There exist many approaches to parallel B&B implementation. An up to date survey on current state of the art in parallel algorithms for Combinatorial Optimization Problems can be found in Ferreira and Pardalos [6]. Initially the implementations were on dedicated parallel machines, for example, Eckstein's [5] code was for CM-5; R. Luling, B. Monien, A. Reinefeld and S. Tschoke [6] have successfully mapped tree structured problems on 1024 processor machine with predefined network topology (i.e. hypercube, mesh, etc.). With advances in cluster computing and emerging relevant software platforms some codes running in such environments have appeared. Bixby et all [4] have presented MILP code which runs on IBM SP2 and cluster of various workstations using TradeMarks (a parallel programming system allowing distributed memory machines to be programmed as shared memory machines) and allowing to treat reasonable size instances. From commercially available MILP solvers OSL and CPLEX include implementations on parallel platforms. For example, CPLEX implementation is exclusively for SGI multiprocessors for shared memory environment while OSL runs under PVM on IBM SP2 machine and RISC6000 workstations.

But the parallel approach requires much effort since **first** certain anomalous behaviour of parallel B&B is observed and **second** sophisticated communication and load balancing strategies are needed to improve processor utilization because of the fast increase of the communication time with the increased number of processors in parallel implementations [7, 9].

In our approach the algorithms are developed for problems with **sparse** constraint matrices, they are scalable and run on a MIMD distributed memory environment under PVM. Our experiments are made, for example, on a heterogeneous cluster of SUN - Unix workstations and on 64 processor SCI Cluster at PC2 in Paderborn Centre for Parallel Computing. We employ *master/slave* approach. We use *depth-first* strategy because it minimises the memory requirements. In contrast with previous approaches our parallel algorithm follows at the beginning the Branch-and-Bound tree. We also apply preprocessing prior to B&B implementation in order to improve problem formulation and we implement B&B algorithms with dynamic distribution of tasks among the processors in order to improve the work load of processors and to minimize the communication.

2. Global placement by quadratic optimization

Now if we assume that we have a number of sources of pollution (or simply sources) that we have to place in a certain area, we can consider the global placement of the sources using a modified force model as we do in case of placement of components on the PCB [2, 11] Based on this force model the quadratic optimization problem with linear constraints is formulated. The objective function of this optimization task is the minimization of the sum of all net lengths. In this paper we assume that the sources are connected by a net weighted in accordance to the number of chemical reactions possible between the pollutants from these sources. The less the possibilities the stronger the connection is. Therefore for pairs of sources, which are connected by nets, attractive forces are defined.

In the force model the sources are considered as dimensionless points. Therefore their overlapping is not taken into account in this step. A linear constraint fixes the center of gravity of all movable points to the center of the region. Additional requirements are incorporated efficiently in the model by defining artificial nets and by weighting of the attractive forces:

- Among the critical sources a strong attractive force is defined to ensure that they will be placed close together.

- To separate different sources or group of sources artificial nets are introduced.

The formulated problem can be transformed to an unconstrained quadratic programming problem and is solvable using the conjugate-gradient method [8].

3. Producing the basic MILP

After finishing the process of global placement we can cut the area considered in several subareas if necessary. For each subarea we can formulate a MILP problem. For this reason we need to define number of constraints and the objective function.

3.1. NON-OVERLAPPING CONSTRAINTS

If we assume that we can approximate the source with a rectangular shape, e.g. depending on some average atmospheric conditions, then we can generate the relevant constraints making analogy with the placement of components. Common to all forms of a feasible solution is the requirement that sources (components) do not overlap with one another, hence the need for non-overlapping constraints. To position a source we assign it a lower left coordinate (x, y). It should be noted that at this initial stage the possibility of rotating a source has not been considered so that the source width (w) is always the difference between the source's left and right x coordinate and the source's height (h) the difference between the lower and upper y coordinate. Non-overlapping constraints are required only for every possible pair of sources within the current area being solved as sources in a different area are protected by constraints, see section 3.2. Hence given two sources c_i and c_j in the current area (room), it is necessary, to prevent overlapping, that at least one of the following linear inequalities holds:-

$$
\begin{array}{llll}
& x_i + w_i & \leq & x_j & c_i \text{ is to the left of } c_j \\
\vee & x_i - w_j & \geq & x_j & c_i \text{ is to the right of } c_j \\
\vee & y_i + h_i & \leq & y_j & c_i \text{ is below } c_j \\
\vee & y_i - h_j & \geq & y_j & c_i \text{ is above } c_j
\end{array}
\tag{1}
$$

As only one of the above inequalities needs to be active, yet in a MILP all constraints must be met at the same time, a way of deactivating the constraints is needed. This is done by introducing two boolean variables x_{ij} and y_{ij} which are produced for each pair of sources and a value W is defined as $W = max P_w, P_h$, where P_w and P_h are the width and height respectively of the whole area. Then (1) above can be written as:-

$$
\begin{array}{llll}
x_i + w_i & \leq & x_j + W(x_{ij} + y_{ij}) & c_i \text{ is to the left of } c_j \\
x_i - w_j & \geq & x_j - W(1 - x_{ij} + y_{ij}) & c_i \text{ is tothe right of } c_j \\
y_i + h_i & \leq & y_j + W(1 + x_{ij} - y_{ij}) & c_i \text{ is below } c_j \\
y_i - h_j & \geq & y_j - W(2 - x_{ij} - y_{ij}) & c_i \text{ is above } c_j
\end{array}
\tag{2}
$$

So for each possible choice of values for (x_{ij}, y_{ij}) only one inequality is active, hence the or conditions have been generated.

3.2. WITHIN AREA CONSTRAINTS

In addition to the nonoverlapping constraints the sources (components) must be kept within the area of the subarea being solved, i.e. within the room (area). Hence the within area constraints are produced:

$$x_i \geq room\_left_x$$
$$y_i \geq room\_lower_y$$

$$x_i + (1 - z_i)w_i + z_ih_i \leq room\_right_x$$
$$y_i + (1 - z_i)h_i + z_iw_i \leq room\_upper_y$$

(3)

Where $room\_left_x$ and $room\_right_x$ are the left and right x coordinates of the possible placement area. This would be equivalent to the whole area if no partitioning had taken place.

3.3. FIXED COMPONENTS

In the description of the area it is possible that some sources (components) are already given a fixed position. As implemented in the MILP production these sources can be fixed by bounding the relevant variables, i.e. lower left coordinates, rotation and side, to produce an equality. So for example to bound the variable x_i to the value 300 the following two inequalities would be used:
$x_i \leq 300$, $x_i \geq 300$.

A better solution would be to remove the variables all together and replace them with constants on the RHS of the constraints. This is done in the presolving phase.

Note: Further constraints are added if we consider some restricted areas, keep some sources within some area etc.

3.4. NETLENGTH MINIMISATION

As in the case of final placement of components on the PCB, we assume that components (sources) are connected with a net [2, 11]. In our case the net is constructed according to the chemical reactions, e.g. two sources are connected with a net if a chemical reaction between some pollutants is possible according to some of the accepted chemical schemes. We assume also that in accordance with global placement phase the sources with less possible chemical reactions are most attracted. We also may assume that the whole area is divided in subareas after the global placement and we can solve the problem only in the relevant subareas. (In case we do not run a global placement beforehand and we wish to consider whole area we can define the relevant artificial nets so we can solve directly the problem in MILP form.)

So we need to minimize the netlength and therefore to minimize the number of possible reactions and respectively the air pollution. In this case in analogy with the objective function in [2, 11] we minimize the netlength and place sources as close as possible within the subarea (area).

Note: One can consider the problem in maximization form assuming that if the sources are further away then less chemical reactions are available according to atmospheric conditions.

Note: The following if-then strategies can investigated, for example:

- placing optimaly a new source(s) of pollution in a given area.

- moving/removing sources of pollution and investigating their influence on the global picture.

- creating an artificial sources of pollution and placing them optimaly under sertain constraints in order to see the possible improvements in the global picture of the area considered. (We can refer here to investigate the case of improving levels of transboundary pollution for example.)

4. The MILP

As a result of the constraints and objective function formulation in the previous section we obtain Integer Linear Programming problems [10, 1] of the form:-

$$min \{cx : Ax \leq b, x \in Z_+^n\} \tag{4}$$

and when relaxing the integrality constraint on some of the variables, we obtain MILPs.

4.1. PREPROCESSING

Most of the known solvers perform some preprocessing in order to improve ILP and MILP formulations of the problem [4]. In our case preprocessing involves: searching for implicit bounds in the matrix, reduction of variables [3], removing of empty rows and columns, scaling rows and variables.

4.2. NODE AND VARIABLE SELECTION

To obtain a lower bound for the value of the objective function at a node of the branching tree we solve the corresponding LP relaxation.

If the solution is integer, if its objective function value is worse than the value of the best known integer solution so far, or if the LP relaxation is infeasible, the node processing is complete. Otherwise a branching variable is selected and two new nodes are created. We use several rules to select the branching variable: first non-integer variable, variable whose smallest degradation is largest [10], penalties method [10]or randomly. At the moment the best results are obtained with the penalties method. The advantages are: low number of branches with 0 degradation, branches which degradation infinity could be avoided The disadvantages are: In dual degenerate problems, where few information is given which variable to branch, and the penalty value becomes 0 for both of the penalties. It is known that calculating penalties is an expensive procedure equivalent to half pivot operation for every integer variable that is checked. But we observed, that in most of the problems the number of nodes searched has reduced up to 20 times in comparison with the rule of choosing to branch the variable with max(min(fractional value)). Thus in each branch, this procedure will ensure that the problem is not dual degenerate, that the duality gap is reduced for each branch.

To select the *next node* to be processed we apply *depth-first* strategy, the advantages being: finding one optimal or good near optimal integer solution faster; requiring less memory, in contrast with *best first* strategy.

5. Parallel Implementation

We have implemented parallel B&B algorithms on a cluster of workstations under PVM (i.e. distributed memory environment). We assume virtual *tree* topology (see figure 1) and we apply *master/slave* approach. The tree has the following structure: Master at the root. Each slave is enrolled as early as possible. The tree parallel architecture follows the the depth-first B&B tree. The master, after the LP relaxation of the first node, produces two nodes following the branching rule. Then it sends one of the nodes to the 2^N-th slave (see figure 1). Where N is a maximal N such that $2^N \leq P$ and P is the number of processors.

In the next iteration the master solves the next node and sends this node to the $2^{(N-1)}$-th processor. The master continues in this way until the number of nodes sent is equal to N. Each of the slaves, after receiving a node and solving it, continues with the expansion of B&B tree by sending problems to its descendants (see Figure 2).

Notice that the diameter of such network is only $O(log_2(P)) = O(N)$. In this way we solve and branch only N nodes in the master

8

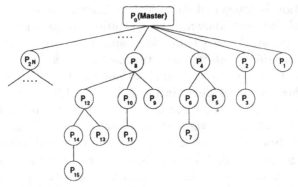

Figure 1. Communication structure of the processors

Branch and Bound tree

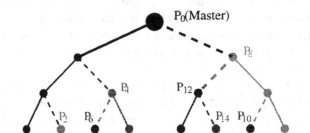

Figure 2. master/slave computation of the nodes in depth-first tree and the tree architecture used

and we are able to spawn the computations on 2^N machines for $O(N)$ steps.

Distribution of the *load* is dynamic. On each processor we have a pool of M nodes. When a process A, for example, runs out of nodes in its pool, it sends a request to its neighbouring processors. For example, processor 4 will send requests to 5, after that to 6, and then to 0; 12 to 13,14 and 8. A process receiving a request, B for example, answers it, only if its own pool of nodes is big enough. The size of the pool is a parameter ($LIST\_LONG\_ENOUGH$), which depends on the Communication/Computation time ratio of the node. If processor B, has not at this time enough nodes in the pool it will keep a record of the request in a queue. When the size of the pool is large enough B answers the request in FIFO order. This strategy allows us to achieve good load balancing and minimize the communication.

The *communication* is organized in the following way: Once one of the processors has found a better solution, it multicasts this better solution so far, to the rest of the processors and checks if another better solution so far has been found by another processor. The rest of the processors are not checking for an update in the value of the best solution so far on every iteration, (even a non blocking receive, costs time, and not always is going to be an update), so only after every K (parameter set to 100 by default) nodes, they check for the best solution so far. This parameter is a compromise, between communication and computation, and it is changed depending on the problem size. This produces anomalies, see *cracb1* where with two processors the number of nodes searched has increased, because the processor 2 did not check for an update until node 100. But in the others cases, with bigger B&B trees, this does not happen.

The algorithm *terminates* when the master runs out of problems (nodes) and checks if its N neighbouring processors, (i.e. processors with number 1,2,4,8,...), have already sent a request for nodes to the master (so the master knowns that the 2^N process of the parallel machine do not have any problem left.) In this way only N of total P processors are checked by the master.

This type of architecture also tackles efficiently *fault-tollerant* issue and algorithm continues successfully if some of the machines become nonoperational during the algorithm execution.

6. Experimental Results

Here we give examples of experiments on problems from industry arising in PCB/MCM layout and having similar structure to the optimisation problems arising in placing sources in air pollution modelling described above. Some experiments on well known ILP and MILP problems are also made. Due to the space constraints we present only some of the results obtained. The examples we consider correspond to area or netlength minimization of subrooms, after the partitioning process, e.g. Comps5 corresponds to a subroom with 5 components, room5 has 6 and room6 - 8 components. Note that *Prb. Send* is the number of nodes exchanged between processors, *Nodes* is the total number of nodes expanded.

Even if the solution time for small problems 6-8 components is high a feasible integer solution for a problem with 20-25 components, (2000-3000 variables) could be available in reasonable time (in some cases a question of minutes).

Table I. Some problems, from the MIPLIB library, and
comps5 room5 and room6 from "Place Box"

Problem	Rows	Variables	Int. Var	Nonzeros
air01	23	771	771	4215
air02	50	6774	6774	61555
air03	124	10757	10757	91028
bm23	20	27	27	478
cracpb1	143	572	572	4158
egout	98	141	55	282
khb05250	101	1350	24	2700
lp4l	85	1086	1086	4677
mod008	6	319	319	1243
modglob	291	422	98	968
stein27	118	27	27	378
stein45	331	45	45	1034
comps5	180	72	56	1212
room5	98	63	47	434
room6	128	78	60	544

Table 3 present the results where algorithms run on a parallel machine, having as the front end ULTRA SUN workstation. The parallel machine is a 64-processors SCI Cluster (Siemens-Nixdorf/SCALI) 64 processors, 19.2GFlops peak performances, 8GByte SDRAM main memory. This is installed at PC2, Paderborn Center for Parallel Computing.

Experimental results show that the algorithm scales well and we can obtain very good efficiency for MILP problems arising in PCB/MCM layout. On the other hand for some of MIPLIB problems we obtain anomalies partially due to the pool size. For example, *air2* problem is too small and there are no subproblems sent when the number of processors increases. On the other hand we observe some superlinear speedup for problem *room6*.

7. Conclusion

In this paper we made an analogy of some optimization problems arising in air pollution modelling while placing sources of pollution with a

Table II. *Comps5 run on Sun10s, the others on Ultra Sparc's. Problems (2) run on heterogeneous network of workstations, ULTRA, SUN20, SUN10.

Problem	Proc.	Nodes	Prb. Send	Optimal Found at Node in %	Efficiency	Time (sec)
comps5 *	1	532673	.	0.024%	.	3498
	2	538749	15	0.064%	108%	1609
	3	537503	63	0.16%	84.1%	1386
	4	540528	91	0.46%	87.8%	995
	5	537931	156	0.38%	84.2%	830
	18(2)	540324	632	1.2%	54.8% (2)	354
	25(2)	542611	641	2.1%	56.4% (2)	248
room5	1	896468	.	28%	.	2828
	2	661198	35	4.7%	112%	1257
	3	671232	90	12%	116 %	808
	4	667924	97	12.7%	105 %	669
	5	672819	242	0.8%	94%	598
	6	672949	191	1.0%	87.1 %	541
	15(2)	674954	521	0.77%	89.7% (2)	210
	22(2)	675157	600	1.0%	62.8% (2)	180
room6	1	103569056	.	1.1%	.	126473
	2	104885771	19	4.0%	83.7%	75513
	3	105168195	115	7.6%	81.0%	52030
	4	105145489	106	12.1%	75.1%	42064
	5	105191704	791	15%	72 %	35131

placement of components on a Printed Circuit Board (PCB) geometrical and technological constraints. We have formulated the placement task as Quadratic Optimization and MILP problems. We have presented a parallel B&B algorithms for ILP and MILP problems with sparse constraint matrices and with dynamic load balancing running in MIMD distributed memory environment under PVM. It is shown that by employing virtual tree topology and dynamic load balancing we have been able to minimize communication in parallel implementation and that the algorithms scale well on a cluster of workstations under PVM. Experiments on known problems from MIPLIB and on real life problems arising in PCM/MCM layout has been carried out and they show that very good efficiency can be achieved in parallel implementation.

Table III. Experiments run in the parallel machine at PC2

Problem	Proc.	Nodes	Prb. Send	Optimal Found at Node in %	Efficiency	Time (sec)
room5	1	896468	.	28%	.	2828
	2	909922	20	14%	132 %	1064
	4	967123	143	.08 %	156 %	453
	8	957771	343	.08 %	164 %	215
	16	976514	615	10 %	149 %	118
comps5	1	532673	.	0.024%	.	3498
	2	2479565	26	0.0001 %	59 %	2917
	4	2467069	146	0.0002 %	37 %	2319
	8	2246969	514	0.0001 %	47 %	913
	16	2480666	1707	0.0001 %	47 %	464
room6	1	16336975	.	1.1%	,	13375
	2	4190909	14	0.9 %	120 %	5573
	4	2929814	57	0.9 %	132 %	2516
	8	1682554	288	0.8 %	290 %	575
	16	2984256	737	0.9 %	179 %	465

Some improvements, such as parallel Interior Point method for sparse matrices, in solving LP relaxation and some fast heuristics on the first stage are required in order to speedup the computations further. On preprocessing and during B&B improvements can be made using Cutting planes [4], Knapsack cuts [4], some heuristic and stochastic methods to decrease the number of nodes expanded and to prove optimality.

References

1. Alonso, J. L., Schmidt, H. and Alexandrov, V. (1997) Parallel Branch and Bound Algorithms for Integer and Mixed Integer Linear Programming Problems under PVM, in M. Bubak, J. Dongarra and J. Wasniewski (eds), *Recent Advances in Parallel Virtual Machine and Message Passing Interface*, LNCS No 1332, Springer, Berlin, pp. 313-320.
2. Alonso, J. L. , Schmidt, H., Alexandrov, V. and Stube, B. (1998) High Quality PCB Placement using Parallel B&B Algorithms for MILP, submitted to *Parallel Computing*.
3. Babayev, D. A. and Mardanov, S. S., (1994) Reducing the Number of Variables in Integer and Linear Programming Problems, *Computational Optimization and Applications*, **3**, 99-109.

Table IV. Some examples from MIPLIB.

Problem	Proc.	Nodes	Prb. Send	Speedup	Efficiency	Time (sec)
air01	1	6	.	.	.	1.18
	2	6	0	0.802	40.1%	1.47
	3	6	0	0.599	19.9%	1.97
air02	1	25	.	.	.	77.8
	2	31	1	0.964	48.1%	80.73
	3	31	1	0.943	31.4%	82.5
air03	1	3	.	.	.	3328
	2	11	2	0.953	47.6%	3491
bm23	1	1232	.	.	.	2.8
	2	1063	4	1.53	52.6%	2.66
	3	1115	0	1.647	54.9%	1.7
	4	1215	2	1.556	38.8%	1.8
	5	928	4	2.154	43.0%	1.3
cracpb1	1	27	.	.	.	47
	2	125	2	0.398	19.9%	118
	3	390	4	0.315	10.5%	149
	4	123	4	1.306	32.6%	36
	5	128	4	2.83	56.6%	16.6
egout	1	8414	.	.	.	9.1
	2	7451	21	1.596	79.8%	5.7
	3	6621	30	2.136	71.2%	4.26
	4	6137	34	2.177	54.4%	4.18
	5	6685	43	2.585	51.1%	3.52
khb05250	1	4104	.	.	.	59.8
	2	5184	12	1.499	74.9%	39.9
	3	5325	22	1.898	63.2%	31.5
	4	3697	32	2.694	67.3%	22.2
	5	5259	36	2.534	50.6%	23.6

4. Bixby, R. E., Cook, D., Cox, D. D. and Lee, T. (1996) Computational Experience with Parallel Mixed Integer Programming in a Distributed Environment, *Annals of Operations Research*.
5. Eckstein, J. (1993) *Parallel Branch-and-Bound Algorithm for General Integer Programming on the CM-5*, TMC-257, Thinking Machines Corporation.

14

Table V. Some examples from MIPLIB (continue from Table IV).

Problem	Proc.	Nodes	Prb. Send	Speedup	Efficiency	Time (sec)
mod008	1	15912	.	.	.	131.6
	2	31203	7	1.8	90.0%	73.1
	3	35692	18	2.502	83.3%	52.6
	4	43538	38	2.464	61.6%	53.4
	5	33705	73	3.218	64.4%	40.9
modglob	1	9638801	.	.	.	109334
	3	12524334	27	1.486	74.3%	73555
	4	15274068	205	2.320	77.3%	47127
	5	18567228	502	2.106	42.2%	51913
stein45	1	121516	.	.	.	2650
	2	130288	12	1.715	85.7%	1545
	3	129936	44	2.390	79.6%	1109
	4	128877	98	3.581	89.5%	740
	5	128175	123	4.027	80.5%	658

6. Ferreira, A. and Pardalos P. (1996) *Solving Combinatorial Optimization Problems in Parallel,* Lecture Notes in Computer Science, No 1054, Springer, Berlin.
7. Gendron, B. and Crainic, T. G. (1994) Parallel Branch and Bound algorithms: survey and synthesis, *Operations Research,* **42**, 1042-1066.
8. Kleinhans, J. M., Sigl, G., Johannes, F. M. and Antreich K. J. (1991) GORDIAN: VLSI placement by quadratic programming and slicing optimization, *IEEE Transactions on CAD,* **10(3)**, 356–365.
9. Laursen, P. S. (1993) Simple Approaches to Parallel Branch and Bound, *Parallel Computing,* **19**, 143-152.
10. Nemhauser, G. L. and Wolsey, L. A. (1988) *Integer and Combinatorial Optimization,* John Wiley and Sons, 1988.
11. Stube, B. and Schmidt, H. (1997) A Multi–Method Approach to PCB Placement under EMC and Thermal Constraints, in *Proceedings of the Workshop on EMC-Adequate Design of Systems and Components at 12th International Zurich Symposium on Electromagnetic Compatibility (in Supplement),* pp. 205–212.

MODELING OF GLOBAL AND REGIONAL TRANSPORT AND TRANSFORMATION OF AIR POLLUTANTS

A.E. ALOYAN
Institute for Numerical Mathematics, RAS
8 Gubkin str., 117951, Moscow, Russia

Abstract

Several numerical models of atmospheric pollutant transport in different scales are considered taking into account different transformation processes. Results from experiments involving gas- and aqueous-phase chemical processes and interaction between gas particles and cloud droplets are presented. Global transport of lindane is considered in the Northern Hemisphere. Optimization problems, studied by using adjoint equations and linear programming methods, are discussed.

1. Introduction

Some models for studying the transport and transformations of atmospheric pollutants in different spatial and temporal scales (mesoscale, regional, and global) are described. The mesoscale model can be described by the following system of equations ([1], [3], [4]):

$$\left(\frac{\partial \varphi_i}{\partial t}\right)_{chem} + \left(\frac{\partial \varphi_g}{\partial t}\right)_{cond} + \left(\frac{\partial \varphi_\eta}{\partial t}\right)_{coag} =$$

$$F_i + F_g + F_\eta + \left[-\mathrm{div}\,\mathbf{u}\varphi + \frac{\partial}{\partial x}K_x\frac{\partial\varphi}{\partial x} + \frac{\partial}{\partial y}K_y\frac{\partial\varphi}{\partial y} + \frac{\partial}{\partial\sigma}K_\sigma\frac{\partial\varphi}{\partial\sigma}\right] + \quad (1)$$

$$B(\varphi_i,\varphi_j) + P(\varphi_{g_1},\varphi_{g_2}) + M(\varphi_{\eta_1},\varphi_{\eta-\eta_1}).$$

Here $B(\varphi_i,\varphi_j)$, $P(\varphi_{g_1},\varphi_{g_2})$, $M(\varphi_{\eta_1},\varphi_{\eta-\eta_1})$ are the chemistry (gas- and aqueous phase), condensation, and coagulation operators, respectively. F_i, F_g, F_η describe sources of gaseous, condensating, and coagulating species, respectively. The expression in square brackets in equation (1) is common for all three processes (photochemistry, condensation and coagulation). The boundary conditions are given as

$$\varphi_\alpha = \varphi_\alpha^0 \quad \text{for} \quad \sigma = H,$$

15

Z. Zlatev et al. (eds.), Large-Scale Computations in Air Pollution Modelling, 15–24.
© *1999 Kluwer Academic Publishers. Printed in the Netherlands.*

$$\varphi_\alpha = \varphi_\alpha^0 \quad \text{for} \quad x = \pm X, \; y = \pm Y \qquad (2)$$

$$a_3 \frac{\partial \varphi_\alpha}{\partial \sigma} = a_\vartheta \frac{\tilde{\beta}_\alpha \varphi_{\alpha,h} - f_\alpha}{\tilde{\beta}_\alpha + a_\vartheta k_\sigma} \quad \text{for} \quad \sigma = \frac{h - \tilde{\delta}}{H - \tilde{\delta}} H$$

where $\alpha = i; \; g; \; \eta; \; \tilde{\beta}_\alpha = \beta_\alpha u_* - \omega_\alpha; \; h$ and H are heights of the atmospheric surface and boundary layers, respectively; β_α characterizes the aerosol particle interaction with the underlying surface; $f_\alpha(x, y, \sigma)$ describes the pollution sources in the roughness level; $\varphi_{\alpha,h}$ is the concentration on the upper boundary of the surface layer. This is considered jointly with the mesoscale three-dimensional nonhydrostatic atmospheric model. The description of the photochemical model can be found in [1]. We have applied here a chemical mechanism for the butane oxidation [4], [5]. The total number of gaseous species is 50 with about 250 chemical reactions. Experiments were performed for different sets of emissions.

Simulations of 3-D photochemical air pollution in the Saint Petersburg region have been performed by using a hydrodynamical model. Particular attention has been given to studying the regional ozone formation mechanisms. Fig. 1 shows ozone concentration contours in the plane (x, y) at height $z = 250$m for $t=16.30$h. The results shows a great variability of gaseous species in the Saint Petersburg region.

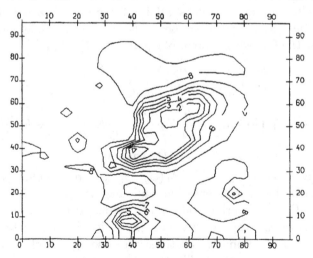

Figure 1. Ozone concentration contours at height 250m

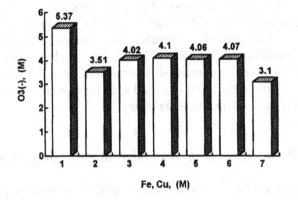

(1) Fe = 10^{-7}, Cu = 0;   (2) Fe = 10^{-5}, Cu = 10^{-5}; (3) Fe = 10^{-5}, Cu=10^{-6};
(4) Fe = 10^{-5}, Cu = 10^{-7}; (5) Fe = 10^{-6}, Cu = 10^{-6}; (6) Fe = 10^{-7}, Cu = 10^{-6};
(7) Fe = 10^{-4}, Cu = 10^{-5}.

Figure 2. Results obtained for different combinations of *Fe* and *Cu*

2. Condensation and coagulation

The numerical experiments for modeling the kinetic processes of coagulation and condensation are described in detail in [1], [3]. Particularly, in [3], comparisons of the numerical results with observation data indicated their qualitative resemblance for the aerosol particle size spectra. Quantitatively, there is some systematic discrepancy between them connected with uncertainties in the emission source specification of the smallest-size particles.

3. Wet convection

To simulate the interaction between gaseous pollutants and cloud drops, a three-dimensional wet convection model (taking into account the long-wave radiation fluxes and other physical mechanisms) is used. However, in this work, we do not consider the hydrodynamical aspects of cloud formation processes which is another problem laying beyond our aims here. On cloudiness modeling, as a first approximation, an average drop size of 10μm is assumed. For these particles, the sensitivity of ozone concentrations to variations of metals *Fe* and *Cu* dissolved in drops is investigated. On fig. 2 the results of a series of numerical experiments are presented for different combinations of metals *Fe* and *Cu*. It is evident from the figure that cloud drops are sinks for ozone with about 40 % decline level.

A photochemical cloud model has been developed for investigating the influence of dissolved Mn and Cu on formation and destruction of ozone [9]. The numerical calculations were performed for the atmospheric conditions of polluted mid-latitude urban areas. The rate of ozone destruction is increased several-fold. In doing so, the metal content in clouds turns to be essential sink for the gas-phase variations of OH, HO_2, and ozone concentrations.

4. Global transport of persistent organic pollutants

A numerical model for the transport of persistent organic pollutant in the Northern hemisphere was developed [2]. Lindane was taken as an example. We formulate the model assuming the Earth's surface to be spherical. A coordinate system (λ, ψ, z) is used where λ is the longitude, ψ is the complement of latitude and z is the altitude measured from the underlying surface. The main equation of pollutant transport on a sphere is expressed in the following form [8]:

$$\frac{\partial \varphi}{\partial t} + \frac{u}{a \sin \psi} \frac{\partial \varphi}{\partial \lambda} + \frac{v}{a} \frac{\partial \varphi}{\partial \psi} + w \frac{\partial \varphi}{\partial z} =$$

$$F - D\varphi + \frac{\partial}{\partial z} \nu \frac{\partial \varphi}{\partial z} + \frac{1}{a^2 \sin^2 \psi} \frac{\partial}{\partial \lambda} \mu \frac{\partial \varphi}{\partial \lambda} + \frac{1}{a^2 \sin \psi} \frac{\partial}{\partial \psi} \mu \sin \psi \frac{\partial \varphi}{\partial \psi}. \quad (3)$$

Here $\varphi = \varphi(\lambda, \psi, z, t)$ is the pollutant concentration; $\mathbf{u} = (u, v, w)$ is the wind velocity vector along λ, ψ, z directions, respectively; μ, ν are the horizontal and vertical turbulence exchange coefficients, respectively; $F = F(\lambda, \psi, z, t)$ is a function standing for emission source magnitudes; D describes the degradation of lindane in the atmosphere; and a is the average radius of the Earth. The boundary conditions are:

$$\varphi(0, \psi, z, t) = \varphi(2\pi, \psi, z, t)$$

$$\varphi(\lambda, -\psi, z, t) = \varphi(\lambda + \pi, \psi, z, t)$$

$$\varphi(\lambda, \pi + \psi, z, t) = \varphi(\lambda + \pi, \pi - \psi, z, t) \quad (4)$$

$$\frac{\partial \varphi}{\partial \psi} = 0 \quad \text{for} \quad \psi = \pi/2$$

The model incorporates the following main parts: planetary boundary layer and surface layer models; flux of lindane in soil and water [6]; wet deposition, degradation in the atmosphere; soil- and see-atmospheric exchange, migration in soil [6].

Calculations were performed for one year period, 1992 using the European Center meteorological information and lindane emission data for Europe. The input parameters were: grid resolution 144×73 corresponding to longitude-latitude step 2.5° × 1.25°; 15 vertical levels up to the tropopause; Time step is 30 min.; meteorological information is from ECMWF; lindane emission data for the European region is taken from the ESQUAD project. The total amount of emissions is 2690 t/year with monthly distribution: February – 10%, March – 15%, April, May, June – 25%, and no emission for the remaining months.

In fig. 3 lindane concentration contours are presented for June and July (z=50 m). Starting from October, lindane concentrations in lower layers of the atmosphere increases though decreasing with altitude because of considerable reemissions from soil. In the course of time, with lindane entering into the atmosphere (reemission) turning noticeable, the maximal concentrations are localized in rather small area. Monthly averaged concentrations at z=50 m were applied. Net gaseous flux and wet deposition to land and water surface in the EMEP modeling area were computed. In fig. 4 the distribution of i net gaseous flux over land and sea by months is demonstrated. The distributions of lindane (in percents) in different media are obtained for different months of 1992. In fig. 5 this is demonstrated for December. Details of numerical experiments can be found in [2].

5. Transboundary transport

It is interesting also to study the sensitivity functions for European countries pollution caused by transboundary transport. Here adjoint functions are used which proved to efficient means in dealing with this kind of problems [7]. [8].

Computations for some European countries were performed jointly with G.I.Marchuk and V.O.Arutyunyan. The problem was considered in the regional scale from Atlantic to Urals in the horizontal and up to the tropopause in the vertical. EMEP emission data for SO_2 were used in the calculations.

With the help of the Lagrange identity we obtain the adjoint equation:

$$-\frac{\partial \varphi^*}{\partial t} - \frac{u}{a\sin\psi}\frac{\partial \varphi^*}{\partial \lambda} - \frac{v}{a}\frac{\partial \varphi^*}{\partial \psi} - (w - w_g)\frac{\partial \varphi^*}{\partial z} -$$
$$\left(\frac{\partial}{\partial z}\nu\frac{\partial \varphi^*}{\partial z}\frac{1}{a^2\sin^2\psi}\frac{\partial}{\partial \lambda}\mu\frac{\partial \varphi^*}{\partial \lambda} + \frac{1}{a^2\sin\psi}\frac{\partial}{\partial \psi}\mu\sin\psi\frac{\partial \varphi^*}{\partial \psi}\right) = p$$

where p is a weight function characterizing the area for which the sensitivity function is determined. The sensitivity function obtained for a part of Former Soviet Union is shown in fig. 6.

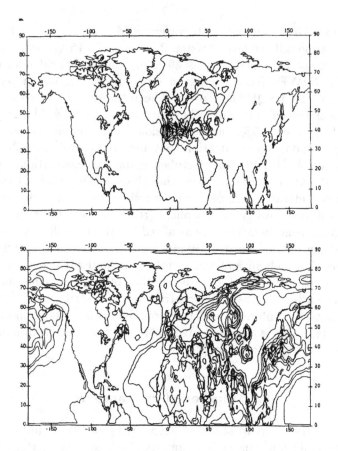

Figure 3. Lindane concentration contours for June and July

The analysis of the experimental results shows that the spatio-temporal structure of the solutions of adjoint problems provide information about the zones that can contribute to the pollution of a chosen region.

6. Optimization problem

Let us now pose an optimization problem for minimizing the economic damage on the environment. An objective function with corresponding constraints is selected to me minimized (this can be damage to agricultural crops, forests, human health, water quality, etc.) Resolving such multicriterion problem brings to specific regularization of emission source magnitudes in the regional scale or in different countries.

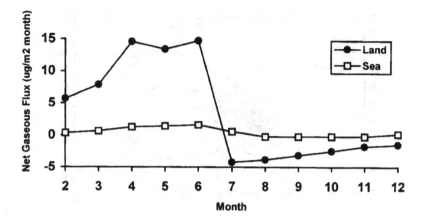

Figure 4. iNet gaseous flux over sea and land

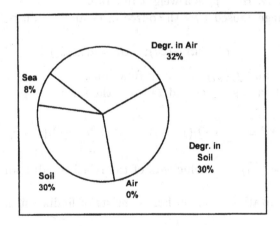

Figure 5. Distribution of lindane in different media for December

Let us have n emission sources at points $x_i \in D$ with concentrations φ_i and source powers Q_i $(i = \overline{1, n})$. Denote by G the environmental control zone $(G \subset D)$.

Consider the functionals

$$a_i^c = \int\limits_0^t \int\limits_G p_i \varphi_i(x, t) \mathrm{d}G; \quad a_i = \int\limits_0^t \int\limits_D p_i \varphi_i(x, t) \mathrm{d}D.$$

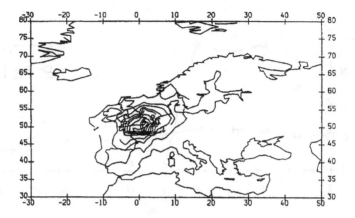

Figure 6. The sensitivity function for a part of the former Soviet Union

representing the total concentrations of a pollutant in regions G and D, respectively. Here p_i is a weight function.

The damage caused by i-th source in G and D can be represented as:

$$\alpha_i^c = a_i^c f_1(\varphi_i); \quad \alpha_i = a_i f_2(\varphi_i)$$

where $f_1(\varphi_i)$ and $f_2(\varphi_i)$ are given functions.

The total damage in G and D takes the form:

$$Y^c = \sum_{i=1}^{n} \alpha_i^c Q_i(1 - e_i), \quad Y_0 = \sum_{i=1}^{n} \alpha_i Q_i(1 - e_i)$$

where e_i $(i = \overline{1, n})$ are coefficients of the relative reduction of emission source powers.

The optimization problem here consists of finding such $\{e_i\}$ that

$$Y^c = \sum_{i=1}^{n} \alpha_i^c Q_i(1 - e_i) \to \min$$

under the following restrictions

$$k_1 Y_0 \leq \sum_{i=1}^{n} \alpha_i Q_i(1 - e_i) \leq k_2 Y_0.$$

Here k_1 and k_2 are coefficients of damage reduction necessary to be achieved.

As an example, we demonstrate the results of numerical calculations for the area of Lake Bajkal. Here the emission control of 6 sources marked by circles is demonstrated for minimizing the damage in region

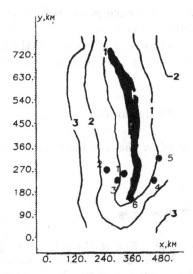

Figure 7. Model results for the area around the lake of Baikal

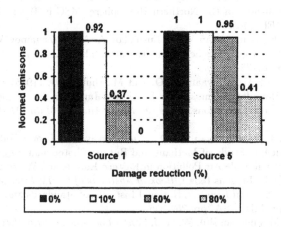

Figure 8. Control of the magnitudes of emission sources

G. In fig. 7 the synthesized contours of φ^* are shown taking into account the rose of winds.

Fig. 8 demonstrates control of emission source magnitudes. When 10%, 50%, 80% damage reduction levels are considered for the whole region, the 1-st and 6-th sources turn to be the most sensitive. Since the 6-th source is also the nearest to the Lake, it makes sense to close it down in general. The 1-st source needs to be regulated always; for example, 50% damage reduction will require 65% decrease of emission and 80% reduction require its total closing. The 2-nd and 3-rd sources are less dangerous though the 3-rd source is located nearby but it's power is weak and so its contribution is small.

It should be noted that this method can be applied to the long-range transport problems replacing the point sources by area sources.

Acknowledgements

This work was partially supported by the Russian Foundation of Basic Research pr. # 960564733.

References

1. Aloyan, A. E. (1997) Numerical Modelling of Minor Gas Constituents and Aerosols in the Atmosphere, in G. Geernaert, A. Walløe Hansen and Z. Zlatev (eds), *Regional Modellin of Air Pollution in Europe*, National Environmental Research Institute, Poskilde, Denmark, pp. 153-176.

2. Aloyan, A. E. and V.O.Arutyunyan, V. O. (1997) Numerical Modeling of Lindane Transport in the Northern Hemisphere, MSC-E Techn. Rep. 12/97, MSC-E EMEP, Moscow.

3. Aloyan, A. E., Arutyunyan V. O., Lushnikov A. A. and Zagainov V. A. (1997) Transport of coagulating aerosol in the atmosphere, *J. Aeros. Sci.*, **28**, pp. 67-8528, pp. 67-85.

4. Aloyan, A. E., Arutyunyaan, V. O. and Marchuk, G. I. (1995) Dynamics of Mesoscale Boundary Atmospheric Layer and Impurity Spreading with Photochemical Transformations, *Russ. J. Num. Anal. & Math. Model.*, **10**, pp. 93-114.

5. Aloyan, A. E., Arutyunyan, V. O., Skubnevskaya, G. I. and Dultseva, G.G. (1997) Modelling Study of Methane and Butane Photooxidation and Application to Regional Scale Air Pollution in Eastern Europe and Western Siberia, in R. Huie and J. Hudgens (eds.), *Book of Abstracts of the IV International Conference on Chemical Kinetics*, National Institute of Standards and Technology, Gethersburg, MD, USA, pp. 254-255.

6. Jacobs, C. M. and Van Pul, W. A. J. (1996) Long-range atmospheric transport of persistent organic pollutants, I: Description of surface-atmosphere exchange modules and implementation in EUROS, RIVM Report 722401013, Bilthoven, Netherlands.

7. Marchuk, G. I. (1986) *Mathematical modeling in the problem of environment*, North-Holland, Amsterdam-New-York.

8. Marchuk, G. I. and Aloyan, A. E. (1995) Global transport of pollutant in the atmosphere, *Izv. AN: Fizika Atm. Okeana*, **31**, 597-606.

9. Mattijsen, J., Builtjes, P. J. and Sedlak, D. L. (1995) Cloud Model Experiments of the Effect of Iron and Copper on Tropospheric Ozone Under Marine and Continental Conditions, *J. Met. & Atm. Physics*, **57**, 43-60.

LONG-TERM CALCULATIONS WITH LARGE AIR POLLUTION MODELS

C. AMBELAS SKJØTH, A. BASTRUP-BIRK,
J. BRANDT AND Z. ZLATEV
National Environmental Research Institute
Frederiksborgvej 399, P.O. Box 358, DK-4000 Roskilde, Denmark.
E-mail: luzz@sun2.dmu.dk, Fax: +45 4630 1214

Abstract

The air pollution levels in a given region depend not only on the emission sources located in it, but also on emission sources located outside the region under consideration, and even on sources that are far away from the studied region. This is due to the transboundary transport of air pollutants. The atmosphere is the major medium where pollutants can be transported over long distances. Harmful effects on plants, animals and humans can also occur in areas which are long away from big emission sources. Chemical reactions take place during the transport. This leads to the creation of secondary pollutants, which can also be harmful. The air pollution levels in densely populated regions of the world, such as Europe, must be studied carefully to find out how the air pollution can be reduced to safe levels and, moreover, to develop reliable control strategies by which the air pollution can be kept under certain prescribed critical levels. This can be done only when large scale mathematical models, in which all physical and chemical processes are adequately described, are used. The numerical treatment of such models leads to huge computational tasks, which can successfully be solved only on high speed computers.

An improved version of the Danish Eulerian Model, which can efficiently be run on several vector and parallel computers, have been prepared and tested. It was demonstrated that the new version runs efficiently on several different high speed computers. The possibility of applying successfully the improved version in a comprehensive study of a wide range of air pollution phenomena in a prescribed European region was demonstrated by carrying out long series of scenarios in order to investigate the influence of different variations of the emission sources to the corresponding concentrations and depositions. Some results from an extensive study, in which the improved version of the Danish Eulerian Model was used, will be presented in this paper.

Z. Zlatev et al. (eds.), Large-Scale Computations in Air Pollution Modelling, 25–38.
© 1999 *Kluwer Academic Publishers. Printed in the Netherlands.*

1. Introduction

High air pollution levels may cause (i) acidification of the soils in some sensitive areas, (ii) damages in certain water areas (especially the closed seas as, for example, the Baltic Sea, the Black Sea and the Mediterranean) and (iii) eutrofication of eco-systems. Moreover, high concentrations of certain pollutants, first and foremost of tropospheric ozone, can directly damage plants, animals and humans. Therefore, systematic studies of the pollution levels are needed in order to answer the following questions:

1. Are certain critical levels exceeded?

2. Will the critical levels be exceeded in the near future?

3. What preventive measures could (and/or should) be taken in the efforts to avoid exceedance of the critical levels?

Such studies have been carried out with an improved version of the Danish Eulerian Model (an air pollution model developed at the National Environmental Research Institute, Roskilde, Denmark). The new high performance computing modules were some of the most important improvements in the new version (because it is impossible to run long series of experiments with a large scale air pollution model, in which all physical and chemical processes are adequately described, without using efficiently high speed computers). A long series of scenarios has been prepared. Then the improved model has been used to study the following issues: (i) different pollution levels in Europe in a nine-year time period (1989-1997), (ii) exceedance of critical levels in certain parts of Europe, (iii) the changes of the air pollution levels in Europe in the period 1989-1997 as a results of the emission changes in Europe and in its surrounding countries in this period, (iv) the effects of systematical reductions of European emissions on the concentrations and/or the deposition of corresponding emissions and (v) representation of the European pollution levels as functions of the variation of the European emissions. These results have been obtained only because the great potential power of the new modern computers has been properly used. As mentioned above, it is impossible to run a model in which all physical and chemical processes are adequately described on computers (even high speed computers) if the code is not optimized so that high computational speed can be achieved. Therefore, we shall start with the presentation of results obtained on such computers, by which it will be demonstrated that the computational speed of the code is sufficiently high when it is run on the available computers.

2. The Danish Eulerian Model

Several physical and chemical processes must be taken into account in the development of a mathematical model for studying long-range transport phenomena. These processes are: (i) horizontal advection, (ii) horizontal diffusion, (iii) deposition, A mathematical model where all these processes are taken into account can be represented by a system of partial differential equations (PDE's) of the following type:

$$\frac{\partial c_s}{\partial t} = -\frac{\partial(uc_s)}{\partial x} - \frac{\partial(vc_s)}{\partial y} - \frac{\partial(wc_s)}{\partial z}$$

$$+\frac{\partial}{\partial x}\left(K_x\frac{\partial c_s}{\partial x}\right) + \frac{\partial}{\partial y}\left(K_y\frac{\partial c_s}{\partial y}\right) + \frac{\partial}{\partial z}\left(K_z\frac{\partial c_s}{\partial z}\right)$$

$$-(\kappa_{1s} + \kappa_{2s})c_s + E_s + Q_s(c_1, c_2, \ldots, c_q), \quad s = 1, 2, \ldots, q.$$

$$(1)$$

where the number of equations q is equal to the number of chemical species studied by the model. The following notation is used: (i) the concentrations of the chemical species are denoted by c_s, (ii) u, v and w are wind velocities, (iii) K_x, K_y and K_z are diffusion coefficients, (iv) the emission sources are described by the terms E_s, (v) κ_{1s} and κ_{2s} are deposition coefficients (for the dry and wet deposition respectively) and (vi) the chemical reactions are denoted by $Q_s(c_1, c_2, \ldots, c_q)$ (it will be assumed that the condensed CBM IV scheme is used in the chemical part; this scheme has been described in [10]; see also [30]).

The space domain of the model (it contains the whole of Europe together with parts of Asia, Africa and the Atlantic Ocean) has been discretized by using, as mentioned above, a (96×96) grid. This means that (i) Europe is divided into 9216 grid-squares and (ii) the grid resolution is approximately $(50\ km \times 50\ km)$. Appropriate initial and boundary conditions are needed. Different initial and boundary conditions are discussed in [30]. Furthermore, large sets of input data, meteorological data and emission data, are also needed in the model. The sets of input data and their reliability are described in [30]. There are several versions of the Danish Eulerian Model. The model is often run as a 2-D model. The mathematical terms describing the vertical transport are removed in this situation, however some vertical exchange between the boundary layer and the free troposphere is still assumed in the 2-D version; see [32]. Much more details about the model can be found in [30], [33], [35] and [36].

3. Splitting procedures

It is difficult to treat the system of **PDE's** (1) directly. This is the reason for using different kinds of splitting. A simple splitting procedure, based on ideas proposed in [19] and [20], can be defined, for $s = 1, 2, \ldots, q$, by five sub-models, representing the physical and chemical processes in (1); the horizontal advection, the horizontal diffusion, the chemistry (together with the emission terms), the deposition and the vertical exchange:

$$\frac{\partial c_s^{(1)}}{\partial t} = -\frac{\partial(uc_s^{(1)})}{\partial x} - \frac{\partial(vc_s^{(1)})}{\partial y} \tag{2}$$

$$\frac{\partial c_s^{(2)}}{\partial t} = \frac{\partial}{\partial x}\left(K_x \frac{\partial c_s^{(2)}}{\partial x}\right) + \frac{\partial}{\partial y}\left(K_y \frac{\partial c_s^{(2)}}{\partial y}\right) \tag{3}$$

$$\frac{dc_s^{(3)}}{dt} = E_s + Q_s(c_1^{(3)}, c_2^{(3)}, \ldots, c_q^{(3)}) \tag{4}$$

$$\frac{dc_s^{(4)}}{dt} = -(\kappa_{1s} + \kappa_{2s})c_s^{(4)} \tag{5}$$

$$\frac{\partial c_s^{(5)}}{\partial t} = -\frac{\partial(wc_s^{(5)})}{\partial z} + \frac{\partial}{\partial z}\left(K_z \frac{\partial c_s^{(5)}}{\partial z}\right) \tag{6}$$

If the model is split into sub-models as shown above, then discretization methods will lead to five ODE systems ($i = 1, 2, 3, 4, 5$):

$$\frac{dg^{(i)}}{dt} = f^{(i)}(t, g^{(i)}). \tag{7}$$

Each system contains $N_x x N_y x N_z x q$ equations (N_x, N_y and N_z are the numbers of grid-points along the coordinate axes). The functions $f^{(i)}$ depend on the particular discretization methods used in the numerical treatment of the different sub-models, while $g^{(i)}$ contain some approximations of the concentrations at the grid-points.

   It is necessary to couple the five ODE systems (7) obtained from (2)-(6). The coupling procedure is connected with the time-integration of these systems. Assume that the values of the concentrations (for all species and at all grid-points) have been found for some $t = t_n$. According to the notation introduced above, these values are components of the vector-function $g(t_n)$. The next time-step, time-step $n + 1$ (at which the concentrations are found at $t_{n+1} = t_n + \Delta t$, where Δt is some increment), can be performed by integrating successively the five systems (7) arising after the discretization of (2)-(6). The values of $g(t_n)$ are used as an initial condition in the solution of the first ODE system, the ODE system (7) obtained from (2). After that the solution of each ODE system (7) (excepting the last one) is used as an initial condition in the solution of the next system. The solution of the last system, the ODE system (7) obtained after the discretization of (6), is used as an approximation to $g(t_{n+1})$. In this way, everything is prepared to start the calculations in the next time-step, step $n + 2$.

The major advantage of using a splitting procedure is the possibility to design the best numerical algorithms for the different sub-models. This has been exploited in the Danish Eulerian Model. Unfortunately, there is also a serious disadvantage of using splitting procedures: it is difficult to control the error caused by the splitting procedure chosen. However, many experiments indicate that the results are rather good when the accuracy required is not very high. This explains why splitting procedures are used in all large-scale air pollution models.

4. Need for high performance computing

The complexity of the problems solved depends on (i) the size of the ODE systems (7), (ii) the number of time-steps that are to be performed in the run under consideration and (iii) the number of runs that are to be carried out in the selected set of scenarios.

The size of the ODE systems is equal to the product of the number of chemical species and the number of grid-points. Some results are given in Table I. The use of more advanced chemical schemes and/or refined spatial resolution lead to a great increase of the number of equations in the ODE systems. The figures are for the 3-D version with ten layers. These figures become considerably smaller (ten times) when the 2-D version is used, but the tasks are still very big. The number of the time-steps depends on the particular task which is to be solved. It varies from several hundreds (pollution forecasts) to several hundred thousands; see Table II. The number of runs can vary in a wide range (depending of the processes that are to be studied). About 720 runs were needed in the simulations carried out in [2] and [37].

Comprehensive studies needed in the efforts to prepare large amount of data which can be useful for biologists (studying different effects), different environmental specialists (both from governmental and non-governmental institutions) and decision makers can successfully be completed only if high performance computing is used. This task will otherwise be impossible unless some mechanisms describing physical and chemical processes are simplified or even neglected.

Several high performance modules for the most time-consuming processes (the advection and the chemistry) were developed and attached to the model. These modules are based on numerical algorithms and ideas discussed in [1], [3], [4], [5], [23], [27], [28], [29], [31], [34]. Some general recommendations from [15], [18], [22], [26], have also been taken into account. Standard parallelization tools (MPI, the Message Passing Interface; see [11]) have been used. This means that good performance should be expected on many parallel architectures. Some results obtained by using this new parallel version of the model are given in Table III.

5. Validation of the output results

The model results must be validated before starting to use the model in long series of scenarios in order to study the variations of the concentrations and depositions caused by variations of certain important parameters the emissions sources). There are two different ways to validate the results:

1. The ability of the numerical algorithms to produce accurate results can be checked by constructing test-examples with known solutions.

2. The ability of the model to calculate reliable concentrations and/or depositions can be checked by comparing the calculated concentrations and/or depositions with measurements

Both checks are needed. If the first series of checks is successfully passed, then this means that the errors due to the numerical algorithms used are small. Thus, if we study the response of the model to the variations of some physical parameter (say, the emissions), then the errors due to the numerical algorithms will not interfere with the changes due to the variation of the parameter chosen, and correct conclusions about the relationship between the model results and the parameter varied can be drawn. If the second series of checks gives good results, then one should expect the model to produce reliable results also in situations where either the measurements needed are not available or measurements are not existent (air pollution forecasts, different control strategies studied by appropriate scenarios, etc.). The second series of checks is often more important. This is why only the main ideas behind the first series of checks will be sketched and then the use of the second series of checks will be discussed in some more detail.

Table I. The number of equations per system of ODE's that are to be treated at every time-step. The number of time-steps for the chemical sub-model is sometimes even larger, because smaller step-sizes have to be used in this sub-model.

Number of species	$(32 \times 32 \times 10)$	$(96 \times 96 \times 10)$	$(192 \times 192 \times 10)$
1	10240	92160	368640
2	20480	184320	737280
10	102400	921600	3686400
35	358400	3225600	12902400
56	573440	5160960	21381120
168	1720320	15482880	61931520

Table II. Approximate numbers of time-steps needed in several important types of tasks.

Time period	Typical task	Time-steps needed
Two-three days	Preparing pollution forecasts	> 300
One month	Studying monthly mean values	> 3400
One season	Studying summer periods for ozone	> 11000
One year	Studying seasonal variations	> 40000
Several years	Studying long-term trends	> 400000

The well-known rotation test is normally used in the first series of checks. For the simple case where pure advection (horizontal transport in the atmosphere) is studied the rotation test has simultaneously been proposed in 1968 in [8] and [21]). After that, this test has been used in several hundred publications. This test has been extended for the more

Table III. Results obtained on three different high speed computers. The computing times are given in seconds. The computational speeds obtained in the runs and the theoretical maxima of the computational speeds for the different architectures, v_{max}, are measured in GFLOPS (billions of floating point operations per second).

Computer	Processors	Time	Speed	v_{max}	Efficiency
CRAY C92A	1	1050	0.460	0.9	51%
POWER CHALLENGE	8	1252	0.384	2.0	19%
IBM SP	8	611	0.791	4.0	20%
IBM SP	32	197	2.452	16.0	15%

complicated case where both advection and chemistry are studied in [16]. This extended version of the rotation test has been used in our experiments. The results indicate that both the algorithms chosen for the advection sub-model and the algorithms chosen for the chemical sub-model perform very well. It should be mentioned that some other tests (discussed in detail in [30]) have also been used in the efforts to evaluate the accuracy of the numerical methods used in the improved version of the Danish Eulerian Model.

The second series of tests is related to use of measurements in the validation procedure. There is a rather big network of measurement stations in Europe, which has been established about 20 years ago within EMEP (the European Monitoring and Evaluation Programme). The results of the measurements taken in all stations, which are located in different European countries, are collected at NILU (the Norwegian Institute for Air Research). After a careful control and calibration of the measured results, these can be used by air pollution modellers to validate the results of their models. Measurements for the period from 1989 to 1995 are at present available at NERI. All available data have been used to compare the calculated by the model concentrations and depositions with measurements. Some other measurements (mainly from Danish stations) have also been used in the comparisons. It should be mentioned here that results from comparisons with measurements have been presented in many publications; [2], [12], [32], [33], [37].

The results of the comparisons of the calculated by the improved version of the Danish Eulerian Model concentrations and depositions with the correspondent measurements taken at stations in different European countries have been presented by using (i) scatter plots and (ii) plots in which the temporal variations of the calculated concentrations and depositions are given together with the temporal variations of the correspondent measurements at the station under consideration.

The conclusion, drawn after a careful investigation of many hundreds of plots, is that the model is able to produce reliable results. The same conclusion can also be drawn by comparing the results obtained by the Danish Eulerian Model by results obtained by some other models (as, for example, the models discussed in [6], [7], [9], [13], [14], [17], [24], [25]). Two illustrations are given here: the first illustration consists of four scatter plots in Fig. 1 - Fig. 4, while the second illustrations shows (in Fig. 5 - Fig. 8) comparisons of calculated and measured ozone concentrations over a long-term period (1989-1995).

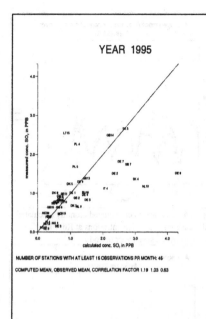

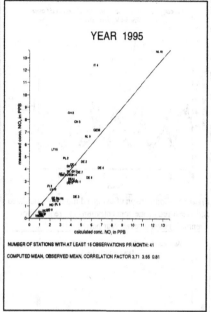

Figure 1. Calculated versus measured SO₂ concentrations

Figure 1. Calculated versus measured SO_2 concentrations

Figure 2. Calculated versus measured NO_2 concentrations

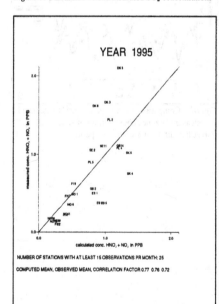

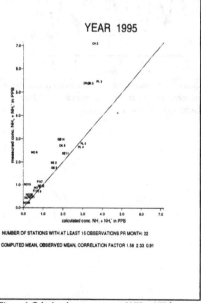

Figure 3. Calculated versus measured $HNO_3 + NO_3^-$ concentrations

Figure 4. Calculated versus measured $NH_3 + NH_4^+$ concentrations

34

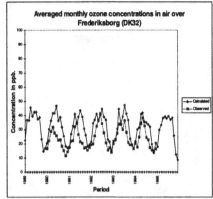

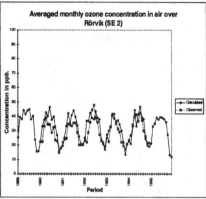

Figure 5. Comparison of averaged monthly calculated and measured ozone concentrations at Frederiksborg (Denmark) over a seven year period.

Figure 6. Comparison of averaged monthly calculated and measured ozone concentrations at Rörvik (Sweden) over a seven year period.

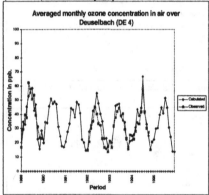

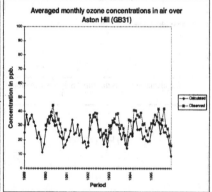

Figure 7. Comparison of averaged monthly calculated and measured ozone concentrations at Deuselbach (Germany) over a seven year period.

Figure 8. Comparison of averaged monthly calculated and measured ozone concentrations at Aston Hill (Great Britain) over a seven year period.

EXPOSURE TO HIGH OZONE CONCENTRATIONS

Numbers of days with 8-hour periods in which
the ozone is higher than 60 ppb
1997

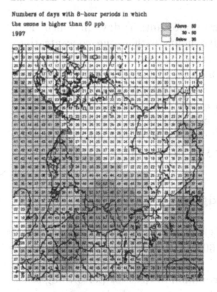

Figure 9

EXPOSURE TO HIGH OZONE CONCENTRATIONS

Numbers of days in which the ozone
is higher than 90 ppb
1997

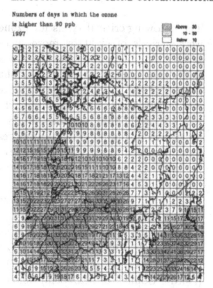

Figure 10

EXPOSURE TO HIGH OZONE CONCENTRATIONS

Numbers of days in which the ozone
is higher than 120 ppb
1997

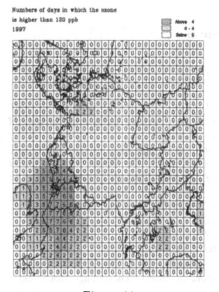

Figure 11

EXPOSURE TO HIGH OZONE CONCENTRATIONS

Percentages: 100*(AOT40)/(3000ppb.hours), showing by how much
the critical value 3000 ppb.hours is exceeded
1997

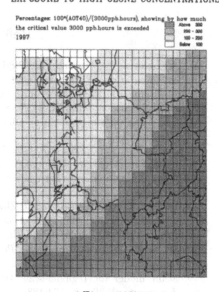

Figure 12

6. Exceeded ozone critical levels in Europe

The following critical levels for ozone are important: (i) days in which there were periods of eight or more successive hours with ozone concentrations greater than 60 ppb, (ii) days in which there were periods of one or more hours with ozone concentrations greater than 90 ppb, (iii) days in which there were periods of one or more hours with ozone concentrations greater than 120 ppb and (iv) AOT40 values for crops which exceed 3000 ppb.hours Some results which indicate that all these values are exceeded in many European regions are shown in Fig. 9 - Fig. 12. The results are for 1997, but similar situations have also been observed in nearly every year in the period 1989-1996.

Acknowledgments

This research was supported by the NATO Scientific Programme under projects ENVIR.CGR.930449 and OUTS.CGR.960312, by the EU ESPRIT Programme under projects WEPTEL (#22727) and EUROAIR (#24618) and by NMR (Nordic Council of Ministers) under a common project for performing sensitivity studies with large-scale air pollution models in which scientific groups from Denmark, Finland, Norway and Sweden are participating. A grant from the Danish Natural Sciences Research Council gave us access to all Danish supercomputers.

References

1. Alexandrov, V., Sameh, A., Siddique, Y. and Zlatev, Z. (1997), Numerical integration of chemical ODE problems arising in air pollution models, *Environmental Modeling & Assessment*, **2**, pp. 365-377.
2. Bastrup-Birk, A., Brandt, J., Uria, I. and Zlatev, Z. (1997), Studying cumulative ozone exposures in Europe during a 7-year period, *Journal of Geophysical Research*. **102**, pp. 23917-23935.
3. Bendtsen, C. and Zlatev, Z. (1997), Running air pollution models on message passing machines, in M. Bubak, J. Dongarra and J. Wasniewski (eds.), *Parallel Virtual Machine and Message Passing Interface*, Springer-Verlag, Berlin, pp. 417-426.
4. Brandt. J., Wasniewski, J. and Zlatev, Z. (1996), Handling the chemical part in large air pollution models, *Appl. Math. and Comp. Sci.*, **6**, pp. 101-121.
5. Brown, J., Wasniewski, J. and Zlatev, Z. (1995), Running air pollution models on massively parallel machines, *Parallel Computing*, **21**, pp. 971-991.
6. Carmichael, G. R., Peters, L. P. and Kitada, T. (1986), A second generation model for regional-scale transport-chemistry-deposition, *Atmospheric Environment*, **20**, pp. 173-188.
7. Chang J. S., Brost, R. A., Isaksen, I. S. A., Madronich, S., Middleton, P., Stockwell, W. R. and Walcek, C. J. (1987), A three dimensional Eulerian acid deposition model: Physical concepts and formulation, *Journal of Geophysical Research*, **92**, pp. 14691-14700.

8. Crowley, W. P. (1968), Numerical advection experiments, *Monthly Weather Review*, **96**, pp. 1-11.

9. Ebel, A., Hass, H., Jacobs, H. J., Laube, M. and Memmesheimer, M. (1991), Transport of atmospheric minor constituents as simulated by the EURAD model, in: J. Pankrath, (ed.), *Proceedings of the EMEP workshop on photooxidant modelling for long-range transport in relation to abatement strategies* Unweltsamt, Berlin, pp. 142-154.

10. Gery, M. W., Whitten, G. Z., Killus, J, P. and Dodge, M. C. (1989), A photochemical kinetics mechanism for urban and regional modeling, *Journal of Geophysical Research*, **94**, pp. 12925-12956.

11. Gropp, W., Lusk, E. and Skjellum, A. (1994), *Using MPI: Portable programming with the message passing interface*, MIT Press, Cambridge, Massachusetts.

12. Harrison, R. M., Zlatev, Z. and Ottley, C. J. (1994), A comparison of the predictions of an Eulerian atmospheric transport chemistry model with measurements over the North Sea, *Atmospheric Environment*, **28**, pp. 497-516.

13. Hass, H., Ebel, A., Feldmann, H., Jacobs, H. J. and Memmesheimer, M. (1993), Evaluation studies with a regional chemical transport model (EURAD) using air quality data from the EMEP monitoring network, *Atmospheric Environment*, **27A**, pp. 867-887.

14. Hass, H, Memmesheimer, M., Geiss, H., Jacobs, H. J., Laube, M. and Ebel, A. (1990), Simulation of the Chernobyl radioactive cloud over Europe using the EURAD model, *Atmospheric Environment*, **24**, pp. 673-692.

15. Hesstvedt, E., Hov, Ø. and Isaksen, I. A. (1978), Quasi-steady-state approximations in air pollution modelling: comparison of two numerical schemes for oxidant prediction. *Internat. J. Chem. Kinetics*, **10**, pp. 971-994.

16. Hov, Ø., Zlatev, Z., Berkowicz, R., Eliassen, A. and Prahm, L. P. (1988), Comparison of numerical techniques for use in air pollution models with nonlinear chemical reactions , *Atmospheric Environment*, **23**, pp. 967-983.

17. Jakobsen, H. A., Jonson, J. E. and Berge, E. (1997), The EMEP regional scale multi-layer Eulerian model, in G. Geernaert, A. Walløe Hansen and Z. Zlatev, (eds), *Regional Modelling of Air Pollution in Europe* National Environmental Research Institute, Frederiksborgvej 399, P. O. 358, DK-4000 Roskilde, Denmark, pp. 63-79.

18. Lambert, J. D. (1991), *Numerical methods for ordinary differential equations*, Wiley, Chichester-New York-Brisbane-Toronto-Singapore.

19. Marchuk, G. I. (1985), *Mathematical modeling for the problem of the environment*, North-Holland, Amsterdam.

20. McRae, G. J., W. R., Goodin, W. R. and Seinfeld, J. H. (1982), Numerical solution of the atmospheric diffusion equations for chemically reacting flows, *Journal of Computational Physics*, **45**, pp. 1-42.

21. Molenkampf, C. R. (1968), Accuracy of finite-difference methods applied to the advection equation, *Journal of Applied Meteorology*, **7**, pp. 160-167.

22. Peters, L. K., Berkowitz, C. M., Carmichael, G. R., Easter, R. C., Fairweather, G., Ghan, S. J., Hales, J. M., Leung, L. R., Pennell, W. R., Potra, F. A., Saylor, R. D. and Tsang, T. T. (1995), The current state and future direction of Eulerian models in simulating tropospheric chemistry and transport of trace species: A review, *Atmospheric Environment*, **29**, pp. 189-222.

23. Skelboe, S. and Zlatev, Z. (1997), Using partitioning in the treatment of the chemical part of air pollution models, in L. Vulkov, J. Wasniewski and P.

Yalamov, (eds), *Numerical Analysis and Its Applications*, Springer, Berlin, pp. 66-77.

24. Stockwell, W. R., Middleton, P., Chang, J. S. and Tang, X. (1990), A second generation regional acid deposition model chemical mechanism for regional air quality modeling, *Journal of Geophysical Research*, **95**, pp. 16343-16367.

25. Venkatram, A., Karamchandani, P. K. and Misra, P. K. (1988), Testing a comprehensive acid deposition model, *Atmospheric Environment*, **22**, pp. 737-747.

26. Verwer, J. G. and Simpson, D. (1995), Explicit methods for stiff ODE's from atmospheric chemistry, *Appl. Numer. Math.*, **18**, pp. 413-430.

27. Zlatev, Z. (1984), Application of predictor-corrector schemes with several correctors in solving air pollution problems, *BIT*, **24**, pp. 700-715.

28. Zlatev, Z. (1988), Treatment of some mathematical models describing long-range transport of air pollutants on vector processors, *Parallel Computing*, **6**, pp. 87-98.

29. Zlatev, Z. (1990), Running large air pollution models on high speed computers, *Math. Comput. Modelling*, **14**, pp. 737-740.

30. Zlatev, Z. (1995), *Computer treatment of large air pollution models*, Kluwer Academic Publishers, Dordrecht-Boston-London.

31. Zlatev, Z., Berkowicz, R. and Prahm, L. P. (1984), Implementation of a variable stepsize variable formula method in the time-integration part of a code for treatment of long-range transport of air pollutants, *Journal of Computational Physics*, **55**, pp. 278-301.

32. Zlatev, Z., Christensen, J. and Eliassen, A., (1993), Studying high ozone concentrations by using the Danish Eulerian model, *Atmospheric Environment*, **27A**, pp. 845-865.

33. Zlatev, Z., Christensen, C. and Hov,Ø. (1992), An Eulerian air pollution model for Europe with nonlinear chemistry, *Journal of Atmospheric Chemistry*, **15**, pp. 1-37.

34. Zlatev, Z., Christensen, J., Moth J. and Wasniewski, J. (1991), Vectorizing codes for studying long-range transport of air pollutants, *Math. Comput. Modelling*, **15**, pp. 37-48.

35. Zlatev, Z., Dimov, I. and Georgiev K. (1994), Studying long-range transport of air pollutants, *Computational Science and Engineering*, **1**, pp. 45-52.

36. Zlatev, Z., Dimov, I. and Georgiev K. (1996), Three-dimensional version of the Danish Eulerian Model, *Zeitschrift für Angewandte Mathematik und Mechanik*, **76**, pp. 473-476.

37. Zlatev, Z., Fenger, J. and Mortensen, L. (1996), Relationships between emission sources and excess ozone concentrations, *Computers and Mathematics with Applications*, **32**, pp. 101-123.

PARALLEL NUMERICAL SIMULATION OF AIR POLLUTION IN SOUTHERN ITALY

G. BARONE, A. MURLI and A. RICCIO
University of Naples "Federico II"and CPS-CNR, Naples, Italy

P. D'AMBRA
Center for Research on Parallel Computing and Supercomputers, (CPS-CNR), Naples, Italy

D. DI SERAFINO
The Second University of Naples, Caserta and CPS-CNR, Naples, Italy

G. GIUNTA
Naval University of Naples and CPS-CNR, Naples, Italy

Abstract

In this paper we present the *Parallel Naples Airshed Model (PNAM)*, a parallel software package for the numerical simulation of air pollution episodes on urban scale domains, using MIMD distributed-memory machines. This is a first result of a research activity aimed at developing a system software to simulate air pollution episodes in the Campania Region, in Southern Italy. PNAM is based on an Eulerian model of the transport and photochemical transformations of air pollutants and uses a time-splitting approach, which separates the advection from the (coupled) diffusion and chemistry phenomena. The parallel implementation is based on grid partitioning and the use of dynamic load balancing techniques is currently under experiment. It is written in Fortran 90 and is based on the parallel Runtime System Library (RSL) to implement domain decomposition, data communication and dynamic load balancing. Numerical experiments have been carried out on a realistic test case, using an IBM SP, to evaluate the parallel performance of PNAM. Execution times, speedup and efficiency have been measured, obtaining a speedup of more than 7 on 12 processors. Preliminary results obtained with a dynamic load balancing strategy have been also analyzed, gaining suggestions for future work.

Z. Zlatev et al. (eds.), Large-Scale Computations in Air Pollution Modelling, 39–52.
© 1999 *Kluwer Academic Publishers. Printed in the Netherlands.*

1. Introduction

The numerical solution of *Air Quality Models* (AQMs) on urban, regional or global domains, is a large-scale Computational Science problem that requires high-performance computational resources. Massively parallel computers and distributed systems appear the most promising platforms to solve these highly computational-demanding problems. Therefore, the porting of existing sequential software on parallel machines and the development of efficient parallel software for AQMs is an active research field [4, 7, 8, 10, 9, 12, 15].

A research activity is carried out at the Center for Research on Parallel Computing and Supercomputers (CPS-CNR) for developing a parallel air quality software system for some areas of Southern Italy, including the Campania Region. The research team consists of a group of computational mathematicians and a group of environmental chemists. The activity involves the individuation of the chemical, physical and meteorological processes that are more significant for the model definition, and the selection and development of efficient and reliable numerical methods, parallel algorithms and software for solution of the problem [1, 2, 11]. In this context, a first version of the *Parallel Naples Airshed Model (PNAM)* has been produced, which is a parallel software for the numerical simulation of air pollution episodes on urban scale domains, developed for MIMD distributed-memory machines.

In this paper, we focus on the computational approach and the parallelism in PNAM. The physical/mathematical model and the numerical solution methods are described in Section 2. The parallel approach is discussed in Section 3, where the problem of load balancing is also addressed. Section 4 deals with software tools used in the development of PNAM. The case study used in numerical experiments is presented in Section 5 and parallel performance results are discussed in Section 6. An analysis of the computed air quality data and a comparison of available experimental data is not performed here; the interested reader is referred to [3]. A few concluding remarks are reported in Section 7.

2. Model Description and Computational Approach

In an Eulerian approach, 3D gas-phase AQMs are modelled by the following *atmospheric advection-diffusion-reaction equations*:

$$\frac{\partial c_i}{\partial t} = -\nabla \cdot (c_i \mathbf{u}) + \nabla \cdot (\mathbf{K} \cdot \nabla c_i) + R_i(t, \mathbf{c}) + D_i(t, \mathbf{c}) + E_i(t, \mathbf{c}) \quad i = 1, \dots, N,$$

$$(1)$$

where $c_i(t, \mathbf{x})$ is the mean concentration of chemical species i, $\mathbf{c}$ is the vector of all the chemical species $(c_1, c_2, \ldots, c_N)$, $\mathbf{u}(t, \mathbf{x})$ is the wind velocity, $\mathbf{K}(t, \mathbf{x})$ is the turbulent diffusivity tensor, and R_i, D_i and E_i are the reaction, dry deposition and emission terms, respectively. These equations are decoupled from the Navier-Stokes equations driving the meteorological phenomena, i.e. any dependence of the turbulent diffusivity, the wind velocity and the radiative properties of the atmosphere on the chemical processes is neglected.

The chemical kinetics and the dry deposition are modeled as in the CIT photechemical model [13]. Therefore, the chemical model is based on the LCC/SAPRC kinetic mechanism [17], that regulate 107 gas-phase reactions among 42 species, and the dry deposition term is based on a *three-resistance approach* to calculate the deposition velocity [18]. The emissions are modeled using data from the EC CorinAir archive; details on the methodology used to realize the emission basecase are given in [3].

The computational approach used to solve system (1) is based on a symmetric time-splitting technique that decouples advection and horizontal diffusion from vertical diffusion and chemistry in the following form:

$$c_i^{t+\Delta t} = L_T^{\Delta t/2} L_{DC}^{\Delta t} L_T^{\Delta t/2} c_i^t \quad i = 1, \ldots, N,$$

where L_T is the advection and horizontal diffusion operator and L_{DC} is the vertical diffusion and chemistry operator. The vertical turbulent diffusion and the chemical kinetics are coupled because they can have similar time scales on urban scale domains, especially for highly turbulent wind fields [16].

A three-dimensional rectangular grid is considered, which is uniformly spaced in the horizontal direction and has variable grid spacing in the vertical direction, so that the region close to the terrain can be resolved accurately. The semi-discretization of the above operators is performed using cell-centered finite-difference schemes.

A third-order upwind positive advection scheme with flux limiters, described in [14], is used for the advective transport operator in the horizontal direction and a first-order upwind in the vertical direction. We do not consider horizontal diffusion, since in most urban-scale flows it is negligibile with respect to the numerical dissipation. Extrapolation techniques are used to obtain boundary conditions. For lateral boundaries of inflow type constant extrapolation is applied, in order to avoid the arising of not real extrema in the solution; in case of outflow a second-order extrapolation is performed. At ground level the wind velocity is set to zero, and hence the concentrations are constant, and a constant extrapolation technique is also used at the top of the domain.

A two-stage second-order explicit Runge-Kutta method (Heun) with time-step stability constraints, taking into account the positivity requirements, is used to solve the system of ODEs arising from the semi-discretization of the advective transport.

The solution of the one-dimensional diffusion and chemistry operator accounts for a large part of the computational work, due to the high stiffness introduced by the chemical reactions. Special-purpose explicit algorithms and software have been largely used in AQMs (see, for example, [13, 24]) for efficiency reasons. In the last few years, the use of implicit methods in the solution of atmospheric chemical kinetic equations has been debated thoroughly [21, 22].

In PNAM, the coupled implicit solution of diffusion and chemistry is implemented, using a modified version of the general-purpose VODE package [2, 6]. The solver, which has been modified to take advantage of the sparsity of the Jacobian matrices arising in the application of the BDF methods used in VODE, has been successfully compared, in the solution of the vertical diffusion-reaction equations, with other special-purpose and general-purpose stiff ODE solvers on different atmospheric chemical models [2]. A *terrain-influenced z*-coordinate transformation, with a one-dimensional stretching function, has been applied to the diffusion operator, and a second-order finite-difference scheme has been used to discretize it. The ground-level boundary condition

$$-K_{zz}\frac{\partial c}{\partial z} = E - v_g c,$$

where K_{zz} is the vertical turbulent diffusivity and v_g is the dry deposition velocity, and the zero-gradient boundary condition at the top of the domain are discretized using first-order approximations. This does not affect significantly the accuracy, since the turbulent diffusion coefficient is very small at the boundaries. For details on the application of the modified version of VODE to the resulting stiff ODE system see [2].

3. Parallelism

The parallelism has been introduced in PNAM using a *grid partitioning* technique, i.e. dividing the computational grid into subgrids and assigning a subgrid to each of the processors available for computation. The computational three-dimensional grid is partitioned along the x and y directions, i.e. the computational domain is decomposed into vertical air columns, and is mapped onto a corresponding logical grid of processors in natural way.

This choice is mainly driven by the problem features. The coupled solution of vertical diffusion and chemistry, which is a very time-consuming task, introduces a global coupling among points in the vertical direction, while requires no interaction between points that lie in different vertical columns. Therefore, no data communication must be performed during the diffusion-chemistry steps. Moreover, this data decomposition allows to use available efficient and reliable sequential software, such as VODE, to solve the stiff ODE systems arising from the semi-discretization of the vertical diffusion and chemistry operators. On the other hand, a nearest-neighbour communication is required in the advection steps, to update the boundary points of each subgrid. Furthermore, a global communication is required each time the meteorological data are read, to compute the maximum advection step allowed by the CFL condition.

We note that a static data decomposition does not lead to load balanced executions. The main source of load imbalance is the stiffness of the chemical kinetics, which varies in time and space, and the different time steps used by the variable step stiff ODE solver in different vertical columns. An additional source of load imbalance can be the use of a non-homogeneous parallel machine, i.e. with processors of different computational power.

To address this problem, the application of dynamic load balancing techniques is currently under experiment. A comparison of different load balancing strategies applied to an atmospheric chemistry transport model can be found in [12].

As a first experience with load balancing, the strategy used in the *MM90* parallel regional weather model [20] has been implemented in PNAM. Following the classification in [12], this algorithm has a *global decision base*, i.e. the decision of remapping the computational grid onto the processor grid is based on a measure of the workload of all the processors, and a *local migration base*, i.e. units of computational work are moved among neighbouring processors. Let $T_{i,j}$ be a measure of the workload of the (i, j)-th vertical column, $T = \sum_{i,j} T_{i,j}$ the corresponding total workload, and m_p and n_p the number of rows and columns in the processor grid. A new mapping of the computational domain is determined by using the following algorithm:

1. divide the computational grid along the y-direction into m_p parts, each containing a number of cells such that their workload sum is as close as possible to T/m_p;

2. divide each part obtained from step 1 along the x-direction into n_p parts, each containing a number of cells such that their workload sum is as close as possible to $T/(m_p \times n_p)$.

The new mapping is applied if the following condition is satisfied:

$$\frac{T}{P \max_p T_p(\Pi^{new})} > \frac{T}{P \max_p T_p(\Pi^{old})} + \epsilon,$$

where $P = m_p \times n_p$ is the total number of processors, $T_p(\Pi^a)$ is the sum of the workload measures of the vertical columns allocated to processor p in the mapping a, and $0 < \epsilon < 1$ indicates the improvement in load balancing that is required from the new mapping.

The workload measure and the time when the check for a new mapping is performed affect the performance of the dynamic load balancing strategy. In the current implementation of PNAM, $T_{i,j}$ is obtained as

$$T_{i,j} = T_{i,j}^A + T_{i,j}^{DC},$$

where $T_{i,j}^A$ and $T_{i,j}^{DC}$ are the computational times of the column (i, j) for solving the horizontal advection and the vertical diffusion and chemistry respectively, accumulated between two successive checks for a new mapping. These times are measured by calling suitable timing routines in the PNAM code. Note that the horizontal coupling of points in the advection steps does not allow to effectively measure the computational time per single column. An estimate of $T_{i,j}^A$ has been obtained measuring the total advection time at the beginning of the simulation and dividing it by the total number of columns. This estimate is based on the assumption that approximately the same work is required to update a generic grid cell during the advection steps. However, this does not hold in general for grid cells along subgrid boundaries, that could wait for data coming from neighbouring processors. Therefore, our estimate of $T_{i,j}^A$ does not take into account idle times due to subgrids of different sizes. Finally, the check for a re-mapping is performed each time the meteorological and emission data are read, just before their input, to avoid redistributions of large amounts of data.

4. Software Tools

PNAM has been written in Fortran 90, using dynamic memory allocation, pointers, modules, and other features enhancing the software flexibility.

The parallel implementation is based on the *Runtime System Library (RSL)*, developed at Argonne National Laboratory, that is specifically tailored for efficient and straightforward implementations of finite-difference regular-grid applications on distributed-memory parallel computers [19].

RSL, developed to parallelize a well-known reagional weather model, the Penn State/NCAR Mesoscale Model, is a software package that provides routines for automatic two-dimensional domain decomposition, automatic local/global index translation, nearest-neighbor and global communication, dynamic load balancing and grid refinement. It is based on the MPI standard message-passing library, which ensures the portability of the package and of the application software based on it.

Domain definition, domain decomposition and data distribution become transparent to the user of RSL. The size of our rectangular domain has been specified by giving the number of horizontal rows and columns, and the RSL default algorithm has been used to divide the domain into subdomains of about the same dimensions. RSL high-level communication routines have been used to perform data exchanges required to update ghost-boundary values and data redistributions needed by the dynamic load balancing strategy.

5. Case Study

The modeling domain is centered at the Naples Gulf and extends about 384×384 Km^2, hence including the Campania Region and some areas of Lazio, Puglia and Basilicata Regions, surrounding Campania (Figure 1). A maximum height of 2360 m is considered. More details on the topography of this domain and on the related air pollution situation can be found in [3].

A uniform grid with 64×64 grid cells of dimension 6×6 Km^2 has been used in the horizontal directions, while the grid used in the vertical direction includes 16 variably spaced grid levels, with a spacing ranging from 10 m at the ground level to 640 m at the upper level.

A photochemical smog episode of 24 hours has been simulated, starting at 0:00 of July 26, 1995, and ending at 0:00 of July 27, 1995. From the available information, it results that during this day the ozone concentration reached 180 ppb, i.e. reached the ozone alarm level fixed by the Italian air quality standards. A startup time of 24 hours has been used to obtain reasonable initial data. A splitting interval of 15 minutes has been considered, and the meteorological data have been read every 60 minutes.

Topographic and land-use data have been obtained from the National Oceanographic and Atmospheric Administration (NOAA) and from the Italian Military Geographic Institute (IGM), respectively. Meteorological data have been generated by using the Topographic Vorticity-mode Mesoscale (TVM) model, which is a three-dimensional

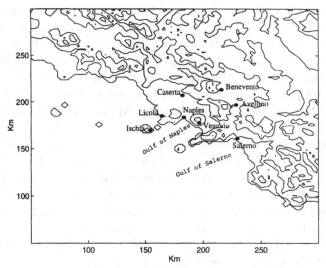

Figure 1. Topography of the modeling domain.

and non-hydrostatic atmospheric circulation model. Details on TVM can be found in [5, 23]. To initialize TVM, synoptic data from the European Centre for Medium Range Weather Forecasting (ECMWF) of Reading (UK) have been used. Finally, initial conditions have been estimated from the data of the monitoring network MARC (Monitoraggio Ambientale Regione Campania), operating by the Campania Regional Board. A few details on the processing of this data are given in [3].

6. Performance Results

Numerical experiments have been carried out on the test case described in Section 5, using an IBM SP machine at CPS-CNR. This machine has 12 Power2 Super Chip thin nodes (160 MHz), each with 512 MBytes of memory, which are connected via an SP switch, with a peak bi-directional bandwidth of 110 MBytes/sec. The available Fortan 90 compiler is XL Fortran, version 4.1. The RSL parallel environment runs on the top of the IBM proprietary version 2.3 of MPI.

   First experiments have been performed without using any dynamic load balancing strategy, i.e. using a static grid partitioning where each processor has the same number of grid cells. Table I shows the execution times, in seconds, the speedup and the efficiency of PNAM on 1, 2, 4, 8 and 12 processors. We note that, in going from 1 to 12 processors, the execution time reduces from 13.17 hours to 1.87 hours; the corresponding speedup is about 7, with a parallel efficiency of about 60%. Higher efficiency values are obtained on 2, 4 and 8 processors.

Table I. Execution times (sec.), speedup, and efficiency of PNAM without dynamic load balancing.

Procs	1	2	4	8	12
Time	47415	27385	14694	8232	6737
Speedup	—	1.73	3.23	5.76	7.04
Efficiency	—	0.87	0.81	0.72	0.59

An analysis of the computational load balance has been performed to see the effects of using a static data decomposition. Since the main source of load imbalance is the use of a variable time step ODE solvers in the solution of the chemical kinetics, the following measure of load imbalance has been considered:

$$LB_P = \frac{\max_p T_p^{DC} - \frac{1}{P} \sum_p T_p^{DC}}{\max_p T_p^{DC}},$$

where T_p^{DC} is the total execution time spent by processor p in the coupled solution of diffusion and chemistry, when P processors are used. The values of LB_P, reported in Table II, show that different processors spent different times in the diffusion/chemistry stages and that the relative difference between the measured maximum time and the ideal one increases with the number of processors.

Table II. Load imbalance with a static data decomposition.

Procs	2	4	8	12
LB_p	0.19	0.22	0.33	0.48

More insight on the load imbalance in the execution of PNAM can be gained looking at Figure 2. On the left, the execution times (sec.) required by the diffusion/chemistry stage, by the first and by the second advection stage (i.e. the advection in the first half and second half splitting interval, respectively) are reported for each splitting interval on one processor. On the right, the same times are reported for a run on four processors, for each of the processors used.[1] We see that the diffusion/chemistry workload is not evenly distributed among the four processors; in particular, around noon, when the diffusion/chemistry process is more significant, the time spent by processor 0 is about 20% of that spent by processor 1. The workload of the first advection stage is evenly distributed among the four processor, each performing about a quarter of the sequential workload.[2]. This is not the case of the second advection stage; in particular, one of the processors spends a time which is greater than the corresponding sequential time. This is due to the differences in the diffusion/chemistry times. The processor with the largest advection time has the smallest diffusion/chemistry time and hence must wait for the other processors to perform advection steps on its subgrid, since each of these steps requires the exchange of ghost-boundary data. Therefore, most of this advection time is actually idle time.

Preliminary experiments have been carried out using the dynamic load balancing strategy described in Section 3. The corresponding execution times (sec.), speedup and efficiency are reported in Table III. There are no sensible performance improvements over the execution with a static data decomposition (the increase in efficiency is at most 4%), and there is also a reduction of efficiency on eight processors.

Table III. Execution times (sec.), speedup, and efficiency of PNAM with dynamic load balancing.

Procs	1	2	4	8	12
Time	47415	25987	14387	8639	6479
Speedup	—	1.82	3.30	5.49	7.32
Efficiency	—	0.91	0.82	0.69	0.61

To analyze this behaviour, we consider the execution times, at each splitting interval, of the diffusion/chemistry and advection stages, that

---

[1] The execution on four processors has been chosen as an example. Analogous conclusions can be drawn analyzing the diffusion/chemistry and the advection times obtained on 2, 8 and 12 processors.

[2] The time spent by each of the four processors is actually a bit larger than the sequential time, due to the communications required by the parallel execution.

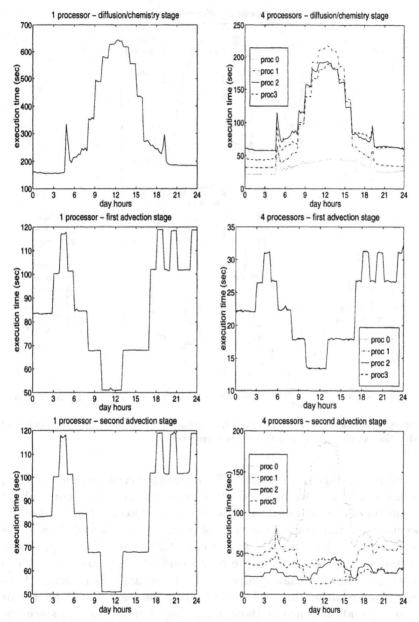

Figure 2. Static data decomposition. Execution times of diffusion/chemistry and advection at each splitting interval, on 1 and 4 processors (left and right, respectively).

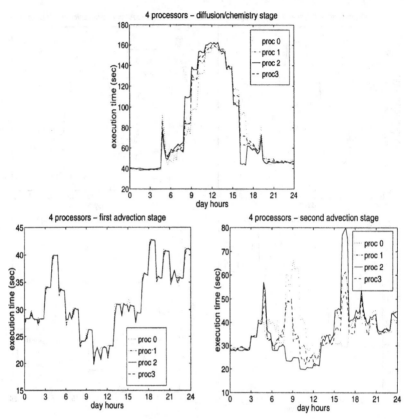

Figure 3. Dynamic load balancing. Execution times of diffusion/chemistry and advection at each splitting interval, on 4 processors.

are shown in Figure 3.[3] In the diffusion/chemistry stage, the workload is now much more balanced; the time spent by each processor is about a quarter of the time spent on one processor. Also the workload of the second advection stage appears to be much more balanced than in the previous case, but the sum of the execution times of each processor is still sensibly larger than the corresponding execution time on one processor, showing that significant idle times are still present. Moreover, the execution times of the first advection stage have a behaviour close to that obtained without balancing, but are larger, since the computational grid is not equally distributed among the processors. Some reasons for the above results can be found in the choice of the workload measure, which does not take into account idle times arising when different processors perform the advection steps on subgrids of

[3] The execution on four processors has been chosen only to perform comparisons with the plots shown in Figure 2.

different sizes (see Section 3). Moreover, different choices of the time between two successive remappings must also be considered.

7. Concluding Remarks

The results reported in the previous section show that PNAM achieves a satisfactory performance on a realistic test case, in going from 1 to 12 processors of an IBM SP, since the execution time decreases from 13.17 hours to 1.80 hours, with a speedup of more than 7 and an efficiency of about 0.60. However, the dynamic load balancing strategy currently implemented in PNAM does not give sensible improvements over the use of a static data decomposition. This can be attributed to the choice of the workload measure and probably also of the time when to check for a new mapping. Future work will be devoted to address these problems and to test PNAM on other parallel machines.

References

1. Barone, G., D'Ambra, P., di Serafino, D., Giunta, G. and Riccio, A. (1998), Numerical simulation of air pollution phenomena in the Neapolitan urban area (Southern Italy): First Experiences, in M. Griebel, O. P. Iliev, P. O. Margenov, P. S. Vassilevski (eds.), *Large-Scale Computations of Engineering and Environmental Problems*, Vieweg & Sons, Braunschweig, Wiesbaden, pp. 128–135.

2. Barone, G., D'Ambra, P., di Serafino, D., Giunta, G. and Riccio, A. (1998), A Comparison of Numerical Methods for Solving Diffusion-Reaction Equations in Air Quality Models, *CPS Tech. Rep. TR98-3*, Center for Research on Parallel Computing and Supercomputers (CPS-CNR), Naples, Italy.

3. Barone, G., D'Ambra, P., di Serafino, D., Giunta, G., Murli, A. and Riccio, A. (1998), Application of a Parallel Photochemical Air Quality Model to the Campania Region (Southern Italy), in B. Sportisse (ed.), *Proceedings of the International Conference on Air Pollution Modelling and Simulation APMS'98, Champs-sur-Marne (Paris), October 26-29, 1998*, ENPS and INRIA, Paris, pp. 57-70.

4. Blom, J. G., Lioen, W. M. and Verwer, J. G. (1998), HPCN and Air Quality Modelling, *CWI Tech. Rep. MAS-R9801*, Centrum voor Wiskunde en Informatica, Amsterdam, The Netherlands.

5. Bornstein, R., Thunis, P., Grossi, P. and Schayes, P. (1996), Topographic Vorticity-Mode Mesoscale$-\beta$ (TVM) Model. Part II: Evaluation, *J. Appl. Meteor.*, 35, pp. 1824-1834.

6. Brown, P. N., Byrne, G. D. and Hindmarsh, A. C. (1989), VODE: a Variable Coefficient ODE Solver, *SIAM J. Sci. Stat. Comput.*, 10, pp. 1038-1051.

7. Brown, J., Wasniewski, J. and Zlatev, Z. (1995), Running Air Pollution Models on Massively Parallel Machines, *Parallel Computing*, 21, pp. 971-991.

8. Bruegge, B., Riedel, E., Russell, A., Segall, E. and Steenkiste, P. (1995), Heterogeneous Distributed Environmental Modeling, *SIAM News*, 28.

52

9. Dabdub, D. and Manohar, R. (1997), Performance and Portability of an Air Quality Model., *Parallel Computing*, **23**, pp. 2187–2200.

10. Dabdub, D. and Seinfeld, J. H.(1996), Parallel Computation in Atmospheric Chemical Modeling, *Parallel Computing*, **22**, pp. 111–130.

11. D' Ambra, P., di Serafino, D., Giunta, G. and Riccio, A. (1997), Parallel Numerical Simulations of Reacting Flows in Air Quality Models, in P. Schiano, A. Ecer, J. Periaux, N. Safotuka (eds.), *Parallel Computational Fluid Dynamics, Algorithms and Results Using Advanced Computers*, Elsevier, Barking, England, pp. 116–123.

12. Elbern, H. (1997), Parallelization and Load Balancing of a Comprehensive Atmospheric Chemistry Transport Model, *Atmos. Environ.*, **31**, pp. 3561–3574.

13. Harley, R. A., Russell, A. G., McRae, G. J., Cass, G. R. and Seinfeld, J. H (1993), Photochemical Modeling of the Southern California Air Quality Study, *Environ. Sci. Technol.*, **27**, pp. 378–388.

14. Hundsdorfer, W., Koren, B., van Loon, M. and Verwer, J. G. (1995), A Positive finite-Difference Advection Scheme, *J. of Comput. Phys.*, **117**, pp. 35–46.

15. Kessler, Ch., Blom, J. G. and Verwer, J. G. (1995), Porting a 3D-Model for the Transport of Reactive Air Polluttants to the Parallel Machine T3D, *CWI Tech. Rep. NM-R9519*, Centrum voor Wiskunde en Informatica, Amsterdam, The Netherlands.

16. Kim, J. and Cho, S. Y. (1997), Computation Accuracy and Efficiency of the Time-Splitting Method in Solving Atmospheric Transport/Chemistry Equations, *Atmos. Environ.*, **31**, pp. 2215–2214.

17. Lurmann, F. W., Carter, W. P. L. and Coyner, L. A. (1987), A Surrogate Species Chemical Reaction Mechanism for Urban-Scale Air Quality Simulation Models. I. Adaptation of the Mechanism. II. Guidelines for Using the Mechanism, *Tech. Rep. EPA/600/3-87/014*, U.S. Environmental Protection Agency and Statewide Air Pollution Research Center.

18. McRae, G. J., Goodin, W. R. and Seinfeld, J. H. (1982), Mathematical Modeling of Photochemical Air Pollution, *Environmental Quality Laboratory 18*, California Institute of Technology, Pasadena, CA, USA.

19. Michalakes, J. (1994), A Parallel Runtime System Library for Regular Grid Finite Difference Models Using Multiple Nests, *Tech. Rep. ANL/MCS-TM-197*, Mathematics and Computer Science Division, Argonne National Laboratory, Argonne, IL, USA.

20. Michalakes, J. (1997), MM90: A Scalable Parallel Implementation of the Penn State/NCAR Mesoscale Model (MM5), *Parallel Computing*, **23**, pp. 2173–2186.

21. Sandu, A., Verwer, J. G., van Loon, M., Carmichael, G. R., Potra, F., Dabdub, D. and Seinfeld, J. H. (1997), Benchmarking Stiff ODE Solvers for Atmospheric Chemistry Problems I: Implicit versus Explicit, *Atmos. Environ.*, **31**, pp. 3151–3166.

22. Sandu, A., Verwer, J. G., van Loon, M., Carmichael, G. R., Potra, F., Dabdub, D. and Seinfeld, J. H. (1997), Benchmarking Stiff ODE Solvers for Atmospheric Chemistry Problems II: Rosenbrock Solvers, *Atmos. Environ.*, **31**, pp. 3459–3472.

23. Schayes, G., Thunis, P. and Bornstein, R. (1996), Topographic Vorticity-Mode Mesoscale—β (TVM) Model. Part I: Formulation, *J. Appl. Meteor.*, **35**, pp. 1815–1823.

24. Scheffe, R. D. and Morris, R. E. (1993), A Review of the Development and Application of the Urban Airshed Model *Atmos. Environ.*, **27B**, pp. 23-29.

REAL TIME PREDICTIONS OF TRANSPORT, DISPERSION AND DEPOSITION FROM NUCLEAR ACCIDENTS

A. BASTRUP-BIRK, J. BRANDT AND Z. ZLATEV
National Environmental Research Institute,
Department of Atmospheric Environment,
Frederiksborgvej 399, P.O. Box 358,
DK-4000 Roskilde, Denmark.
E-mail: lujbr@sun4.dmu.dk
Fax: +45 3630 1214

Abstract

A tracer model, DREAM, has been developed for studying transport, dispersion, and deposition of air pollution caused by a single but strong source. The model is based on a combination of a Lagrangian short-scale puff model and an Eulerian long-range transport model. The Lagrangian model is used in the area near the source to calculate the initial transport and dispersion of the release and the Eulerian model is used for long-range transport calculations in the whole model domain. The model has been run and validated against measurements from the two ETEX releases and from the Chernobyl accident. DREAM will be applied in the WEPTEL project (Innovative Weather Presentation on TELevision, EU-Esprit 22727) to calculate real time predictions of the transport, dispersion and deposition of radioactive material from an accidental release (as e.g. the Chernobyl accident). Such predictions are important tools for emergency planning and for the information of the public in case of a major accident. DREAM, and its application in the WEPTEL project will be described.

1. Introduction

The Chernobyl accident, April 25, 1986, emphasized the need for fast and reliable forecasts of transport, dispersion and deposition of radioactive air pollutants. Following the Chernobyl accident, many national and international activities have therefore been initiated to develop reliable models that can be used in connection with accidental releases. A comprehensive, high-resolution, 3-D tracer model has been developed.

Z. Žlatev et al. (eds.), *Large-Scale Computations in Air Pollution Modelling*, 53–62.
© 1999 *Kluwer Academic Publishers. Printed in the Netherlands.*

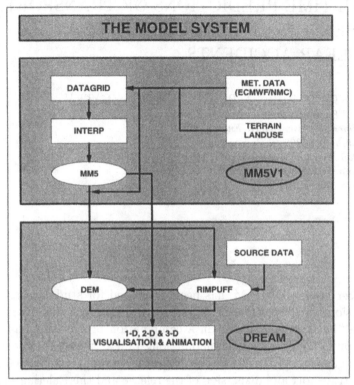

Figure 1. Schematic diagram showing main modules for MM5V1 and DREAM and flow chart for the whole model system. TERRAIN/LANDUSE, DATAGRID and INTERP modules are preprocessors for interpolation of terrestrial and meteorological input data. Emission data are given as input to the Lagrangian model. Advanced visualization and animation techniques have been implemented in the model system.

The combined model has been tested against numerical experiments, using a revised version of the Molenkamp-Crowley rotation test [6], and against measurements from the two ETEX releases in the autumn of 1994 [11], and the Chernobyl accident [2] [3] [4].

2. The DREAM

The tracer model is based on a combination of a Lagrangian short-scale puff model and an Eulerian long-range transport model. The combined 3-dimensional tracer model is called the DREAM (the Danish Rimpuff and Eulerian Accidental release Model) [2] [3] [4] [5] [6] [7]. The Lagrangian model is used to calculate the initial transport, dispersion and deposition of the plume in an area close to the source and the Eulerian model is used to calculate transport, dispersion and deposition on long range. By coupling a Lagrangian model with an Eulerian model, the idea is to gain the advantages of both kind of models [2]. For a more precise description of the mean meteorological fields, the meteorological meso-scale model MM5V1 [9] is used as a meteorological driver for the transport model. A schematic diagram for the whole model system is shown in figure 1.

In the Eulerian modelling framework, the advective transport, dispersion, emission, dry- and wet deposition and radioactive decay in the atmosphere is described by the equation (see, e.g. [12])

$$\frac{\partial C}{\partial t} = -u\frac{\partial C}{\partial x} - v\frac{\partial C}{\partial y} - \dot{\sigma}\frac{\partial C}{\partial \sigma} \tag{1}$$

$$+K_x\frac{\partial^2 C}{\partial x^2} + K_y\frac{\partial^2 C}{\partial y^2} + \frac{\partial}{\partial \sigma}\left(K_\sigma\frac{\partial C}{\partial \sigma}\right)$$

$$+E(x, y, \sigma, t)$$

$$-v_d C - \Lambda C - k_r C$$

where C is the tracer mixing ratio (the concentration c divided by the air density ρ), $u, v, \dot{\sigma}$ are the wind speed components in the x, y, σ directions, respectively, K_x, K_y and K_σ are dispersion coefficients, $E(x, y, \sigma, t)$ is the emission, v_d is the dry deposition velocity, (which in practice is applied as a lower boundary condition in the vertical dispersion), Λ is the scavenging coefficient for wet deposition and k_r is representing the radioactive decay. The vertical coordinate in the model is in σ-coordinates:

$$\sigma = \frac{P - P_t}{P_s - P_t} \tag{2}$$

where P is the pressure, P_s is the surface pressure and P_t is the pressure at the top of the model.

The Lagrangian model is a puff-model which simulates a continuous release changing in time by sequentially releasing a series of puffs which are advected and dispersed individually along trajectories. The puffs are Gaussian shaped in the horizontal direction but have been transformed into σ-coordinates in the vertical direction by using the hydrostatic approximation and the ideal gas law [2]. Assuming total reflection from the ground and from the mixing height, the contribution from the total number of puffs N to the mixing ratio $C_{x,y,\sigma}$ at x, y, σ is given by

$$C_{x,y,\sigma} = \sum_{i=1}^{N} \frac{M_i g (\sigma_{c_i} + \phi)}{(2\pi)^{3/2} \tilde{\sigma}_{xy_i} \tilde{\sigma}_{\sigma_i} \rho R \bar{T}_v} \tag{3}$$

$$\times \; exp \left(-\frac{1}{2} \left(\left(\frac{x - x_{c_i}}{\tilde{\sigma}_{xy_i}} \right)^2 + \left(\frac{y - y_{c_i}}{\tilde{\sigma}_{xy_i}} \right)^2 \right) \right)$$

$$\times \; \left(exp \left(-\frac{1}{2} \left(\frac{\sigma_{c_i} + \phi}{\tilde{\sigma}_{\sigma_i}} ln \left(\frac{\Psi_{c_i}}{\Psi} \right) \right)^2 \right) \right.$$

$$+ exp \left(-\frac{1}{2} \left(-\frac{\sigma_{c_i} + \phi}{\tilde{\sigma}_{\sigma_i}} ln \left(\Psi \Psi_{c_i} \right) \right)^2 \right)$$

$$+ exp \left. \left(-\frac{1}{2} \left(\frac{\sigma_{c_i} + \phi}{\tilde{\sigma}_{\sigma_i}} ln \left(\frac{\Psi \Psi_{c_i}}{2 \Psi_{H_{mix}}} \right) \right)^2 \right) \right)$$

where M_i is the mass of the individual puff i, g is the acceleration due to gravity, R is the gas constant for dry air, $\bar{T}_v$ is the mean virtual temperature, $x_{c_i}, y_{c_i}, \sigma_{c_i}$ is the location of the center of the puffs, $\tilde{\sigma}_{xy_i}, \tilde{\sigma}_{\sigma_i}$ are the horizontal and vertical standard deviations of the puffs, Ψ is given by

$$\Psi = \frac{\sigma(P_s - P_t) + P_t}{P_s} \tag{4}$$

If the pressure at the top of the model $P_t = 0$ then $\Psi = \sigma$. ϕ is given by

$$\phi = \frac{P_t}{P_s - P_t} \tag{5}$$

Many different parameterizations for vertical dispersion, dry and wet deposition, and the mixing height have been implemented and tested in the model and the sensibility to the accuracy of the meteorological input data [2] [3]. The coupling of the Lagrangian model and the Eulerian model is carried out using the puff-couplig [6].

It is difficult to treat the partial differential equation (PDE) (1) directly in the Eulerian model. The Eulerian model has therefore been splitted according to rules given in McRae [10] (see also [5]) into 3 sub-models containing:

1. three-dimensional advection, horizontal dispersion and emissions,

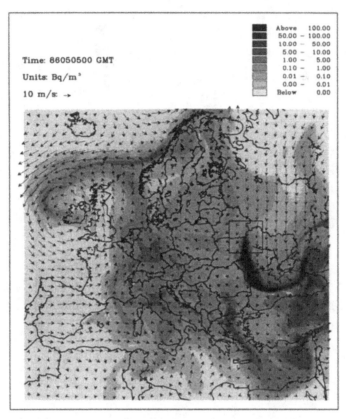

Figure 2. [131]I surface concentrations in Europe, 9 days after start of the Chernobyl accident.

2. vertical dispersion and dry deposition, and

3. wet deposition and radioactive decay.

Spatial discretization of submodels 1) and 2) is performed by using a finite element algorithm with piecewise linear independent functions; the chapeau functions [2]. This algorithm has been tested together with other schemes in [5]. Time integration of sub-model 1) is carried out using a predictor-corrector scheme with up to three correctors [13]. Especially the three dimensional advection is a very important process in transport modelling, so this submodel requires an accurate numerical method. Sub-model 2) is solved using the less expensive, but more stable implicit θ-method [8] [2]. The third sub-model is very simple and is solved directly. The combined model is applied on a polar stereographic projection, using an Arakawa A grid. The spatial grid-resolution in the Eulerian model is $25km \times 25km$ and in the Lagrangian

58

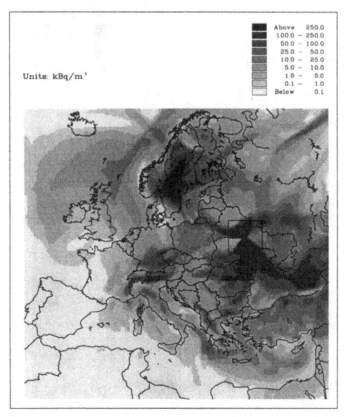

Figure 3. [131]I accumulated total (wet and dry) depositions in Europe, 12 days after start of the Chernobyl accident.

model $5km \times 5km$. The model is presently running on a CRAY C92A with an efficiency with respect to peak performance of 50%.

3. Some examples of model results

In Fig. 2 and Fig. 3 some examples of model results calculated by using DREAM are given. Fig. 2 shows the surface concentrations of ^{131}I, 9 days after start of the release of the Chernobyl accident. The release period was 10 days. In Fig. 3 the accumulated total (wet and dry) deposition of ^{131}I is given. The calculated concentrations and total deposition have been compared with measurement showing good agreement [2].

4. Implementation in WEPTEL

The DREAM model together with the Danish Eulerian model (DEM) [1] [13] is presently being applied in WEPTEL (Innovative Weather Presentation on TELevision). One of the aims of this project is to be able to calculate real time forecasts of air pollution episodes as e.g. major nuclear accidents or episodes of high levels of air pollution as e.g. O_3 for the use in TV broadcasting. The meteorological data and the air pollution data are calculated at different locations and transferred to the TV company, see Fig 4. Fig 5 shows a list of the participants in this project and the www-site for further information.

An example of a photochemical episode over the Netherlands in July, 1994, is given in Fig. 6. The figure shows the O_3 concentrations calculated by using DEM. According to the new EU regulatives, the population should be informed when the O_3 concentrations exceed 90 ppb. Furthermore the population should be warned when the concentrations exceed 120 ppb. As seen in Fig. 6 the O_3 concentrations exceeded both critical levels considerable in this period.

Table I. Turn around time for a 3 days forecast.

Task	Turn around time
Meteorological model	20 min
Data transfer (met to air)	2-20 min
Air pollution model	< 20 min
Data transfer (met+air to TV)	< 1 min
Total (approximately)	**< 1 hour**

When calculating air pollution forecasts in real time, all major processes (including data transfers and data processing) must be carried out fast. In table I the estimated turn around time in the WEPTEL project, using the present computers, for these processes are given. A certain time is necessary for data transfer of 3-D meteorological data.

5. Conclusions and plans for future research

The DREAM model has been described in this paper. the model was developed for handling transport, dispersion and deposition, both on short- and long-range. The model has been validated against measurements from ETEX (the European Tracer EXperiment) and the

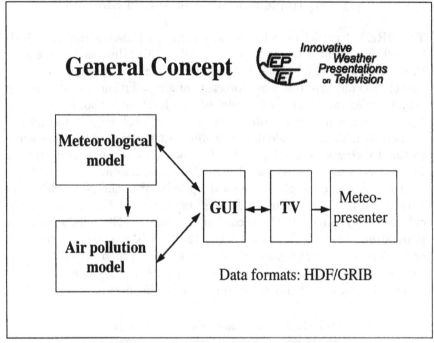

Figure 4. General concept and data flowchart for the WEPTEL system

Figure 5. The participants in WEPTEL

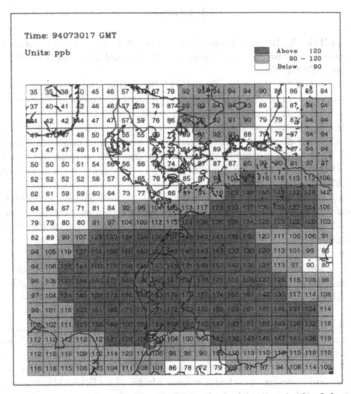

Time: 94073017 GMT

Units: ppb

Above 120
90 – 120
Below 90

Figure 6. O₃ concentraions in The Netherlands during an episode, July 30, 1994.

Chernobyl accident with good results (see, [2]). DREAM together with the Danish Eulerian model are presently being applied in a real time forecast system, WEPTEL.

Some of the plans for future developments of the model is implementation on super computers with parallel architecture in order to improve real time performance. Improvements of the parameterizations which are used in the model will also be carried out.

6. Acknowledgments

Meteorological data have been kindly provided by EURAD, University of Cologne, Germany, and EMEP (European Monitoring and Evaluation Programme). The Danish Research Academy, Denmark, the Danish Science Research council, Denmark, EU-Esprit (projects WEPTEL, Esprit-22727 and EUROAIR, Esprit-24618), and NATO (ENVIR.CRG.930449, OUTS.CRG.960312) are greatly acknowledged for financial support.

References

1. Bastrup-Birk, A., Brandt, J. and Zlatev, Z. (1997) Studying cumulative ozone exposures in Europe during a 7-year period, *Journal of Geophysical Research.* **102**, pp. 23917-23935.

2. Brandt, J (1998) *Modelling Transport, Dispersion and Deposition of Passive Tracers from Accidental Releases*, PhD Thesis, National Environmental Research Institute, Roskilde, Denmark.

3. Brandt, J., Bastrup-Birk, A., Christensen, J. H., Mikkelsen, T., Thykier-Nielsen, S. and Zlatev, Z. (1998), Testing the importance of accurate meteorological input fields and parameterizations in atmospheric transport modelling using DREAM - validation against ETEX-1. *Atmospheric Environment*, **32**, pp. 4167-4186.

4. Brandt, J., Bastrup-Birk, A., Christensen, J. H. and Zlatev, Z. (1998), Numerical modelling of transport, dispersion and deposition - validation against ETEX-1, ETEX-2 and Chernobyl, submitted to *Environmental Modelling and Software*.

5. Brandt, J., Dimov, I, Georgiev, K., Wasniewski, J. and Zlatev, Z. (1996), Coupling the advection and the chemical parts of large air pollution models, in J. Wasniewski, J. Dongarra, K. Madsen and D. Olesen (eds.), *Applied Parallel Computing, Industrial Computation and Optimization*, Springer, Berlin, pp. 65-76.

6. Brandt, J., Mikkelsen, T., Thykier-Nielsen, S. Zlatev, Z. (1996), Using a combination of two models in tracer simulations, *Mathematical and Computer Modelling.* **23**, pp. 99-115.

7. Brandt, J., Mikkelsen, T., Thykier-Nielsen, S. Zlatev, Z. (1996), The Danish Rimpuff and Eulerian Accidental release Model (The DREAM), *Physics and Chemistry of the Earth*, **21**, pp. 441-444.

8. Christensen, J. H. (1995), *Transport of Air Pollution in the Troposphere to the Arctic*, PhD Thesis. National Environmental Research Institute, Roskilde, Denmark.

9. Grell, G. A., Dudhia, J. and Stauffer, D. R. (1995), A Description of the Fifth-Generation Penn State/NCAR Mesoscale Model (MM5), NCAR/TN-398+STR, NCAR Technical Note, Mesoscale and Microscale Meteorology Division, National Center for Atmospheric Research, Boulder, Colorado.

10. McRae, G. J., Goodin, W. R. and Seinfeld, J. H. (1982), Numerical solution of the atmospheric diffusion equation for chemically reacting flows. i *Journal of Computational Physics*, **45**, pp. 1-42.

11. Nodop, K. (Ed.) (1997), *ETEX symposium on long-range atmospheric transport, modelverification and emergency response, 13-16 May, 1997, Vienna (Austria)*, Office for Official Publications of the European Communities, Luxembourg.

12. Pudykiewicz, J. (1989), Simulation of the Chernobyl dispersion with a 3-D hemispheric tracer model. *Tellus*, **41B**, pp. 391-412.

13. Zlatev, Z. (1995), *Computer Treatment of Large Air Pollution Models*. Kluwer Academic Publishers, Dordrecht-Boston-London.

MIXING HEIGHT IN COASTAL AREAS - EXPERIMENTAL RESULTS AND MODELLING, A REVIEW

E. BATCHVAROVA

National Institute of Meteorology and Hydrology
66 Blvd. "Tzarigradsko chaussee", 1784 Sofia, Bulgaria

1. Introduction

The mixing height (MH), or the height above the surface where pollutants become vertically dispersed, is related to the atmospheric boundary layer (ABL) in homogeneous regions. This layer is characterised by turbulence generated mechanically (from the friction drag of the air moving across the rough and rigid surface of the Earth) and thermally (from the convection process or "bubbling-up" of air parcels from the heated surface). The height of the atmospheric boundary layer (or the depth of surface related influence) is changing in time for a given location depending on the strength of the surface generated mixing and reaches about 1 to 2 km during the day due to heating of the surface by the Sun and about 100 m during the night in conditions of radiative cooling. The boundary layer height changes also depending on the geographical, terrain and vegetation characteristics of a given area [1].

When the Earth surface is non-homogeneous and especially in regions with abrupt change of surface characteristics (sea or lake coast, field-forest, rural-urban) , the air flow starts to transform over the new surface and (within the ABL) an internal boundary layer (IBL) starts to develop. This internal boundary layer grows downwind with the distance from the change of surface and far from it (if no other changes of surface occur) may reach the height of the ABL [2]. The internal boundary layer at coastal regions is one of the reasons for air pollution episodes, as its depth is only about 500-700 m at a distance of 15 km from the shore. In such regions the mixing height should be connected with the height of the internal boundary layer.

Z. Zlatev et al. (eds.), Large-Scale Computations in Air Pollution Modelling, 63–68.
© *1999 Kluwer Academic Publishers. Printed in the Netherlands.*

2. Mixing height over uniform surface (ABL)

The depth of the ABL depends from the strength of the turbulence generated at the ground surface. But to define the top of the ABL is not simple, as it is dependent on the choice of the turbulence characteristic to consider (model or measure) [1], [2], [3]. The level at which changes in the vertical profile of a given parameter (wind, temperature, humidity, aerosol concentrations, heat flux, turbulent kinetic energy (TKE) dissipation rate, etc.) is observed, is connected to the ABL depth. These levels for the different parameters generally do not coincide. Consequently, when discussing the mixing height or the atmospheric boundary layer depth, in order to be precise it is always necessary to clarify which parameter is considered. From another side, different measuring techniques reflect different parameters - free and tethered balloon sounding represent profiles of mean wind, temperature and humidity; sodars provide profiles of the vertical wind velocity and its variances and a number of other characteristics [4]. Furthermore, when comparing modelled and measured ABL depths, one should be aware of the above described differences.

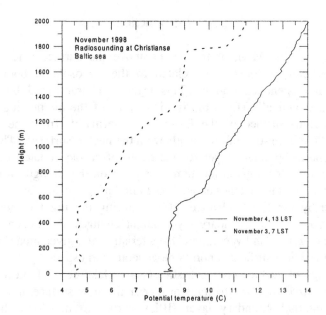

Figure 1. Potential temperature profiles from Christiansø

Most often the depth of the atmospheric boundary layer during daytime (convective conditions) is connected with the base of the elevated inversion [1] in the vertical profile of the potential temperature (the air temperature corrected for the pressure changes with height), about 550m in Figure 1 (solid line). When including the importance of the entrainment zone between the ABL and the free atmosphere (FA) aloft for the mixing processes, the depth of the ABL can be defined by the level within the entrainment zone (or capping inversion) at which conservation is obtained for the total heat deficit between the ABL and the FA [5], about 620 m on the solid line in Figure 1. The depth of the entrainment zone is typically 30 % of the convective ABL height [6], but can become a bigger fraction of it under specific conditions (like the morning growth of the convective ABL) [7]. Thus, if considering the mixing that occur within the entrainment zone, the mixing height chosen in dispersion models should be 15 to 30 % higher than the base of the elevated inversion.

It can be pointed out that typically over sea the ABL is lower than over land, but rather deep ABL can form when cold air is advected over the warmer sea. Another feature typical for the boundary layer over sea is that a convective boundary layer with depth comparable to daytime conditions can develop during night and early morning, Figure 1, dashed line.

Over land when radiative cooling takes place in night and early morning a stable layer is formed within which mixing is due to mechanical turbulence [8]. The height of this layer is typically 100 m to 200 m and is not trivial to identify and model [3]. As consequence the mixing height over land differs significantly from day to night and in the morning there is a graduate process of its growth [9]. The depth of the convective ABL over land reaches in middle latitudes 1500 m to 2000 m , 1400 m for example, for the Wangara Experiment day 33 after Stull [2].

In tropical and equatorial regions moisture content in the air is significant and is rather important in controlling stability and energy balance in the atmosphere, thus leading to increase of the depth of the daytime ABL [10]. The usual (midlatitude) parametrizations of both the convective and the stable ABL are not applicable.

3. Mixing height in regions of transition from one to another type of surface (IBL)

In coastal regions, when the air flows from one to another type of surface, a transformation starts over the new surface and the so called internal boundary layer (IBL) is formed downwind [2], [5]. The top of the IBL is also marked by an inversion layer in the potential temperature profile and with a sharp decay of atmospheric turbulence. The air aloft is stable stratified (conditions of suppression of turbulence), while the air within the IBL is well mixed by

66

convective and mechanical turbulence. Thus, in coastal regions the mixing height is connected to the height of the internal boundary layer which is significantly less deep than the ABL (that could exist at the place if no change of surface was present) and air pollution episodes are typical for big coastal cities [11], [12], [13], [14], [15], [17].

In Figure 2 the measured (by tethered and free balloon radiosoundinhs and lidar observations) IBL height at different places is presented. The data used are published by Batchvarova and Gryning [11] for Athens, Kerman et al. [16] for Nanticoke, Gryning and Batchvarova [5] for Øresund, Raynor et al. [17] for Brookhaven, Steyn and Oke [14] for Vancouver in 1978, Chen and Oke [16] for Vancouver 1986.

The meteorological, synoptic and orographic conditions for the different experiments differ significantly. Still, the IBL height at 10 km downwind inland is 300 to 400 m typically and becomes higher under conditions of low winds and intensive heating of the surface (Vancouver, 1986).

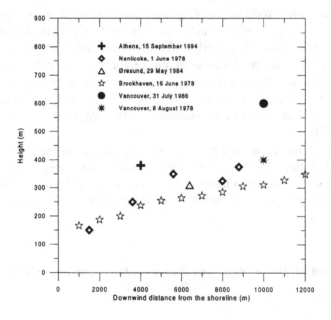

Figure 2. Internal boundary layer height measured at different sites

A number of models of one-dimensional models of the internal boundary layer height and their validation against experimental data are presented by

Stunder and SethuRaman [19] and Gryning and Batchvarova [5]. The variation of the internal boundary layer height due to irregular coastline, changing wind direction and spatial variation of the surface parameters (or landuse variation) require 3-D modelling as performed by Gryning and Batchvarova [15] and Batchvarova *et al.* [20].

4. Concluding remarks

The main goal of this review was to point out that the dispersion parameter mixing height is connected to the complex structure of the atmospheric boundary layer. Different values and different parametrizations for the mixing height are to be used in dispersion modelling for homogeneous conditions over land and over water; in middle and in equatorial latitudes; at coastal regions.

Acknowledgements

Thanks are due to Sven-Erik Gryning from Risø National Laboratory for supplying the radiosounding data in Figure 1.

References

1. Oke, T. R. (1987), *Boundary Layer Climates*, Methuen, London.
2. Beyrich, F., Gryning, S.-E., Joffre, S., Rasmussen, A., Seibert, P. and Tercier, P. (1996), On the Determination of the Mixing Height - A Review, in J. G. Kretzschmar and G. Cosemans (Eds), *4th Workshop on Harmonization within Atmospheric Dispersion Modelling for Regulatory Purposes, 6-9 May 1996, Oostende, Belgium, E&M.RA9603,* VITO, Belgium, pp. 155-162.
3. Stull, R. B. (1988), *Introduction to Boundary Layer Meteorology,* Kluwer Academic Publishers, Dordrecht.
4. Beyrich, F. (1997), Mixing Height Determination Using Remote Sensing Systems - General Remarks, in S.-E. Gryning, F. Beyrich and E. Batchvarova (Eds), *The Determination of the Mixing Height - Current Progress and Problems, EURASAP Workshop Proceedings, 1-3 October 1997, Risø National Laboratory,* Risø-R-997(EN), pp. 71-74.
5. Gryning, S.-E. and Batchvarova, E. (1990), Analytical model for the growth of the coastal internal boundary layer during onshore flow, *Q. J. R. Meteorol. Soc.,* **116,** pp. 187-203.
6. Gryning, S.-E. and Batchvarova, E. (1994), Parametrization of the Depth of the Entrainment Zone above the Dytime Mixed Layer, *Q. J. R. Meteorol. Soc.,* **120,** pp. 47-58.

7. Batchvarova, E. and Gryning, S.-E. (1994), An Applied Model for the Daytime Mixed Layer and the Entrainment Zone, *Boundary-layer Meteorol.*, **71**, pp. 311-323.

8. Gryning, S.-E. (1998), Some aspects of dispersion in the stratified atmospheric boundary layer over homogeneous terrain, *Boundary-layer Meteorol.*, in print.

9. Batchvarova, E. and Gryning, S.-E. (1991), Applied model for the growth of the daytime mixed layer, *Boundary-Layer Meteorol.*, **56**, 261-274.

10. Samah, A. A. (1997), Modelling the development of mixing height in near equatorial region, in S.-E. Gryning, F. Beyrich and E. Batchvarova (Eds), *The Determination of the Mixing Height - Current Progress and Problems, EURASAP Workshop Proceedings, 1-3 October 1997, Risø National Laboratory*, Risø-R-997(EN), pp. 31-34.

11. Batchvarova, E. and Gryning, S.-E. (1998), Wind Climatology, Atmospheric Turbulence and Internal Boundary Layer Development in Athens during the MEDCAPHOT-TRACE Experiment, *Atmos. Environ.*, **32**, pp. 2055-2069.

12. Kallos, G., Kossemenos, P. and Pielke, R. (1993), Synoptic and Mesoscale Circulation Assocoated with Air Pollution Episodes in Athens, Greece, *Boundary-Layer Meteorol.*, **62**, pp. 163-184.

13. Lalas, D. P., Tombru-Tsella, M., Petrakis, M., Assimakopoulos, D. N. and Helmis, C. G. (1983), Sea-Breeze Circulation and Photochemical Pollution in Athens, Greece, *Atmos. Environ.*, **17**, pp. 1621-1632.

14. Steyn, D. and Oke, T. R. (1982), The Depth of the Daytime Mixed Layer at Two Coastal Sites: A Model and its Validation, *Boundary-Layer Meteorol.*, **24**, pp. 151-180.

15. Gryning, S.-E. and Batchvarova, E. (1996), A Model for the Height of the Internal Boundary Layer over an Area with an Irregular Coastline, *Boundary-Layer Meteorol.*, **78**, pp. 405-413.

16. Kerman, B. R., Mickle, R. E., Portelli, R. V., Trivett, N. B. and Misra, P. K. (1982), The Nanticoke Shoreline Diffusion Experiment - II. Internal Boundary Layer Structure, *Atmos. Environ.*, **16**, pp. 423-437.

17. Raynor, G. S., SethuRaman, S. and Brown, R. M. (1979), Formation and Characteristics of Coastal Internal Boundary Layers During Onshore Flow, *Boundary-Layer Meteorol.*, **16**, pp. 487-514.

18. Chen, J. M. and Oke, T. R. (1994), Mixed-layer Heat Advection and Entrainment during the Sea Breeze, *Boundary-Layer Meteorol.*, **68**, pp. 139-158.

19. Stunder, M. and SethuRaman, S. (1985), A Comparative Evaluation of the Coastal Internal Boundary-Layer Height Equations, *Boundary-Layer Meteorol.*, **32**, pp. 177-204.

20. Batchvarova, E., Cai, X.-M., Gryning, S.-E. and Steyn, D. (1998), Modelling Internal-Boundary-Layer Development in a region with Complex Coastline, *Boundary-layer Meteorol.*, in print.

ON SOME ADAPTIVE NUMERICAL METHODS
FOR LINEAR ADVECTION EQUATIONS

R. BOCHORISHVILI
Vekua Institute of Applied Mathematics
Tbilisi State University, University str. 2, 380040, Georgia
e-mail: rdboch@viam.hepi.edu.ge

1. Introduction

The most important feature of adaptive numerical methods can be characterized as their capability to concentrate computational efforts where they are most needed. As usual, in adaptive methods redistribution of the efforts happens automatically in dynamic manner with account of properties of derived in the process of computations approximate solution of the problem under consideration. Such flexibility allow adaptive methods to achieve comparable accuracy and at the same time substantially to reduce computational costs in comparison with non adaptive ones. Example of such successful adaptive method is given in [7] . Namely, for calculation of steady state flow over aircraft's bow shape in 2D comparable numerical results are achieved by adaptive numerical method with 700-1000 equations and by non adaptive one with 100 000 equations. Because of such significant advantage development of good adaptive numerical methods is and will be important despite increasing the power of computers. The following 3 basic approaches for achieving adaptation are often used:

(i) The moving grid method allows the user to control spacing of the computational mesh while keeping total amount of nodes unchanged. This is achieved by moving nodal points in order to concentrate them in regions where solution of a problem under consideration may have complex character and higher numerical errors are expected.

(ii) The *p−method* allows us to control automatically the truncation errors. The accuracy of numerical scheme may vary from node to node in order to apply more accurate schemes in regions where this is needed.

(iii) Mesh refinement and coarsening allows us to control amount of nodal points and their distribution in computational domain. The mesh is refined by adding new points in sub-domains where finer resolution is necessary for ensuring the desired accuracy. The points are reduced, if the accuracy needed can be achieved on a coarse mesh.

Z. Zlatev et al. (eds.), Large-Scale Computations in Air Pollution Modelling, 69–78.
© *1999 Kluwer Academic Publishers. Printed in the Netherlands.*

In all these approaches adaptation is usually made by estimating the expected error by applying the approximate solution obtained in the process of the solution of the problem under consideration. There exist several particular mechanisms for implementing the above general approaches as adaptive numerical schemes, see [7]-[3]. Three new adaptive approaches for linear advection equations in one space dimension

$$\frac{\partial u}{\partial t} + a(t, x)\frac{\partial u}{\partial x} = 0. \tag{1}$$

are discussed in this paper. These adaptive approaches are designed in such a way that they can be easily incorporated into existing numerical schemes for linear advection equation and corresponding software of solution. The possibilities of extension of these approaches to higher space dimensions and other equations are discussed as well. The aim is to develop adaptive methods suitable for using in air pollution models.

2. Moving grid scheme

For constructing moving grid schemes we follow the approach developed in [3]. The resulting moving grid scheme can be formulated in 3 steps: (i) rewrite the equation in equivalent form for the cells of the grid, (ii) discretize it by taking into account the stability requirement and (iii) update the adaptation conception. This is a general algorithm, that can be used together with many discretization methods. An application is given below. We develop a semidiscrete moving grid scheme for (1) based on a Courant-Isaacson-Rees type discretization in space, see also the split coefficient matrix method [5]. The results of each steps of the algorithm are:

(i) equation for the cell of moving grid:

$$\frac{d}{dt}(h_j(t)u_j(t)) - (x'_{j+\frac{1}{2}}(t)u_{j+\frac{1}{2}}(t) - x'_{j-\frac{1}{2}}(t)u_{j-\frac{1}{2}}(t)) +$$

$$+ \int_{x_{j-\frac{1}{2}}(t)}^{x_{j+\frac{1}{2}}(t)} a(t, x)\frac{\partial u}{\partial x}dx = 0, \tag{2}$$

where $u_j(t)$ is approximate solution of (1) in time moment t and nodal point $x_j(t)$, $h_j(t) = x_{j+\frac{1}{2}}(t) - x_{j-\frac{1}{2}}(t)$, $x_{j\pm\frac{1}{2}}(t)$ are nodal points of cell interfaces in time moment t,

$x_{j+\frac{1}{2}}(t) = \frac{1}{2}(x_j(t) + x_{j+1}(t))$, $u_{j\pm\frac{1}{2}}(t)$ are values of approximate solutions at cell interfaces, but primes stand for denoting derivatives;

(ii) semidiscrete scheme:

$$\frac{d}{dt}(h_j(t)\, u_j(t)) + a^+(t, x_j(t))(u_j(t) - u_{j-1}(t)) +$$

$$+ a^-(t, x_j(t))(u_{j+1}(t) - u_j(t)) - (x'^{+}_{j+\frac{1}{2}}(t) u_{j+1}(t) + x'^{-}_{j+\frac{1}{2}}(t) u_j(t)) +$$

$$+ (x'^{+}_{j-\frac{1}{2}}(t)\, u_j(t) + x'^{-}_{j-\frac{1}{2}}\, u_{j-1}(t)) = 0,$$

(3)

where

$$a^\pm(t, x_j(t)) = \frac{1}{2}(a(t, x_j(t)) \pm |a(t, x_j(t)|),$$

$$x'_{j\pm\frac{1}{2}}(t) = \frac{1}{2}\left(x'_{j\pm\frac{1}{2}}(t) \pm \left|x'_{j\pm\frac{1}{2}}(t)\right|\right);$$

(iii) adaptation conception

$$\frac{dx_{j+\frac{1}{2}}}{dt} = \frac{(x_{j+\frac{3}{2}} - x_{j+\frac{1}{2}})e_{j+1}(t) - (x_{j+\frac{1}{2}} - x_{j-\frac{1}{2}})e_j(t)}{h},$$

(4)

where h is an average discretization step in space, $h_j(0) = h$, $h_j(t) \succ 0$, and it is somehow measured error in $x_j(t)$, e.g. truncation error calculated by using finite difference derivatives of numerical solution.

Thus, in order to accomplish construction of moving grid scheme it remains only to update corresponding initial and boundary conditions for semidiscrete equations (3) and for equations (4) defining the motion of nodal points $x_j(t)$.

Clearly, in order to have completely discrete version of the above moving grid scheme discretization of (3),(4) with respect to time variable should be performed.

Analysis of equations (4)show that nodal points maintain the same ordering as at was in starting moment. So if $x_{j+\frac{1}{2}}(0) \succ x_{j-\frac{1}{2}}(0)$ then $x_{j+\frac{1}{2}}(t) \succ x_{j-\frac{1}{2}}(t)$ for any $t \succ 0$. Another important property of equations (4) is the providing of concentrating of nodal points in those sub domains where higher errors expected:

$x_{j+\frac{1}{2}}$ moves toward $x_{j+\frac{3}{2}}$, if $(x_{j+\frac{3}{2}} - x_{j+\frac{1}{2}})e_{j+1} \succ (x_{j+\frac{1}{2}} - x_{j-\frac{1}{2}})e_j$;

$x_{j+\frac{1}{2}}$ moves toward $x_{j-\frac{1}{2}}$, if $(x_{j+\frac{3}{2}} - x_{j+\frac{1}{2}})e_{j+1} \prec (x_{j+\frac{1}{2}} - x_{j-\frac{1}{2}})e_j$;

So equations (4) regulate the distance between nodal points in such a way that errors are redistributed uniformly between cells of the grid thus achieving "maximum" accuracy with given amount of nodes .

Theorem 1 justifies the developed moving grid method for equation (1) with initial condition

$$u(0, x) = u_0(x), x \in \Re, u_0 \in C^2. \tag{5}$$

THEOREM 1. *Approximate solutions constructed by means of semidiscrete moving grid scheme (3),(4) with initial data consistent with initial condition (5) converge to exact solution of (1),(5) as $h \to 0$.*

A similar result is valid for two-level implicit schemes which are obtained from (3)-(4) and are unconditionally stable. The semidiscrete moving grid analogue of (1) differs from the similar one on a fixed grid by last two terms in left hand side of (3). Thus, the extension of the above moving grid method to other partial differential equations with first order derivative in time is easy. The generalizations to higher space dimensions is straightforward as well, if a suitable discretization method for space derivatives is selected.

3. Method of filtering

In the context of modern control theory the word filtering refers to deducing the best in some sense estimate of measured quantities that are corrupted by noise. Below the same word is used in the same sense but with respect to numerical solutions of nonlinear advection equations. The idea of our method of filtering is the following: evidently numerical solutions of the equations contain errors; so it is reasonable to apply suitable filtering in order to improve the accuracy of the numerical solution. The most suitable for our purposes approach seems to be the one that has a common origin with the dynamic programming approach, [2]. Let $u_j(t)$ be numerical solution in x_j, $v_j(t)$ be corresponding filtered one that is looking for. Then the error in x_j on time interval (t_{n-1}, t_n) in the sense of L^2 can be defined as

$$\int_{t_{n-1}}^{t_n} |u_j(t) - v_j(t)|^2 \, dt. \tag{6}$$

Let us introduce the function $c_j^n(t, z)$ value of the error in nodal point x_j and time moment $t, t \succ t_{n-1}$,

$$c_j^n(t, z) = \int_{t_{n-1}}^{t} |u_j(\zeta) - z(\zeta)|^2 \, d\zeta, z \in \Re. \tag{7}$$

Then the best filtering is defined as

$$c_j(t, z^*) = \min c_j^n(t, z), \quad u_j(t) = z^*. \tag{8}$$

Following Bellman's approach and modifying it correspondingly for c_j^n the following approximate model is suggested:

$$\frac{\theta}{\theta t} c_j^n + k_j^n \frac{\theta}{\theta z} c_j^n = (u_j - z)^2, \tag{9}$$

where

$$k_j^n = a(t_n, x_j) \left(\frac{\theta u}{\theta x} \right)_{j,apr}^n,$$

but via $\left(\frac{\theta u}{\theta x} \right)_{j,apr}^n$ some approximation of $\frac{\theta u}{\theta x}$ in (t_n, x_j) is defined, e.g. the quantity calculated by using finite differences of the numerical solution $\{ u_j^n \}$.

It is easy to verify that, if $c_0(z)$ is initial value of $t = t_{n-1}$ then solution of corresponding Cauchy problem for equation (9) is given by the formulae:

$$c_j^n(t, z) = c_0(z - k_j^n(t - t_{n-1})) + \int_{t_{n-1}}^{t} (u_j(\zeta) - z + k_j^n(t - z)) d\zeta. \tag{10}$$

One can derive simple formulae for calculation of filtered solution by using (6)-(8) and (10). The filtered solution is the best one in the sense defined above, if the stencil of $\left(\frac{\theta u}{\theta x} \right)_{j,apr}^n$ does not contain nodal point (t_n, x_j). If this is not so then the approximate model (9) and formulae (10) are valid only for those z that are sufficiently close to u_j^n. For the remaining z another model is needed which can be constructed in a similar way with different k_j^n. The accuracy of the developed approach coincides with the accuracy of the approximate model. A variety of approximate models can be developed.

The method of filtering can also be used as a correction step for improving the accuracy of the numerical solution. The method can be combined with any numerical method, but a further theoretical investigation is needed to prove the convergence and to select numerical schemes which are most efficient when used together with filtering. The method of filtering can be applied in conjunction with adaptation. It can be used for some selected nodal points where higher errors are observed in numerical solution. The same approaches as those in the previous paragraph can be used to compute an a posteriori error estimation. An attractive features of the developed method of filtering is its simplicity. Its extension to the multidimensional case and for other equations with first order derivatives in time including systems of equation is straightforward.

4. Wavelet like mesh refinement.

In the last years wavelets have been used in different fields because of their good properties, [6]. One such good property is their adaptability. If sequence of spaces is given

$$V_k \subset V_{k+1} \subset V_{k+2} \subset \ldots \subset L^2$$

then each space $V_{k+m;}$, $m \in N$, can be represented as

$$V_{k+m} = V_k \oplus W_k \oplus \ldots \oplus W_{k+m-1}, \tag{11}$$

where W_k is the complement of V_k to V_{k+1}. In multiresolution analysis the bases in spaces V_k are defined by using the so called scaling function but wavelets create corresponding basis for the complement spaces W_k. Refinement by factor 2 is made when going from V_k to V_{k+1}, i.e. roughly speaking, if $x_{k_j} = 2^{-k}j$ are nodal points corresponding to V_k then $x_{kj} = 2^{-k-1}j$ are the same ones for V_{k+1}. Using standard notations and denoting via φ_{kj} basis functions in V_k and via ψ_{kj} corresponding vawelets then any function $f \in V_{k+m}$ can be represented as

$$f(x) = \sum_j s_{kj}\varphi_{kj}(x) + \sum_j d_{kj}\psi_{kj}(x) + .. + \sum_j d_{k+m-1,j}\psi_{k+m-1,j}(x).$$
$$\tag{12}$$

By using some ideas from wavelets theory and related topics, wavelet like mesh refinement is developed below. The approach is suitable for mesh adaptation in variety of numerical methods, e.g. finite differences or finite elements. This ideas can be applied to: (i) mesh refinement (one step of the refinement is defined by analogy with formulae (11) as space plus complement to next; this allows us to use the representation (12) and to work only in complement spaces after the refinement) and (ii) thresholding (that is usually used for compression of information, for example, in image processing; in our approach we do thresholding of initial data and the operator as well). We use the same notations as in (12) but $\psi_{kj}(x)$ and $\varphi_{kj}(x)$ are arbitrary basis functions, e.g. corresponding with finite elements. Thus, in fact we do not use wavelets and that is why have wavelet like mesh refinement.

The key procedures of the approach are related to the approximation of initial data, thresholding of the operator and the composition of the solution.

Approximation of initial data is made as follows:

$$u_0(x) = u_{0k}(x) + w_{0k}(x) + w_{0k+1}(x) + \ldots, \tag{13}$$

$$u_{0k}(x) = \sum_j u_0(x_{kj})\varphi_{kj}(x), \tag{14}$$

$$w_{0k}(x) = \sum_j d_{kj}\psi_{kj}(x), \quad d_{kj} = u_0(x_{k+1j}) - u_{0k}(x_{k+1j}), \tag{15}$$

$$w_{0k+1}(x) = \sum_j d_{k+1j}\psi_{k+1j}(x), \quad d_{k+1j} = \\ u_0(x_{k+2j}) - (v_{0k}(x_{k+2j}) + w_{0k}(x_{k+2j})), \tag{16}$$

Formulas (15),(16) are valid, if $\psi_{kj}(x_{ki}) = \delta_{ki}$ but development of similar formulas are possible in other cases as well.

Thresholding of initial data is made in standard way by means of assigning zero to all coefficients d_{kj} less then given threshold ε. As a result we have the approximation of initial data with given accuracy and compression of it in selected basis.

For thresholding of the operator let define:

$$\Omega_{kj} = \{x_{kj+l} : d_{kj} \neq 0, \ d_{kj+1} \neq 0, \ d_{kj-1} = 0, \ 0 \prec l \prec m, \ d_{kj+m} = 0\},$$

$$\Omega_{kj}^0 = \{x_{kj+l} : d_{kj} = 0, \ d_{kj+1} = 0, \ d_{kj-1} \neq 0, \ 0 \prec l \prec m, \ d_{kj+m} \neq 0\}.$$

It is expected that Ω_{kj}^0 will contain more elements then Ω_{kj}, especially, on finer scales. This fact allows to make thresholding of the operator and to arrive at natural domain decomposition leading to several initial value problems but on smaller domains. In particular thresholding procedure can be formulated into following two steps: (i) add $[\tau a_{\max}/h]$ nodes to each of sets Ω_{kj} and denote by Ω_{kj}^+; the result and (ii) update the homogeneous boundary conditions in a suitable way for a corresponding numerical scheme on Ω_{kj}^+.

The discretization step in time is denoted by τ in (i), while the notation h_k is used for the step corresponding to W_k. Formulas similar to (13),(14) are used for the composition of the solution. Finally the algorithm for wavelet like mesh refinement is formulated as follows:

1. select bases, define the numerical scheme and input data;
2. approximate the initial data;
3. threshold the data for current time level;
4. threshold the operator;

5. solve the problems on Ω_{kj}^+;

6. compose the solution;

7. go to 3 until next time level is less then $T_{\max}$.

The first step in the algorithm above is necessary to ensure consistency of accuracies of the numerical scheme and bases used. The numerical scheme is not specified in step 5. Thus, the above algorithm can be combined with many finite difference and finite element schemes.

Theorem 2 provides a theoretical justification of the developed approach for a two-level implicit scheme with Courant-Isaacson-Rees type approximation in space and for the so called hat basis functions.

THEOREM 2. *The approximate solution constructed by the wavelet like mesh refinement algorithm using hat basis functions and the implicit scheme with Courant-Isaacson-Rees type approximation as solver converges to solution of (1)-(5) as discretization step tends to zero, if the ratio between discretization steps in time and in space is kept constant and the threshold ε is defined by*

$\varepsilon =$ *(discretization step on finest scale)*(discretization step in time).*

The rate of the convergence is the the same as the convergence rate obtained by using the discretization step on finest scale.

It should be emphasized that the extension of this approach to higher dimensions is straightforward and the extension is possible not only for rectangular but also for triangular meshes. Wavelet like mesh refinement for other hyperbolic and parabolic equations can be performed in a similar way but a special investigation of the thresholding of operators is needed in each particular case.

5. Conclusions.

In order to make conclusions and to compare the developed approaches, the following criteria are used:

(i) accuracy of the approach;

(ii) capabilities to be incorporated into existing numerical methods and software;

(iii) capabilities of speeding up computations;

(iv) capabilities to be extended for air pollution models.

Numerical calculations of test problems were performed using each of the developed approaches. As basis solver in each case two level implicit scheme weighted with $\sigma = 1/2$ was applied. For approximation of space derivatives, as it was mentioned above, split-coefficients approach was used. Calculations were performed with 64 and 128 nodes in space.

Numerical results show that using of adaptive approaches leads to improving of accuracy of approximate solution in comparison with the one calculated by means of corresponding non adaptive scheme. By the moving grid scheme good results are obtained, if solution of the problem has complex behavior in some isolated sub domains e.g. like shock wave. If there are several sub domains where moving grid scheme should ensure concentrating of nodes then numerical results are not so good. The method of filtering gives very good numerical results in some nodal points while in other ones there are no improvements of accuracy. We do not know exactly why this happens and suppose that further investigations are needed, especially, with the purpose of selection of suitable approximate models for filtering. The wavelet like mesh refinement gave good results for all test problems.

One of the attractive features of each of developed approaches is their capability relatively easily to be incorporated into existing numerical methods and software. From this point of view the method of filtering is the most attractive one among the 3 approaches presented above. The wavelet like mesh refinement is also very suitable because it can be incorporated without any need to change the basic solvers. Most efforts are required by the developed moving grid schemes because this requires modification of existing basic solvers by introduction of additional terms to take into account the motion of nodes.1 Furthermore the resulting equations that should be solved are nonlinear despite the original problem, see for example the semidiscrete equations (3),(4).

In moving grid schemes speeding up of the computations is achieved at the expense of optimal distribution of nodes. The results from test problems have shown that this is possible, if solution of the problem under consideration has only isolated complex behavior. Preliminary theoretical estimation indicates that the method of filtering can speed up the computations very much because the approximate models, low or higher order ones, require comparable amount of arithmetic operations. Wavelet like mesh refinement can undoubtedly speed up computations. It incorporates in itself best features of moving grid and multigrid methods, thus, enabling us to locate complex from computational point of view regions and to use finer meshes only in those sub domains where necessary. This method allows us to obtain good approximations if there are different scales and to have the sub-problems of smaller size to be solved numerically because of thresholding.

The solution of advanced air pollution models requires huge amount of computational resources and this need is increased very quickly when the number of grid points is increased [4]. Thus for air pollution modelling the numerical schemes are most suitable if they allow us to achieve the desired accuracy at minimum amount of nodal points. Taking into consideration that air pollution models represent large systems of partial differential equations, the method of filtering and wavelet like mesh refinement seem to be more suitable for this purpose. Note that in case of moving grids for systems of equations difficulties arise in selection of efficient adaptation concepts. Thus, we arrive at following conclusions:

(i) moving grid schemes are not suitable for air pollution problems,

(ii) the method of filtering is very attractive from many points of view but needs further investigation/modification,

(iii) the wavelet like mesh refinement can give good results and speed up computations several times as well.

References

1. Anderson, D. A., Tannenhill, J. C. and Pletcher, R. H. (1990) *Computational Fluid Mechanics and Heat Transfer*, Publishers MYR, Moscow.
2. Bellman, R. (1960), *Dynamic Programming*, Inostrannaja Literatura, Moscow.
3. Bochorishvili, R. (1997), Three point semidiscrete moving grid schemes for nonlinear scalar conservation laws in one space dimension, *Reports of enlarged Sessions of the Seminar of Vekua Institute of Applied Mathematics*, 13 pp. 1-5.
4. Brandt, J., Dimov, I., Georgiev, K., Uria, I., Zlatev, Z. (1997), Numerical algorithms for the three-dimensional version of the Danish Eulerian Model, in G.Geernaert, A.Walloe Hansen and Z.Zlatev (eds), *Regional Modelling of Air Pollution in Europe*, National Environmental Research Institute, Roskilde, Denmark, pp. 249-262.
5. Chakravarthy, S. R., Anderson, D. A. and Salas, M. D. (1980), The split-coefficient matrix method for hyperbolic systems of gas dynamics equations, *AIAA*, 1 Paper 80-0268, Pasadena, California.
6. Daubechies, I. (1995) Wavelets and other phase space localization methods, in *Proceedings of the International Congress of Mathematicians, Zurich, Switzerland 1994*, Birkhauser Verlag, Basel, Switzerland, pp.57-74.
7. Research Directions in Computational Mechanics (1991), National Academy Press, Washington D.C.
8. Thompson, J. F. (1985), Grid generation techniques in computational fluid dynamics, *Aero/cosmicheskaya Techniqua*, 3, 141-171 (translation from AIAA Journal, Vol.22, No.11, 1984).

A PARALLEL ITERATIVE SCHEME FOR SOLVING THE CONVECTION DIFFUSION EQUATION ON DISTRIBUTED MEMORY PROCESSORS

L. A. BOUKAS and N. M. MISSIRLIS
Department of Informatics, University of Athens,
Panepistimiopolis 15710,Athens, Greece. Email: nmis@di.uoa.gr

Abstract

In this paper we introduce an iterative scheme for solving the Convection Diffusion equation. In fact, the method can be applied to any Partial Differential Equation which uses iterative schemes for their numerical solution. Parallelism is introduced by decoupling the mesh points with the use of red–black ordering for the 5–point stencil. The optimum set of values for the parameters involved, when the Jacobi iteration operator possesses real or imaginary eigenvalues, is determined. The performance of the method is illustrated by its application to the numerical solution of the convection diffusion equation. It is found that the proposed method is significantly more efficient than local SOR when the absolute value of the smallest eigenvalue of the Jacobi operator is larger than unity. Finally, the parallel implementation of the method is discussed and results are presented for distributed memory processors with a mesh topology.

1. Introduction

The model problem considered here is that of solving the second order convection diffusion equation

$$\Delta u - f(x,y)\frac{\partial u}{\partial x} - g(x,y)\frac{\partial u}{\partial y} = 0 \tag{1}$$

on a domain $\Omega = \{(x,y)\}|0 \leq x \leq \ell_1, 0 \leq y \leq \ell_2\}$, where $u = u(x,y)$ is prescribed on the boundary $\partial\Omega$. The discretization of (1) on a rectangular grid $M_1 \times M_2 = N$ unknowns within Ω leads to

$$u_{ij} = \ell_{ij}u_{i-1,j} + r_{ij}u_{i+1,j} + t_{ij}u_{i,j+1} + b_{ij}u_{i,j-1}, \tag{2}$$
$$i = 1, 2, \ldots, M_1 \ , \ j = 1, 2, \ldots, M_2$$

Z. Zlatev et al. (eds.), Large-Scale Computations in Air Pollution Modelling, 79–88.
© *1999 Kluwer Academic Publishers. Printed in the Netherlands.*

with

$$\ell_{ij} = \frac{k^2}{2(k^2 + h^2)}\left(1 + \frac{1}{2}hf_{ij}\right), \quad r_{ij} = \frac{k^2}{2(k^2 + h^2)}\left(1 - \frac{1}{2}hf_{ij}\right)$$

$$\tag{3}$$

$$t_{ij} = \frac{h^2}{2(k^2 + h^2)}\left(1 - \frac{1}{2}kg_{ij}\right), \quad b_{ij} = \frac{h^2}{2(k^2 + h^2)}\left(1 + \frac{1}{2}kg_{ij}\right),$$

where $h = \ell_1/(M_1 + 1)$, $k = \ell_2/(M_2 + 1)$, $f_{ij} = f(ih, jk)$ and $g_{ij} = g(ih, jk)$. For a particular ordering of the grid points, (2) yield a large, sparse linear system of equations of order N of the form

$$Ax = b. \tag{4}$$

We consider iterative methods for the numerical solution of (4) on a mesh connected processor array. In order to use the Successive Over-relaxation (SOR) method [15], [16] we have to color the grid points red-black [1], [14] so that sets of points of the same color can be computed in parallel. However, the parameter ω which accelerates the rate of convergence of SOR is computed adaptively in terms of $u^{(n+1)}$ and $u^{(n)}$ when the method is applied to (2) [10]. This computation requires global communication between the processors for each iteration which means $O(\sqrt{N})$ communication complexity for a mesh connected $\sqrt{N} \times \sqrt{N}$ array of processors. As the number of iterations for SOR is proportional to $O(\sqrt{N})$, it follows that its execution time is $O(N)$.

In order to avoid global communication, which increases the execution time, we use local relaxation methods [2], [8], [9].

2. The local Modified SOR method

The local SOR method was introduced by Ehrlich [8], [9] and Botta and Veldman [2] in an attempt to further increase the rate of convergence of SOR. The idea is based on letting the relaxation factor ω vary from equation to equation. This means that each equation of (2) has its own relaxation parameter denoted by ω_{ij}. Kuo et. al [11] combined local SOR with red black ordering and showed that is suitable for parallel implementation on mesh connected processor arrays. In the present study we generalize local SOR by letting two different sets of parameters $\omega_{ij}, \omega'_{ij}$ to be used for the red $(i + j$ even) and black $(i + j$ odd) points, respectively. An application of our method to (2) can be written as follows:

$$u_{ij}^{(n+1)} = (1 - \omega_{ij})u_{ij}^{(n)} + \omega_{ij}J_{ij}u_{ij}^{(n)}, \quad \text{red points} \tag{5}$$

$$u_{ij}^{(n+1)} = (1 - \omega_{ij}')u_{ij}^{(n)} + \omega_{ij}'J_{ij}u_{ij}^{(n+1)}, \quad \text{black points} \qquad (6)$$

where

$$J_{ij}u_{ij}^{(n)} = l_{ij}u_{i-1,j}^{(n)} + r_{ij}u_{i+1,j}^{(n)} + t_{ij}u_{i,j+1}^{(n)} + b_{ij}u_{i,j-1}^{(n)} \qquad (7)$$

and J_{ij} is called the local Jacobi operator. The parameters $\omega_{ij}, \omega_{ij}'$ are called local relaxation parameters and (5)-(8) will be referred to as the local Modified SOR (MSOR) method. Note that if $\omega_{ij} = \omega_{ij}'$, then (5), (6) reduce to the local SOR method studied in [11]. Moreover, if $\omega_{ij} = \omega_{ij}' = \omega$ (5), (6) degenerate into the classical SOR method with red black ordering.

Next, we apply the local Fourier analysis to find an eigenvalue relationship between the eigenvalues of the local MSOR iteration operator and J_{ij} the local Jacobi operator. Writting (5), (6) in terms of the error vector $e^{(n)} = u_{ij}^{(n)} - u_{ij}$, we have

$$e_R^{(n+1)} = (1 - \omega_{ij})e_R^{(n)} + \omega_{ij}J_{ij}e_B^{(n)} \quad \text{for red points } (i+j \text{ even}), \quad (8)$$

$$e_B^{(n+1)} = (1 - \omega_{ij}')e_B^{(n)} + \omega_{ij}'J_{ij}e_R^{(n+1)} \quad \text{for even points } (i+j \text{ odd})(9)$$

where e_R and e_B represent the errors at the red and black points, respectively. By eliminating $e_R^{(n+1)}$, (9) is written as

$$e_B^{(n+1)} = \omega_{ij}'(1 - \omega_{ij})J_{ij}e_R^{(n)} + (1 - \omega_{ij}' + \omega_{ij}\omega_{ij}'J_{ij}^2)e_B^{(n)}. \qquad (10)$$

Equations (8) and (10) can be written as

$$\begin{pmatrix} e_R^{(n+1)} \\ e_B^{(n+1)} \end{pmatrix} = \mathcal{L}_{\omega_{ij},\omega_{ij}'}(J_{ij}) \begin{pmatrix} e_R^{(n)} \\ e_B^{(n)} \end{pmatrix}, \qquad (11)$$

where

$$\mathcal{L}_{\omega_{ij},\omega_{ij}'}(J_{ij}) = \begin{bmatrix} 1 - \omega_{ij} & \omega_{ij}J_{ij} \\ \omega_{ij}'(1 - \omega_{ij})J_{ij} & 1 - \omega_{ij}' + \omega_{ij}\omega_{ij}'J_{ij}^2 \end{bmatrix} \qquad (12)$$

is called the local MSOR iteration operator. By assuming that an eigenfunction of $\mathcal{L}_{\omega_{ij},\omega_{ij}'}(J_{ij})$ possesses the form $(c_1 e^{i(k_1 x + k_2 y)}, c_2 e^{i(k_1 x + k_2 y)})^T$ and that the corresponding eigenvalue is λ_{ij}, or

$$\mathcal{L}_{\omega_{ij},\omega_{ij}'}(J_{ij}) \begin{pmatrix} c_1 e^{i(k_1 x + k_2 y)} \\ c_2 e^{i(k_1 x + k_2 y)} \end{pmatrix} = \lambda_{ij} \begin{pmatrix} c_1 e^{i(k_1 x + k_2 y)} \\ c_2 e^{i(k_1 x + k_2 y)} \end{pmatrix}, \qquad (13)$$

it follows that

$$\mathcal{L}_{\omega_{ij},\omega_{ij}'}(\mu_{ij}) \begin{pmatrix} c_1 \\ c_2 \end{pmatrix} = \lambda_{ij} \begin{pmatrix} c_1 \\ c_2 \end{pmatrix} \qquad (14)$$

since

$$J_{ij}e^{i(k_1x+k_2y)} = \mu_{ij}(k_1,k_2)e^{i(k_1x+k_2y)}, \tag{15}$$

where

$$\mu_{ij}(k_1,k_2) = l_{ij}e^{-ik_1h} + r_{ij}e^{ik_1h} + t_{ij}e^{ik_2k} + b_{ij}e^{-ik_2k}. \tag{16}$$

Furthermore, from (14) it follows that

$$det\,(\mathcal{L}_{\omega_{ij},\omega'_{ij}}(\mu_{ij}) - \lambda_{ij}) = 0 \tag{17}$$

or, because of (12)

$$\lambda_{ij}^2 - (2 - \omega_{ij} - \omega'_{ij} + \omega_{ij}\omega'_{ij}\mu_{ij}^2) + (1-\omega_{ij})(1-\omega'_{ij}) = 0. \tag{18}$$

The above relationship between the eigenvalues of the local MSOR iteration operator and the local Jacobi operator is a generalization of the SOR relationship [16]. Indeed, if $\omega_{ij} = \omega'_{ij} = \omega$ and $\mu = \mu_{ij}$, then (18) degenerates into

$$\lambda^2 - (\omega^2\mu^2 - 2(\omega-1))\lambda + (1-\omega)^2 = 0$$

which is the well known Young's relationship. The above approach is known as the local Fourier analysis [6], [7] and holds for linear constant coefficient PDEs on an infinite or on a rectangular domain with Dirichlet or periodic boundary conditions. Under these assumptions the eigenfunctions of the Jacobi iteration matrix are sinusoidal functions. If the PDE under consideration has space-varying coefficients with general boundary conditions, the sinusoidal functions are not eigenfunctions. In this case Fourier analysis holds approximately. The case for $\mu_{ij}(k_1,k_2)$ real was studied in [11], [3] and [4] for the local SOR and local MSOR methods, respectively. Next, we consider the case where the eigenvalues of the Jacobi iteration operator are imaginary and proceed to select ω_{ij} and ω'_{ij} such that to minimize the spectral radius of the local MSOR iteration operator $\mathcal{L}_{\omega_{ij},\omega'_{ij}}(J_{ij})$. The whole analysis is presented in [5]. Here we state our results which are the following.

THEOREM 1. *Let the local Jacobi operator J_{ij} have imaginary eigenvalues $i\mu_{ij}$ (μ_{ij} real). Then $S(\mathcal{L}_{\omega_{ij},\omega'_{ij}})$ the spectral radius of the local MSOR iteration operator for the 5–point stencil with red–black ordering is minimised at*

$$\omega_{ij} = \widehat{\omega}_{1,i,j} = \frac{2}{1 - \overline{\mu}_{ij}\underline{\mu}_{ij} + \sqrt{(1+\overline{\mu}_{ij}^2)(1+\underline{\mu}_{ij}^2)}}$$

and $\qquad\qquad\qquad\qquad\qquad\qquad\qquad\qquad\qquad\qquad\qquad$ (19)

$$\omega'_{ij} = \widehat{\omega}_{2,i,j} = \frac{2}{1 + \overline{\mu}_{ij}\underline{\mu}_{ij} + \sqrt{(1+\overline{\mu}_{ij}^2)(1+\underline{\mu}_{ij}^2)}}$$

and its corresponding value is given by

$$S(\mathcal{L}_{\widehat{\omega}_{1,i,j},\widehat{\omega}_{2,i,j}}) = \sqrt{(1 - \widehat{\omega}_{1,i,j})(1 - \widehat{\omega}_{2,i,j})} = \frac{\sqrt{1 + \overline{\mu}_{ij}^2} - \sqrt{1 + \underline{\mu}_{ij}^2}}{\sqrt{1 + \overline{\mu}_{ij}^2} + \sqrt{1 + \underline{\mu}_{ij}^2}}.$$

(20)

The following corollaries are easily proved from the previous theorem.

COROLLARY 2. *Under the hypothesis of the previous theorem, if* $\underline{\mu}_{ij} = 0$, *then* $\widehat{\omega}_{1,i,j} = \widehat{\omega}_{2,i,j}$ *and*

$$S(\mathcal{L}_{\omega_{ij}^*,\omega_{ij}^*}) = 1 - \omega_{ij}^* = \frac{\sqrt{1 + \overline{\mu}_{ij}^2} - 1}{\sqrt{1 + \overline{\mu}_{ij}^2} + 1},$$

(21)

where

$$\omega_{ij}^* = \frac{2}{1 + \sqrt{1 + \overline{\mu}_{ij}^2}}.$$

(22)

COROLLARY 3. *Under the hypothesis of the previous theorem, if* $\underline{\mu}_{ij} = \overline{\mu}_{ij}$, *then*

$$\omega_{ij} = 1 \quad, \quad \omega'_{ij} = \frac{1}{1 + \overline{\mu}_{ij}^2}$$

and

$$S(\mathcal{L}_{\omega_{ij},\omega'_{ij}}) = 0.$$

Note that for linear constant coefficient PDEs defined on a unit square with Dirichlet boundary conditions, the spectral radii of all local Jacobi operators are the same which in turn means that $\omega_{ij} = \omega$ and $\omega'_{ij} = \omega'$ indicating that local MSOR degenerates into the classical MSOR method. A simple comparison of (20) and (22) reveals that

$$S(\mathcal{L}_{\widehat{\omega}_{1,i,j},\widehat{\omega}_{2,i,j}}) < S(\mathcal{L}_{\omega_{ij}^*,\omega_{ij}^*}).$$

Moreover, a study of the behaviour of $S(\mathcal{L}_{\widehat{\omega}_{1,i,j},\widehat{\omega}_{2,i,j}})$ with respect to $\underline{\mu}_{ij}^2$ reveals that $S(\mathcal{L}_{\widehat{\omega}_{1,i,j},\widehat{\omega}_{2,i,j}})$ is a decreasing function of $\underline{\mu}_{ij}^2$. In conclusion, local MSOR will attain at least the convergence rate of local SOR, whereas its convergence rate will increase as $\underline{\mu}_{ij}$ increases.

3. Numerical results

If the PDE has space-varying coefficients with Dirichlet boundary conditions, then for the 5–point stencil the quantities $\overline{\mu}_{ij}$ and $\underline{\mu}_{ij}$ are determined by [8]

$$\mu_{ij} = 2 \left(\sqrt{\ell_{ij} r_{ij}} \cos \frac{k_1 \pi}{M_1 + 1} + \sqrt{t_{ij} b_{ij}} \cos \frac{k_2 \pi}{M_2 + 1} \right), \qquad (23)$$

where $k_1 = 1, 2, \ldots, M_1$ and $k_2 = 1, 2, \ldots, M_2$. From (23) we find

$$\overline{\mu}_{ij} = 2 \left(\sqrt{\ell_{ij} r_{ij}} \cos \pi h + \sqrt{t_{ij} b_{ij}} \cos \pi k \right) \qquad (24)$$

and

$$\underline{\mu}_{ij} = 2 \left(\sqrt{\ell_{ij} r_{ij}} \cos \frac{\pi(1-h)}{2} + \sqrt{t_{ij} b_{ij}} \cos \frac{\pi(1-k)}{2} \right). \qquad (25)$$

From (23) we see that the eigenvalues μ_{ij} may be real, imaginary or complex. In the sequel we distinguish these cases.

Case 1: μ_{ij} are real. This case applies when $\ell_{ij} r_{ij} \geq 0$ and $t_{ij} b_{ij} \geq 0$. The optimum values of the local MSOR parameters are given by [3]

$$\omega_{1,i,j} = \frac{2}{1 - \overline{\mu}_{ij}\underline{\mu}_{ij} + \sqrt{(1 - \overline{\mu}_{ij})(1 - \underline{\mu}_{ij})}}$$

and $\qquad (26)$

$$\omega_{2,i,j} = \frac{2}{1 + \overline{\mu}_{ij}\underline{\mu}_{ij} + \sqrt{(1 - \overline{\mu}_{ij})(1 - \underline{\mu}_{ij})}}$$

where $\overline{\mu}_{ij}$ and $\underline{\mu}_{ij}$ are computed by (24) and (25), respectively.

Case 2: μ_{ij} are imaginary. This case applies when $\ell_{ij} r_{ij} \leq 0$ and $t_{ij} b_{ij} \leq 0$. The optimum values for the local relaxation parameters are given by Theorem 1 and $\overline{\mu}_{ij}, \underline{\mu}_{ij}$ are computed by (24) and (25), respectively.

Case 3: μ_{ij} are complex. This case applies when (i) $\ell_{ij} r_{ij} > 0$ and $t_{ij} b_{ij} < 0$ or (ii) $\ell_{ij} r_{ij} < 0$ and $t_{ij} b_{ij} > 0$. Letting $\mu_{ij} = \mu_R + i\mu_I$, Botta and Veldman [2] suggested for the local SOR the following 'good' (near the optimum) value for ω_{ij},

$$\omega_{i,j,opt} = \frac{2}{1 + \sqrt{1 - \overline{\mu}_{ij}^2}}, \qquad (27)$$

where

$$\overline{\mu}_{ij}^2 = \overline{\mu}_R^2 - \overline{\mu}_I^2 (1 - \overline{\mu}_R^{2/3})^{-1} \qquad (28)$$

with

$$\underline{\mu}_R \leq |\mu_R| \leq \overline{\mu}_R \text{ and } \underline{\mu}_I \leq |\mu_I| \leq \overline{\mu}_I. \tag{29}$$

The derivation of optimum values for the parameter sets ω_{ij} and ω'_{ij}, when the eigenvalues of J_{ij} are complex is an open problem.

Table I. Comparison of local iterative methods for $h = 1/81$, * indicates no convergence after $5 \cdot 10^4$ iterations

#	Method	$Re = 1$	$Re = 10$	$Re = 10^2$	$Re = 10^3$	$Re = 10^4$
	(R,I)	(0, 6400)	(0, 6400)	(0, 6400)	(0, 6400)	(0, 6400)
	$\underline{\mu}_{min}$	0.581E-01	0.613E+00	0.613E+01	0.613E+02	0.613E+03
	$\underline{\mu}_{max}$	0.118E+00	0.120E+01	0.120E+02	0.120E+03	0.120E+04
	LSOR Vel	274	572	5137	*	*
	LSOR Tak	188	327	2573	25419	*
	LSOR Rus	167	321	2570	24660	*
1	LSOR Str	167	321	2570	24660	*
	LSOR Bot	167	321	2568	25413	*
	LSOR Nat	167	321	2566	25394	*
	LSOR R/B	89	264	2501	24289	*
	LMSOR	89	278	452	519	733
	(R,I)	(6400, 0)	(6400, 0)	(0, 6400)	(0, 6400)	(0, 6400)
	$\underline{\mu}_{min}$	0.194E-01	0.153E-01	0.938E-01	0.958E+00	0.958E+01
	$\underline{\mu}_{max}$	0.194E-01	0.167E-01	0.119E+00	0.120E+01	0.120E+02
	LSOR Vel	3390	476	278	613	6437
	LSOR Tak	3222	368	186	333	3082
	LSOR Rus	183	161	170	329	3078
2	LSOR Str	171	162	170	329	3078
	LSOR Bot	183	161	170	329	3077
	LSOR Nat	183	161	170	329	3074
	LSOR R/B	132	82	91	307	3145
	LMSOR	132	82	98	317	459

In order to test our theoretical results we considered the numerical solution of (1) with $u = 0$ on the boundary of the unit square. The initial vector was chosen as $u^{(0)}(x, y) = xy(1-x)(1-y)$. The solution of the above problem is zero. For comparison purposes we considered the application of local SOR (LSOR) with natural (LSOR$_{nat}$) and red black ordering (LSOR$_{rb}$), local MSOR (LMSOR with red-black ordering) as well as all the local SOR methods with different set of parameters as proposed by Veldman, Takemitsu, Russell, Strikwerda and Botta [2]. In all cases the iterative process was terminated when the criterion $||u^{(n)}||_\infty \leq 10^{-6}$ was satisfied. Various functions for the coefficients $f(x, y)$ and $g(x, y)$ were chosen such that the eigenvalues μ_{ij} to be mainly imaginary. Complex eigenvalues do not appear in our examples, whereas real eigenvalues appear in a few cases. The type of eigenvalues for each case is indicated by the tuple (# real, # imaginary) in the second row of each table. The coefficients used in each problem are:

1. $f(x, y) = Re(2x - 10)^3$, $g(x, y) = Re(2y - 10)^3$

2. $f(x, y) = Re(2x - 10)$, $g(x, y) = Re(2y - 10)$

3. $f(x, y) = g(x, y) = Re \cdot 10^4$

4. $f(x, y) = Re(2x - 10)^5$, $g(x, y) = Re(2y - 10)^5$.

The number of iterations for the problems considered are presented in Tables I, II for $Re = 10^m$, $m = 0, 1, 2, 3$ and 4. From the results of Tables I and II we can distinguish three cases depending upon the bounds on $\underline{\mu}$, $\underline{\mu}_{min}$ and $\underline{\mu}_{max}$: (i) $\underline{\mu}_{max} < 0.1$, (ii) $(\underline{\mu}_{min}, \underline{\mu}_{max}) \subset [0.1, 1.2]$ and (iii) $\underline{\mu}_{min} \geq 1$. If case (i) holds, then we see that the rate of convergence of both local SOR and local MSOR methods is approximately the same (see prob. # 1 for $Re = 1$, prob. # 2 for $Re = 1, 10$). If case (ii) holds, then local SOR has a better rate of convergence than local MSOR (see prob. # 1 for $Re = 10$, prob. # 2 for $Re = 10^2, 10^3$). As can be seen case (iii) is the most common case and local MSOR performs better than any other local iterative method. In fact, as $\underline{\mu}$ increases the number of iterations of local MSOR is significantly smaller compared with the other methods. In conclusion, when $\underline{\mu}_{min} \geq 1$, then local MSOR outperforms all the considered iterative methods. Note that in problem 3, since the coefficients are constants both local SOR and local MSOR degenerate into their classic counterparts. Even in this problem, the improvement in the convergence rate for using MSOR over SOR is significant for large values of $\underline{\mu}$.

Finally we considered the implementation of the local MSOR method on a distributed memory MIMD computer with a square mesh topology.

In particular, the 5-point local MSOR was implemented on the Parsytec GCel 3/512 machine using T805 transputers with 4MB of external memory. Fig. 1 show the speedup and the efficiency, respectively obtained by the 5-point local MSOR method, for various sizes of the mesh (h) and the block size of grid points associated with each transputer.

Table II. Comparison of local iterative methods for $h = 1/81$, * indicates no convergence after $5 \cdot 10^4$ iterations

#	Method	$Re = 1$	$Re = 10$	$Re = 10^2$	$Re = 10^3$	$Re = 10^4$
	(R,I)	(0, 6400)	(0, 6400)	(0, 6400)	(0, 6400)	(0, 6400)
	$\underline{\mu}_{min}$	0.120E+01	0.120E+02	0.120E+03	0.120E+04	0.120E+05
	$\underline{\mu}_{max}$	0.120E+01	0.120E+02	0.120E+03	0.120E+04	0.120E+05
	LSOR Vel	744	6943	*	*	*
	LSOR Tak	339	3206	32207	*	*
	LSOR Rus	331	3246	32086	*	*
3	LSOR Str	331	3246	32086	*	*
	LSOR Bot	331	3246	32086	*	*
	LSOR Nat	342	3413	34221	*	*
	LSOR R/B	397	3473	34653	*	*
	LMSOR	331	459	690	1894	5757
	(R,I)	(6400, 0)	(6400, 0)	(0, 6400)	(0, 6400)	(0, 6400)
	$\underline{\mu}_{min}$	0.392E+01	0.392E+02	0.392E+03	0.392E+04	0.392E+05
	$\underline{\mu}_{max}$	0.120E+02	0.120E+03	0.120E+04	0.120E+05	0.120E+06
	LSOR Vel	3856	39192	*	*	*
	LSOR Tak	1933	19748	*	*	*
	LSOR Rus	1959	20132	*	*	*
4	LSOR Str	1959	20132	*	*	*
	LSOR Bot	1927	19742	*	*	*
	LSOR Nat	1926	19727	*	*	*
	LSOR R/B	1929	19103	*	*	*
	LMSOR	475	558	692	926	1175

Acknowledgment

We would like to thank the staff of the Athens High Performance Computing Laboratory for allowing access to their parallel machine.

References

1. Adams, L. M., Leveque, R. J. and Young, D. (1988), Analysis of the SOR iteration for the 9–point Laplacian, *SIAM J. Num. Anal.*, 9, pp. 1156–1180.
2. Botta, E. F. and Veldman, A. P. (1981), On local relaxation methods and their application to convection-diffusion equations, *J. Comput. Phys.*, **48**, pp. 127-149.

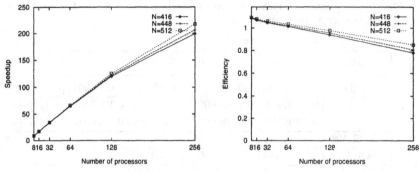

Figure 1. Speed up and Efficiency graphs of the local MSOR for different mesh sizes.

3. Boukas, L. A. (1998), *Parallel Iterative Methods for Solving Partial Differential Equations on MIMD Multiprocessors*, Ph.D Thesis, Department of Informatics, University of Athens, Athens.

4. Boukas, L. A. and Missirlis, N. M. (1998), *Convergence theory of Extrapolated Iterative Methods for Solving the Convection Diffusion Equation*, Technical Report, Department of Informatics, University of Athens, Athens.

5. Boukas, L. A. and Missirlis, N. M. (1998), A parallel local modified SOR for nonsymmetric linear systems, *Intern. J. of Comput. Math.*, **68**, pp.153-174.

6. Brandt, A. (1977), Multi–level adaptive solutions to boundary–value problems, *Math. Comput.*, **31**, pp. 333–390.

7. Chan, T. F. and Elman, H. C. (1989), Fourier analysis of iterative methods for solving elliptic problems, SIAM Review, **31**, pp. 20–49.

8. Ehrlich, L. W. (1981), An Ad-Hoc SOR method, *J. Comput. Phys.*, **42**, pp. 31-45.

9. Ehrlich, L. W. (1984), The Ad-Hoc SOR method: A local relaxation scheme, in Elliptic Problem Solvers II, Academic Press, New York, pp. 257-269.

10. Hageman, L. A. and Young, D. M. (1981), *Applied Iterative Methods*, Academic Press, New York.

11. Kuo, C.-C. J., Levy B. C. and Musicus, B. R. (1987) A local relaxation method for solving elliptic PDE's on mesh-connected arrays, *SIAM J. Sci. Statist. Comput.*, **8**, pp. 530-573.

12. Missirlis, N. M. (1984), Convergence theory of Extrapolated Iterative methods for a certain class of non–symmetric linear systems, *Numer. Math.*, **45**, pp. 447–458.

13. Niethammer, W. (1979), On different splittings and the associated iteration method, *SIAM J. Numer. Anal.* **16**, pp. 186–200.

14. Ortega, J. M. and Voight, R. G. (1985), *Solution of Partial Differential Equations on Vector and Parallel Computers*, SIAM, Philadelphia.

15. Varga, R. S. (1962), *Matrix Iterative Analysis*, Prentice-Hall, Englewood Cliffs, NJ.

16. Young, D. M. (1971), *Iterative Solution of Large Linear Systems*, Academic Press, New York.

DATA - ASSIMILATION AND HPCN-EXAMPLES OF THE LOTOS-MODEL

P. J. H. BUILTJES

TNO Institute of Environmental Sciences, Energy Research and Process Innovation
P.O.Box 342, 7300 AH Apeldoorn, The Netherlands

1. Introduction

Photo-oxidants in the troposphere over Europe, with as main trace gas ozone, have been studied extensively over the last three decades. The key scientific question is the understanding of the processes which govern the photo-oxidant formation. The determination of these processes is also of policy relevance; ozone levels over Europe exceed regularly limit values. This fact requires the determination of abatement strategies to reduce ozone concentration. Such an abatement can only be effective, and cost-effective, in case the reasons for these high ozone levels are known, also in a quantitative way.

Studies concerning photo-oxidant formation are performed by means of dedicated field experiments, data analysis of results from air quality networks, laboratory studies, fundamental chemistry research and atmospheric chemistry modelling.

In view of the increasing knowledge, including information about precursor emissions and meteorology, atmospheric chemistry-air pollution modelling is currently performed with well advanced models, requiring large scale computations.

First, some examples will be presented of both scientific studies as well as policy applications with the so-called LOTOS-model, Long Term Ozone Simulation.

It will be shown that the current computer resources put a limitation to the modelling studies, and the need for High Performance Computing and Networking (HPCN) will be clarified. Special attention will subsequently be given to the development of specific data-assimilation techniques using the LOTOS-model, and the paper will be ended by giving some conclusions and remarks concerning future developments.

Z. Zlatev et al. (eds.), Large-Scale Computations in Air Pollution Modelling, 89–98.
© 1999 *Kluwer Academic Publishers. Printed in the Netherlands.*

2. Examples of the LOTOS-model

The LOTOS-model is an Eulerian 3-D grid model which calculates in an hour-by-hour way photo-oxidant formation in the lower troposphere over Europe for longer time periods [1]. The sophistication of the LOTOS-model is between the one-layer EMEP-trajectory model and the 3-D Eulerian grid model EURAD, which covers the whole troposphere [2]. The horizontal grid resolution is 1.0 x 0.5 latlong (optional 0.5x0.25), there are 4 layers in the vertical upto about 3 km.

The chemical scheme used is CBM IV. Boundary conditions are derived from the TNO-version of the 2-D global model by Isaksen.

The antropogenic and biogenic emissions are ,as far as available, taken from CORINAIR, with time and temperature dependencies according to the LOTOS-structure. Meteorological data are determined using ECMWF-data and a diagnostic scheme from the Free Univ. Berlin. Calculations have been performed for the complete year 1990 and 1994, and the summer of 1997. A full year calculation takes about 100 hours at a HP 9000/740 workstation.

Recent science-oriented studies have focussed on fundamental questions concerning the differences in ozone pattterns over Europe.

Field measurements show that for stations close to the western coast of Europe, the ozone concentrations show a peak in spring, whereas more inland a peak in the summer is observed. LOTOS-model results, as shown in Figure 1 have the same tendency, but not so outspoken. The real reason for the spring time maximum is not fully clear yet. [3].

Budget studies have been performed with the LOTOS-model over different areas in Europe. Clear differences in the strenght of the different terms, ozone production, destruction, deposition, horizontal and vertical transport, are calculated. The results show that the differences in ozone-behaviour between different parts in the south of Europe are as large as the differences between the northern and the southern part of Europe as such. [4]

An analysis of the trends in ozone measurements over the last 2 decades show large differences over Europe, which are not really understood yet. LOTOS-model calculations indicate that it is unlikely that these differences are the direct consequence of the change in emission quantities of the precursor emissions of NO_x and VOC. [5]

LOTOS-studies with a mixture of science and application have been made concerning model validation and the so-called AOT 40.

Model validation covers a wide range of aspects, multi-component validation, analysis of observed and modelled phenomena, process and overall validation, model intercomparison, modelling of trends [6]. Both for science studies and for policy applications it is of utmost importance to know the reliability of the model used in the study, and the recent results show that an overall model validation methodology is not available yet.

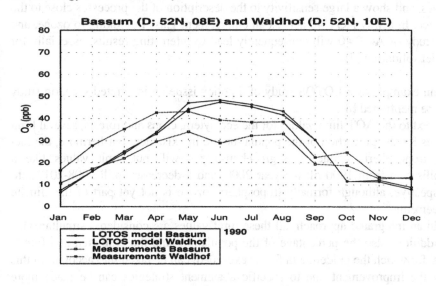

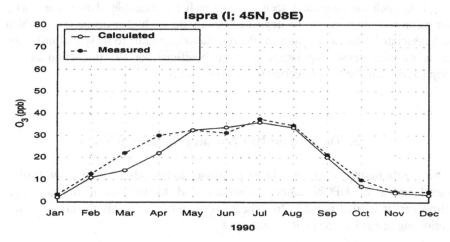

Figure 1. Annual variations of calculated and measured ozone concentrations at Bassum and Waldhof(Germany), and Ispra (Italy)

The limit value for the protection of the eco-system is the so-called accumulated ozone over threshold, with a threshold of 40 ppb, AOT 40.

The threshold is valid for daytime (global radiation over 50 W/M2), and for the 3 month period june, july, august for vegetation and the 6 month period april-september for forest(over 24 hours). LOTOS-model calculations, and also results of other models like EMEP, show substantial differences with measured AOT 40 values, and show a large sensitivity to the description of the processes close to the surface. In a way AOT 40 acts as an enlargement glass in looking to ozone, and the study of AOT 40 will consequently lead to interesting results, especially for model validation [7].

As an example of a LOTOS study of a policy issue an integrated scenario study can be mentioned [8].

Next to the AOT limit values for the eco-sytem, crops and forests, several limit values for the protection of the population are in existence. Furthermore, the fact that tropospheric ozone is a greenhouse gas will put limit values like a stabilisation from 1900 till the year 2000, and a decrease uptill 2008-2012 into perspective, although formally tropospheric ozone is not yet part of the climate agreements.

In an integrated approach, all these limit values are considered simultaneous. In addition, also the percentage of the population, and the crop-area and forest-area for which the guideline in fact is exceeded, can be taken into account. In this way the improvement due to specific abatement strategies can be made more explicit. Such an integrated scenario approach for example shows that traffic emission reduction, due to its impact especially on the higher ozone levels which are associated with limit values for the population, and due to its spatial influence in areas with dense population, is more benificial for the protection of the population than for the protection of the eco-system.

3. The need for HPCN in air pollution studies

There are a number of aspects in air pollution studies where the use and further development of HPCN-techniques would lead to substantial improvement. However, it is also possible to indicate some pittfalls in this direction. In the following several aspects will be adressed.

1) The use of chemical schemes
At the moment in atmospheric chemistry transport models only strongly condensed chemical schemes can be used. As such, there is an extensive

knowledge concerning chemical transformation in the atmosphere, contained in so-called master-mechanisms. However, these master-mechanisms are much too large to be used at the moment, due to the limitation in computer capacity. Even substantial research is performed, like in the subproject CMD (chemical mechanism development) of EUROTRAC-II to optimise condensed chemical schemes.

In the future, with increasing computer capacity, 'full' chemical schemes can be used combining gas phase, aerosols and cloud chemistry.

The warning is that the use of more complete schemes should be in balance with the information available in emission data bases. It would be useless to incorporate a chemical scheme with say 500 seperated VOC-species, when no knowledge would be available concerning their emission rate.

2) The determination of model sensitivity

At the moment to determine the sensitivity of model results to input data or specific model parameters, a limited number of model sensitivity runs is performed. This leads to a first estimate about the model's sensitivity.

In the future, a complete statistical analysis using all relevant parameters, seperated and in all possible combinations, will be feasable. This leads in principle to a complete picture of the sensitivity.

The warning is that the amount of information as a result of such a complete analysis will be enormous, leading to inherent difficulties in the coherent handling of all these data.

3) The use of models in scenario studies.

At the moment scenario studies to determine for example blame matrices between different countries are normally performed with relatively simple models to be able to perform the required scenario runs (1000-2000 runs) in a reasonable time.

In the future such scenario studies can also be performed with state-of -the-art models, even including the determination of the overall uncertainty.

The warning is that the use of more complex models including uncertainty analysis in stead of the use of more simple concensus models has to be incorporated, and even more important, accepted in policy decisions.

4) The presentations of model results

At the moment model results are often presented by several coloured contour plots, 2-D graphical presentations and sometimes videos.

In the future interactive data presentation will be possible, where model results are presented in a virtual reality environment, a cave. Researchers will be able to walk through the modelled results, like an ozone cloud.

The warning is that there is no garantee that such a sophisticated presentation will always lead to a better understanding of the calculated phenomena. Even more important, although a trivial remark, is that more sophisticated presentations are no garantee at all that the model results are more correct.

5) The accuracy of model calculations.
At the moment, model calculations still contain errors, in general small, due to the used numerics which are limited in sophistication due to restricted computer power. An example is the use of operator splitting which unavoidable leads to numerical errors.

In the future there will be a substantial improvement in a further minimalisation of these numerical errors.

In fact, in this case there is no warning, except for the higher computer costs.

6) Model validation
At the moment, model validation is often restricted to just a graphical comparison between model results and observations, in the case of photo-oxidants also even restricted to only ozone.

In the future, given that also field data of other trace gases are available, the model validation will be multi-component, and linked with a full statistical analysis, including sensitivity calculations.

The warning is that at the moment, a general accepted model validation methodology is not available, and should be developed before full use of modern HPCN techniques can be made in this respect.

7) Real time ozone forecasting
At the moment, real time ozone forecasting is in general done with empirical models or expert systems, based on an analysis of previous ozone episodes.

In the future, real time ozone forecasting will be performed based on the use of on-line weather forecasting in combination with state-of-the-art chemical schemes/models and optimal use of on-line ozone measurements in combination with data-assimilation

The warning is that the on-line availability of ozone measurements might be problematic, and that the availability of weather forecast is at the moment in Europe restricted to the official meteorological institutes.

8) The use of satellite data
At the moment, the use of tropospheric air chemistry satellite data is just starting and in its infancy.

In the future, the combination of tropospheric satellite data, with state-of-the-art models and ground level observations and vertical soundings, all combined with data-assimilation techniques will lead to an enormous enlargement of the availble experimental/model derived information concerning air quality.

The warning is that there will be always an inherent, and often quite substantial uncertainty of the satellite results, which should be taken into account to avoid not justified results.

4. Data-assimilation techniques using the LOTOS-model

Data-assimilation has been mentioned in the previous paragraph in combination with ozone forecast and the optimal use of satellite data.

Data-assimilation has been used, especially in meteorological models, regularly over the last decade. The normal practice is to use measurements as initial conditions for the prognostic meteorological model, followed by model calculations over a certain period, say 48 hours.

Then, at the end of the period, the model results are compared with new measurements, and calculated and measured data are nudged, combined, and the model calculations are continued.

Another approach used in air pollution modelling is the determination of the accuracy of emission data used as input to the model by comparing calculated results with field observations, and performing inverse calculations. Uptill now, this method has been applied mainly to lineair problems, for example CH4-emissions.

Recently, new developments have been made in the field of data-assimilation focussing on photo-oxidant formation. The difference with the methods mentioned before is that in this case data-assimilation has to be developed for a non-lineair system like ozone- formation, instead of the lineair approaches, as for the inverse modelling for CH4.

Two approaches are possible in the non-lineair case, the so-called 4-D var method, where an adjoint model is made which replaces the original model, and the extended Kalman filter approach, where a model independent system is build around the original model.

Concerning the LOTOS-model, an extended Kalman filter approach has been developed [9]. In Table 1 the parameters are listed, with their assumed deviations, which are used in the data-assimilation scheme. By introducing measurements in the model, the scheme will try to decrease the differences between the measurements and the observations using the freedom given by the introduced noise. An example is given in Figure 2, were on purpose the deviation between the base run results and the observations has been made large. The results presented in Figure 2 show that the extended Kalman filter built around the LOTOS model is functioning properly.

Table 1. Introduction of noise in LOTOS

Module	Parameter	Deviations
chemistry	reaction rates	± 10%
meteo	cloud cover	± 25%
	wind field	additional
	mixing height	± 25%
	temperature	± 1°C
boundary concentrations	O_3	± 20%
emissions	NO_x	± 20%
	VOC antropogeen	± 30%
	VOC biogeen	-50%, +200%
	CO	± 30%
land use	deposition factor	± 30%

In the future, a major application of these data-assimilation techniques will be the analysis and use of troposheric trace gas observations from satellites. As an example tropospheric ozone can be mentioned, were first results are available from the GOME-instrument, and more accurate results are expected from Sciamachy, which will be launched in the middle of the year 2000.

The data coming from the satellites will be made only once a day, or once every 3 days, at a specific moment. Furthermore, there will be no data in case of clouds (average global cloud coverage is about 50-60 %), and there will be interference when the aerosol content is too high. This means that, although there will be a substantial increase in experimental data of ozone, there will be also many data-gaps in the results. These gaps can only be filled in -and so-called added value can be given to the satellite data- by using adequate models with data-assimilation.

Because of the large computer time required for data-assimilation, and the large amount of satellite data, this can only be handled in the framework of HPCN.

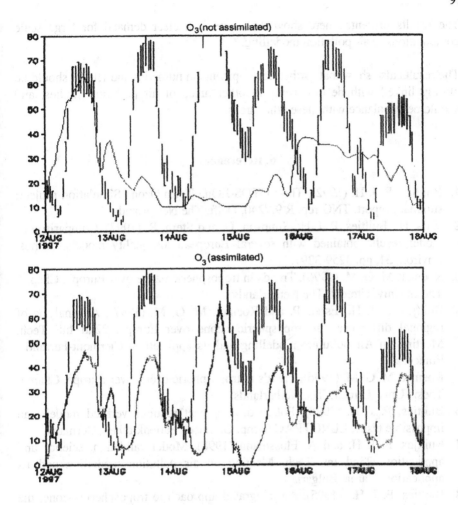

Figure 2. Measured (with assumed error bars) and calculated ozone concentrations for a station in Germany using LOTOS without and with data-assimilation technique.

5. Conclusions and remarks

Atmospheric Transport and Chemistry modelling is a part from the overall research into atmospheric chemistry.

ATC-modelling as such covers the items input data, model performance, assessment studies, and numerical aspects. Numerical aspects are important, but forms only a part of ATC-modelling.

The results presented here show that there is a clear demand for large scale computations in air pollution modelling.

The results also show that further developments in numerics and HPCN should be closely linked with developments in other areas of air pollution studies, and should be in balance with these other areas.

6. References

1. Builtjes, P. J. H. (1992), The LOTOS-LOng Term Ozone Simulation-project; summary report. TNO-rep. R 92/240, Delft, The Netherlands.
2. Hass, H., Builtjes, P. J. H., Simpson, D. and Stern, R. (1997), Comparison of model results obtained with several European air quality models. Atmos. Environ. **31**, pp. 3259-3297.
3. Roemer, M. G. M. (1996), Trends in tropospheric ozone over Europe. Chap. 5 Thesis Univ. Utrecht, The Netherlands.
4. Builtjes, P. J. H., Esser, P. and Roemer, M. G. M. (1997), An analysis of regional differences in tropospheric ozone over Europe. 22nd Int. Tech. Meeting on Air pollution modelling and its application, Clermont-Ferrand , France.
5. Roemer, M. G. M. (1996), Trends in tropospheric ozone over Europe, Chap. 6 Thesis Univ. Utrecht, The Netherlands.
6. Builtjes, P. J. H. (1998), Policy development requires verified models: an impossible task?, EUROTRAC Symp. Garmisch Partenkirchen, Germany.
7. Builtjes, P. J. H. and A. Flossmann (1998), Model validation, science and application. 23nd Int. Tech. Meeting on air pollution modelling and its application, Varna, Bulgaria.
8. Builtjes, P. J. H. (1995), An integrated approach to tropospheric ozone, the PHOXA and LOTOS experience. EMEP workshop on the control of photochemical oxidants over Europe, St. Gallen, Switzerland.
9. Segers, A., Heemink, A. W., Builtjes, P. J. H. and van Loon, M. (1997), Data assimilation applied to nonlinear atmospheric chemistry models: first experiences. 1st GLOREAM workshop, Aachen, Germany.

COMPUTATIONAL CHALLENGES OF MODELING INTERACTIONS BETWEEN AEROSOL AND GAS PHASE PROCESSES IN LARGE SCALE AIR POLLUTION MODELS

G. R. CARMICHAEL, A. SANDU, C. H. SONG, S. HE, M. J.
PHANDIS, D. DAESCU, V. DAMIAN-IORDACHE, F. A. POTRA
Center for Global & Regional Environmental Research and the
Department of Chemical & Biochemical Engineering University
of Iowa. Iowa City, Iowa 52240, Tel: 319/335-1399
FAX: 319/335-1415 EMAIL: gcarmich@icaen.uiowa.edu

Abstract

In this paper we discuss computational challenges in air quality modeling (as viewed by the authors). The focus of the paper will be on t1he "current" state-of-affairs. Due to limitation of space the discussion will focus on only a few aspects of air quality modeling: i.e., chemical integration, sensitivity analysis and computational framework; with particular emphasis on aerosol issues.

KEY WORDS: Air quality, atmospheric chemistry, sensitivity analysis

1. Introduction

A detailed understanding of the relationships between the emissions and the resulting distribution of primary and secondary species in the atmosphere is a requisite to designing actions for the maintenance of a healthy environment. Scientific efforts to understand the atmospheric processes governing these relationships involve a combination of laboratory experiments, field studies, and modeling analysis. Laboratory experiments provide basic data on individual physical and chemical processes. Field studies are designed to investigate a limited number of processes under conditions in which a few processes dominate. Unlike controlled laboratory experiments, field studies cannot be parametrically controlled. Since laboratory experiments and field studies by themselves cannot fully elucidate complex atmospheric phenomena, comprehensive models that allow multiple processes to occur simultaneously are required for data analysis and scientific inquiry.

99

Z. Zlatev et al. (eds.), Large-Scale Computations in Air Pollution Modelling, 99–136.
© *1999 Kluwer Academic Publishers. Printed in the Netherlands.*

The models from [9], [41], [25] are examples of regional and global scale atmospheric chemistry models in use today. These models treat transport, chemical transformations, emissions and deposition processes in an integrated framework, and serve as representations of our current understanding of the complex atmospheric processes. They provide a means to perform parametric studies for quantitative analysis of the relationships between emissions and the resulting distribution, and can also be used to study the response of the pollutant distributions to system perturbations, and to link pollutant distributions to environmental effects.

As our scientific understanding of atmospheric chemistry and dynamics has expanded in recent years, so has our ability to construct comprehensive models which describe the relevant processes. However, these comprehensive atmospheric chemistry models are computationally intensive because the governing equations are nonlinear, highly coupled, and extremely stiff. As with other computationally intensive problems, the ability to fully utilize these models remain severely limited by today's computer technology.

The scientific issues associated with analysis of our chemically perturbed atmospheres are dominated by a number of underlying considerations. Several of the more important ones are: (a) the anthropogenic sources of trace species are quite localized and occur only over a fraction of the Earth's land area; (b) natural sources of trace species are, for the most part, very disperse and are not in the same areas as the anthropogenic sources (although this trend may be changing in regions such as tropical rain forests and the savannahs); (c) in virtually no case can an individual species be studied in isolation from other species; (d) many of the mechanisms that effect transformation of the species are non-linear (e.g., chemical reactions and nucleation processes); and (e) species of importance have atmospheric lifetimes that range from milliseconds and shorter to years (e.g., OH radical to CH_4). These considerations require: finer grid resolutions than currently existing ones in present-day models; simultaneous treatment of many species; and long simulation times (i.e. months to years) to assess the impacts. These demands present considerable challenges to the air quality modeling community.

Aerosols are an area of increased importance that put additional computational burden on air quality models. Atmospheric particles have various influences on the earth-atmospheric system, including the scattering and absorption of solar and terrestrial radiation ([14]) and visibility impairment ([76]). In order to better quantify the role of aerosols in the chemistry and physics of the atmosphere, it is necessary to improve our ability to predict aerosol composition as a function of size. However, the modeling of aerosol processes is difficult because of the strong spatial and temporal variability in aerosol composition and size distributions, and the complexities and uncertainties in aerosol micro-physical processes, transport characteristics, and chemical inter-actions between the particulate and gas phases.

Atmospheric aerosols are formed in two ways: (i) direct emissions from natural sources, and (ii) gas-to-particle conversion. Aerosol parti-cles from natural sources tend to be in the coarse mode with diameters larger than $1\mu m$. In contrast, particles formed by gas-to-particle conver-sion, such as nucleation and absorption ([77], [80], [24]) usually compose the fine mode with diameters from $0.001\mu m$ to $1\mu m$.

Different particle modes have quite different chemical/physical char-acteristics due to the size-dependency of the formation mechanism. These different aerosol characteristics determine the distribution of volatile species between the gas and aerosol phases. Both the gas/aerosol interaction and the formation mechanism have been important subjects in the field of air pollution modeling, atmospheric physics, and aerosol thermodynamics ([77], [24], [31], [72], [64], [65], [81], [3], [54], [42], [43]). Even though many investigators have studied this topic, it remains a challenge to model the gas-to-particle conversions and thermodynamic processes which control the compositional size-distribution of aerosols.

In this paper we discuss the issues of modeling aerosol processes in air quality models.

2. Model Framework

Modeling aerosol processes is computationally intensive and challeng-ing. Gas-to-particle conversion, coagulation and deposition are the most important dynamic processes of aerosols. Gas-to-particle conversion (or particle-to-gas conversion) is caused by four major pathways: nucle-ation, condensation, absorption and dissolution.

The mathematical description of aerosol dynamics can be described by the general dynamics equation (GDE),

$$\frac{\partial Q_i}{\partial t} + \nabla \bullet (uQ_i) = \nabla \bullet (K\nabla Q_i) + \left(\frac{\partial Q_i}{\partial t}\right)_{cond/evap} + \left(\frac{\partial Q_i}{\partial t}\right)_{coag}$$
$$\left(\frac{\partial Q_i}{\partial t}\right)_{react} + \left(\frac{\partial Q_i}{\partial t}\right)_{source/sink} \tag{1}$$

where $\left(\frac{\partial Q_i}{\partial t}\right)_{source/sink}$ is the rate of change of the aerosol mass of species due to nucleation, primary aerosol emission and removal, while

$\left(\frac{\partial Q_i}{\partial t}\right)_{cond/evap}$, $\left(\frac{\partial Q_i}{\partial t}\right)_{coag}$ and $\left(\frac{\partial Q_i}{\partial t}\right)_{react}$ are rates of changes due to condensation/ evaporation, coagulation and surface reaction.

There are many approaches to implement these equations in air pollution models. In this chapter we illustrate the issues related to aerosol modeling by presenting some of our efforts in this area. We are building aerosol processes into our in-house developed regional scale trace gas model, STEM ([8], [9]). The STEM model is a three-dimensional (3D), Eulerian numerical model which accounts for the transport, chemical transformation and deposition of atmospheric pollutants ([10]). The earlier versions of the model have been used quite extensively for scientific studies and policy evaluations in the eastern United States and the Pacific Region ([13], [50], [51], [78], [12]).

The current 3D version of the model includes the chemical mechanism based on [53] and [2] and modified for explicit treatment of isoprene. This version of the STEM model is a combination of the classical gas and aqueous phase chemistry calculations with an extension to treat aerosol species also. A total of 107 species (90 gas phase and 17 aerosol phase chemical species) and a total of 190 gas and aerosol phase reactions are included. The 90 gas phase species are further divided into 62 long-lived species that are transported and 28 short-lived species such as free radicals which follow a pseudo-steady-state approximation. The transport of the gaseous species is computed using a locally one dimensional (LOD) Crank Nicholson Galerkin finite element method (FEM). More details of this treatment can be found in [10] and [9]. The computational structure of the model is shown in Fig. 1.

2.1. AEROSOL DYNAMICS

The aerosol dynamic component treats several physical aerosol processes, such as nucleation, condensation/ evaporation, coagulation, and absorption/dissolution. In order to describe each physical aerosol process, a mixture of fundamental and empirical formulas are utilized as discussed in [80], [24], and [71].

2.2. THERMODYNAMIC AND KINETIC APPROACHES

Traditionally there have been two approaches for simulating atmospheric aerosols. The first approach can be called the "kinetic approach". Several researchers have performed atmospheric modeling studies in this way ([80], [24]). They have considered the irreversible uptake characteristics from the gas to the aerosol particles using transport parameters such as gas phase diffusion coefficient, accommodation coeffi-

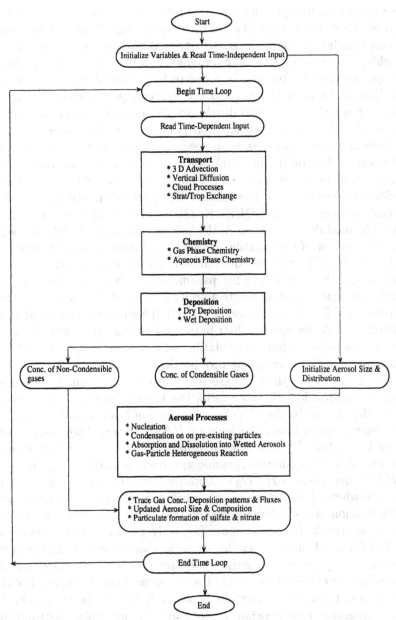

Figure 1. Schematic of the computational structure of STEM-III.

cient and so on. In this procedure, regardless of the volatility, the species deposits irreversibly from the gas to the particles. Thus, this approach cannot take into account the possibility that volatile species such as hydrochloric acid, nitric acid, and ammonia can be redistributed between the gas and aerosol phases. In order to better simulate the reversible distribution between the gas and the fine/coarse particles, thermodynamic approaches have been developed. Technically, this consideration is possible with the assumption that particles are charge-neutral.

Several aerosol modules have been developed to estimate the distribution of the volatile inorganic species between the gas and aerosol phases. Kim et al.[49] have compared the SCAPE (Simulating Composition of Atmospheric Particles at Equilibrium) module with other aerosol modules, such as EQUIL, KEQUIL, MARS, SEQUIL, and AIM. All of the modules other than AIM (Aerosol Inorganic Model) deal with the distribution of the volatile species in a thermodynamic manner.

In the SCAPE module, the Gibbs energy minimization scheme is employed with thermodynamic parameters, such as equilibrium constant, chemical potential, solute activity coefficient, water activity, and relative humidity of deliquescence (RHD). This module can predict the amount of volatile species distribution as well as the existing forms (ions or ion pairs) of both the volatile and non-volatile species (sulfates and mineral cations) in the aerosol particle.

Fig. 2 represents the comparisons of the results using the thermodynamic approach with those using the kinetic approach. This figure shows the change in the HNO_3 concentration level. If only gas phase chemistry is considered, the level of HNO_3 is very high. But, when either the uptake kinetics (kinetic approach) or thermodynamic equilibrium estimation (thermodynamic approach) is carried out, the level of HNO_3 decreases as HNO_3 is deposited onto the aerosol particles. In the case where thermodynamic equilibrium is assumed, the nitric acid concentration is considerably higher than that from the case of the kinetic approach. In kinetic modeling, only gas-to-particle conversion for HNO_3 is taken into account, as is the case of sulfate. In the thermodynamic approach, the most significant fact is that nitrates should neutralize the excessive cations in the coarse particle. Thus, most of the nitrate goes to the coarse particle. However, in the kinetic approach, the most important consideration is the aerosol number-size distribution, as is the case of the sulfate-uptake. Thus, most of the nitrate tends to be in the fine mode because the fine mode has more particles, and hence more available surface area.

This difference between the two approaches may have important implications regarding aerosol modeling. In terms of the fraction of nitrate-deposition in the fine/coarse particles, the results from the ther-

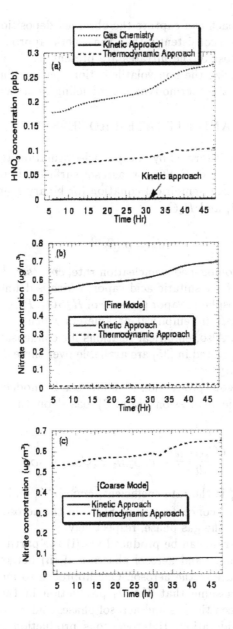

Figure 2. Comparison of the results from the thermodynamic approach with those from the kinetic approach

modynamic approach can capture the observed deposition characteristics of aerosol more consistently than the kinetic approach. As pointed out previously, most of the nitrates are present in the coarse particle in order to neutralize the non-volatile cation species. This fact can be well-explained by the thermodynamic modeling as shown in Fig. 2.

2.3. SULFATE AND NITRATE PROCESSES

Particulate sulfate is formed by new particle formation and by chemical conversion of SO_2 and H_2SO_4 on aerosol surfaces. Peter and Easter [63] introduced a semi-empirical equation for binary homogeneous nucleation of H_2SO_4 and H_2O,

$$J_n = \alpha \left(S_s\right)^\beta \tag{2}$$

where, J_n is the homogeneous nucleation rate, $cm^{-3}sec^{-1}$, and S_s is the saturation ratio of the sulfuric acid vapor, which is defined as the ratio of the actual atmospheric vapor pressure of H_2SO_4, P_a, to its saturation vapor, P_s, and α, β are empirical parameters which are dependent on relative humidity. Also, empirical relations for condensation/ evaporation processes suggested in [35] are available over three different ranges of relative humidity.

The heterogeneous production of sulfate can be modeled as a pseudo-first- order kinetic expression (Eqn. 3) based on the observation in [52],

$$\frac{dC_{sulfate}}{dt} = k_{SO_2,H_2SO_4} C_{SO_2,H_2SO_4} \tag{3}$$

where k_{SO_2,H_2SO_4} is the rate coefficient, cm^3sec^{-1}, and $C_{sulfate}$ and C_{SO_2,H_2SO_4} are the concentration of SO_4^{2-} in the aerosol particles and SO_2 or H_2SO_4 in the gas phase, respectively.

Particulate nitrate can be produced via (i) the partitioning of nitric acid between the gas and aerosol phases and (ii) heterogeneous reactions on the surface of the aerosol. One approach to modeling nitrate formation is to assume that HNO_3 production in the gas phase is partitioned between the gas and aerosol phases and controlled by thermodynamic consideration. Heterogeneous production involves night-time reactions with NO_3 and N_2O_5 on the surface of the aerosol. For heterogeneous production processes, the following equation can be used ([24], [32]):

$$k_j = \int_{r_1}^{r} 2k_{d,j}(r)n(r)dr \tag{4}$$

where $n(r)dr$ is the number-concentration distribution between r and $r+dr$, (often assumed to be log-normal), and $k_{d,j}(r)$ is the size-dependent mass transfer coefficient. The expression for $k_{d,j}(r)$ is given by

$$k_{d,j} = \frac{4\pi D_j V}{1 + K_n \left(\lambda + 4 \left(1 - \frac{\alpha}{3\alpha} \right) \right)} \tag{5}$$

where D_j is the gas phase diffusion coefficient in $cm^3 sec^{-1}$, K_n is the dimensionless Knudsen number $\left(= \frac{l}{r} \right)$, λ is the effective free path of a gas molecule in air, V is the ventilation factor, and α is the accommodation coefficient (or sticking coefficient).

Many previous studies have assumed that the absorbed nighttime species are very quickly converted into nitric acid via fast aqueous phase reactions with hydroxyl radicals ([37]), and thus the overall rate is governed (limited) by mass transport processes with characteristic parameters as described by Eqns. 4 and 5.

Nitric acid may also be produced on aerosol surfaces via reactions involving surface sorbed NO and NO_2. However, the sticking coefficients for NO and NO_2 are very small. The mass accommodation coefficient (α) of NO over sulfuric acid ($H_2SO_4 \, nH_2O$, 96 wt. %H_2SO_4) at 298 K is less than 1 x 10^{-6} ([23]). Also, the mass accommodation coefficient of NO_2 for the same type of surface at 273 K is of the order of 6 x 10^{-4} ([23]). As new experiments are conducted and more is learned about possible surface reactions, such reactions may prove to be important.

2.4. ORGANIC AEROSOL

While the importance of inorganic aerosols (nitrate-ammonium-sulfate mixtures) has been widely appreciated and studied, there is a growing awareness that organic aerosols are also important. Secondary organic aerosols (SOA) are a major component of urban environments ([59], [66]), and have also been found to represent an appreciable component of the fine aerosol mass (comparable to that for sulfate) at the Grand Canyon ([56]), as well as in the marine atmosphere ([55]). SOA are produced as a by-product of the oxidation of many organic molecules, including aromatics ([59]), monoterpenes ([60]), and carbonyls ([34]). These reactions yield semi-volatile compounds, which along with primary semi-volatile organics, partition themselves between the gas and particle phases. Furthermore, the presence of polar organic species on the aerosol surface can make the aerosol more hygroscopic ([70]), and thus increase the light scattering by particles. These organic aerosols may represent an important source of CCN ([58]).

Since aerosols and ozone share common chemical precursors and emissions sources, and are intricately linked through the photochemical oxidant cycle, it is important that they be studied in a simultaneous and coupled manner. Yet it is usual for these topics to be treated as isolated problems. A complicating factor of potentially large importance is the role of aerosols as reactive surfaces. Reactions on aerosol surfaces have been shown to play a significant role in stratospheric chemistry. However, little is known about the role of heterogeneous reactions in the lower regions of the atmosphere. However, there is a growing body of information that suggests that the photochemical oxidant cycle may be perturbed through heterogeneous uptake of volatile organic compounds, nitrogen oxides and free radicals, and subsequent reactions on the surface. Through these processes the formation rates of tropospheric ozone and secondary aerosols, and the sensitivity of O_3 to NO_x and VOC emissions may be altered.

Organic aerosol formation processes can be added to the modeling approach described here. This can be done following [60], with aerosol production potentials/yields taken from that work and the recent work of [59], and partitioning between gas and aerosol phases calculated based on the approach from [62].

2.5. TWO-BIN SCAPE MODEL

In studies of aerosol composition, the size composition is often discretized into size bins ([7]). In each bin, all particles are assumed to have the same composition (i.e., internally mixed). It is widely recognized that aerosols often exhibit bi-modal distributions, and the fine and coarse aerosols are externally mixed ([16]). Based upon this concept, a two-bin (fine and coarse mode) gas/aerosol equilibrium model (two-bin SCAPE model) has been developed. At first, the SCAPE calculates aerosol phase concentrations in the fine section, using the total concentrations of the volatile species and the fine section concentrations of the non-volatile species. Next, after excluding the fine-section volatile species concentrations from the total concentrations, SCAPE calculates the coarse-section concentrations of the volatile species, using the remaining total concentrations of the volatile species and the coarse-section concentrations of the non-volatile species.

2.6. "ARTIFACTS"

An advantage of the thermodynamic equilibrium approach is related to the fact that it can help estimate measurement error, which is unavoidable during the measurement procedure. This type of measurement error is sometimes called an "artifact". Artifacts take place in the case of both cumulative and long time collection of aerosol particles, and simultaneous collection of cation/anion species using cascade impactors or filter packs ([47]). There are many reaction pathways for such measurement error. Reactions between the collected particles on the filters and acidic (HCl, SO_2, HNO_3) or alkaline (NH_3) gases can cause such measurement error. For example, already-deposited volatile species, HCl, on the surface of a filter can be replaced by less volatile species such as SO_4^{2-} and/or NO_3^- mainly due to the difference of vapor pressures between species. When chloride ions are released from the sample, these chloride ions may capture either hydrogen ions or ammonium (NH_4^+). Thus, the released form of chlorides may be in the form of NH_4Cl or HCl. In this case, both HCl or NH_3 concentration in the gas phase and SO_4^{2-} and NO_3^- in the aerosol phase are overestimated, whereas both Cl^- and NH_4^+ in the aerosol phase and SO_2 and HNO_3 in the gas phase are underestimated.

Another possible way for artifact error to take place is under the strong influence of sea-salt aerosol. Sea-salt particles are abundant in $NaCl$. As the sea-salt aerosols deposit onto the surface of filters, chlorine species can be released via the reactions:

$$NaCl(aero) + HNO_3(aero) \longrightarrow NaNO_3(aero) + HCl(g) \quad (6)$$

$$2NaCl(aero) + H_2SO_4(aero) \longrightarrow Na_2SO_4(aero) + 2HCl(g) \quad (7)$$

This phenomenon has been called "chlorine loss" or "chlorine deficiency" ([40], [36]). In this case, the chloride concentration in the aerosol particle is underestimated. This dissipation of volatile species from aerosol filters results in the overestimation of gas phase concentration for HCl.

There are additional ways for artifact error to occur. The following reactions also occur in the coastal marine environment:

$$H_2SO_4(aero) + 2NH_4NO_3(aero) \longrightarrow (NH_4)2SO_4(aero) + 2HNO_3(g)$$
$$(8)$$

$$H_2SO_4(aero) + 2NH_3(g) \longrightarrow (NH_4)2SO_4(aero) \quad (9)$$

These reactions can occur under the conditions of sulfate or SO_2 rich air or strong sea-salt influence. These conditions are created in marine boundary layers under the influence of continental outflows.

Thermodynamic equilibrium modules, such as SCAPE, provide an effective means to estimate such measurement error and to estimate the thermodynamically possible distributions for those species.

Table I. Calculation time required for each simulation.

Model	Calculation time
I) Gas phase chemistry model	11 seconds
II) Gas/aerosol model	
Irreversible kinetic model	265 seconds
Hybrid model ($CASEI$)	6360 seconds
Hybrid model ($CASEII$)	331 seconds
Hybrid model ($CASEIII$)	8100 seconds

2.7. COMPUTATIONAL CONSIDERATIONS

Three-dimensional models without aerosol modules are basically composed of two parts: (i) gas-phase chemistry and (ii) horizontal and vertical transports. Dabdub and Seinfeld [19] and [20] reported that the computation of the rate of gas-phase chemical reactions is the most time-intensive calculation component. Generally, the chemistry part of the calculation takes 75-90% of the total calculation time. Thus, many efforts have been concentrated on the chemistry part for saving the computation time ([20], [11], [68]). These efforts can be categorized into two groups: (i) development of more efficient implicit integrators ([68]) or effective explicit solvers ([20]) and (ii) implementation of parallel computation with multiple processors ([11], [18]).

The situation is changed when an aerosol module is incorporated into three-dimensional models. Table I shows CPU times for coupled aerosol/ chemistry calculation. The CPU times in Table I are obtained from an HP-UNIX A 9000/735 with 160-M RAM Machine. For the gas-phase reaction calculation, only 11 seconds are required. But, if irreversible aerosol kinetics are added into the gas-phase chemistry model, more CPU time is needed (265 seconds). (Here, the irreversible aerosol kinetics implies that the model includes aerosol dynamics such as nucleation, condensation/ evaporation, and irreversible absorption for sulfate and nitrate ([80], [24], [75]).)

When the SCAPE module is incorporated into the Eulerian model and is called at every reaction time step (Hybrid model (Case I) [1]),

---

[1] SCAPE module is called at every reaction time step.

the total computation time drastically increases up to 6360 seconds (Actually, the STEM-box model has two characteristic time scales: one for reaction integration (10 secons) and the other for transport-equation integration (300 seconds)). In this particular case, the number of calls to the SCAPE module is 18,680 for a two-day simulation. However, it is important to note two facts: (i) the current STEM model does not have any reaction for NH_3 and HCl; and (ii) even though the current STEM model has HNO_3-related reactions, HNO_3 is a reservoir species. In other words, the level of HNO_3 does not greatly affect the concentrations of other species. Hence, we can call the SCAPE module at every transport time scale (300 seconds) instead of reaction time scale (10 seconds) (Hybrid model (Case II) [2]). In this particular case, the SCAPE module is only called 576 times for two day simulation, and it only takes 331 seconds. This scheme was employed in the aged sea-salt particle case ([75]). (In Cases I and II, the SCAPE module employs a non-iterative scheme, which is also explained in [75].) If the SCAPE module with the iterative scheme is utilized at every transport time step (300 seconds) and the two-bin SCAPE module is iterated until two modes (fine and coarse modes) share the same gas-phase concentration (hybrid model (Case III) [3]), the computation time drastically increases up to 8100 seconds. Of course, in this case, the computation time is dependent on the accuracy level (i.e. error tolerance).

The iterative scheme used to find the equilibrium concentrations in the two-bin gas-aerosol model (SCAPE) requires expensive computations and it is an extremely time-consuming process. The complex expressions of the activity coefficients, the wide range of the equilibrium constants and the influence of the temperature and humidity, to name only a few, make difficult the implementation of high-order methods to solve the nonlinear equilibrium equations. On the another hand, using a (safe) bisection method requires a large number of function evaluations, involving computations of the ion-pair concentrations from the ion concentrations, the activity and binary activity coefficients, and the water content. On average these computations require 70% to 80% of the CPU time, and when most of the species have non-zero ionic concentratinos in the aqueous phase the time will increase to about 90%.

The performance of the code can be improved considerably in two ways:

a) By reducing the time necessary to solve the equilibrium equations inside of each bin (we will refer at this process as an internal process).

---

[2] SCAPE module is called at every transport time step (aged particle case).

[3] SCAPE module is called at every transport time step and employs iteration scheme (naturally emitted aerosol case).

b) By reducing the number of iterations necessary to equilibrate the gaseous concentrations of the volatile species in the fine-mode with those in the coarse-mode (we will refer at this process as an external process).

The internal process can be improved by combining the safety of the bisection method with the speed of the secant method into a Brent-type algorithm. Taking advantage of the previous predicted concentrations will improuve the iterative process. If the concentrations don't change significantly after few iterations, a fix point method combined with extrapolation is expected to converge very fast. If this fails, we return to the basic method. While the time spent is insignificant, the benefits may be considerable. In all the numerical experiments performed using these techniques in the implementation of the code will reduce the CPU time. Table II shows the timing results obtained with the old version of the code (SCAPE1) and Table III shows the results obtained with the new implementation (SCAPE2), using data from Yakushima, 3/28-31/88, with different methods for activity coefficients[4]. The error tolerance inside the bins was 10^{-3} for the gas-liquid phase computations and 10^{-2} for the solid computations, and the relative error for equilibrium between the bins was 4×10^{-3}. Relative differences between the predicted equilibrium concentrations of the volatile species (HCl, HNO_3, NH_3) are presented in Table IV. It can be seen that the predictions remain within the required error interval, and the CPU time may be reduced up to 10 times.

Improving the external process is defintely the most important part, but also the most delicate. One possible way is to require more accuracy in the internal process. While this will incresase considerable the time necessary to solve the equilibrium equations in each bin, the iteration number in the external process may be reduced dramaticaly. Denote by T_1 and T_2 the time necessary to solve the equilibrium equations in the fine, and coarse modes, by N_1 the number of iterations necessary to equilibrate concentrations between bins. We may assume that the time consumed to update the results from one bin to another is negligible compared with $min(T_1, T_2)$. If reducing the error tolerance will increase T_1 to $c_1 T_1$ and T_2 to $c_2 T_2$, and will reduce the number of external iterations to N_2, then we obtain faster convergence whenever $(T_1 + T_2)N_1 > (c_1 T_1 + c_2 T_2)N_2$. Good results were obtained with data at Cheju and Yakushima. Futher experiments should be done in order to decide for what kind of distribution of the concentrations in each bin is preferable to use smaller errors in the internal process. Once this is

---

[4] P represents Pitzer method to compute the activity coefficients, K-M Kusil-Meissner method and B represents the Bromley method

done, the code will automatically adjust the error to obtain the fastest convergence.

Table II. SCAPE1 results for rh=0.80.

Temp.(K)	CPU time(sec.)			Nr. of iterations			Time(sec.)/iter		
	P	K-M	B	P	K-M	B	P	K-M	B
275	25.3	20.7	38.1	32	9	50	.79	2.3	.76
280	16.7	10.1	18.4	15	4	11	1.11	4.17	1.67
285	9.8	7.1	11.4	8	4	6	1.22	1.77	1.9
290	6.6	6.6	9.8	5	3	5	1.32	2.2	1.96
295	5.5	6.4	6.5	4	3	3	1.37	2.13	2.16
300	0.39	2.12	1.88	3	2	3	0.13	1.06	.62

Table III. SCAPE2 results for rh=0.80.

Temp.(K)	CPU time(sec.)			Nr. of iterations			Time(sec.)/iter		
	P	K-M	B	P	K-M	B	P	K-M	B
275	3.2	5.0	2.7	33	8	20	.097	.625	.135
280	1.7	2.5	1.6	14	4	9	.121	.625	.178
285	1.0	1.3	1.2	8	3	6	.125	.433	.2
290	0.7	0.9	0.8	5	3	4	.14	.3	.2
295	0.6	0.79	0.67	4	3	3	.15	.263	.223
300	0.37	1.52	1.58	3	2	3	.123	.76	.527

These results were obtained on a HP-UX 9000 workstation, model B180L.

Table IV. Relative differences between the predicted results ($\times 10^{-3}$).

Temp.(K)	HCl			NH_3			HNO_3		
	P	K-M	B	P	K-M	B	P	K-M	B
275	4.4	3.0	4.8	5.2	3.9	1.0	4.7	1.9	4.7
280	1.0	1.3	3.9	2.8	6.6	5.4	0.9	1.5	4.3
285	2.9	0.9	7.2	1.4	2.8	8.7	2.9	8.3	7.7
290	5.9	3.7	4.8	9.7	0.3	6.3	6.2	3.7	5.4
295	0.2	0.7	4.1	1.8	1.1	1.0	0.3	0.5	4.4
300	0.3	8.4	3.5	3.0	0.3	4.3	0.4	10.2	3.9

2.8. IMPROVED ALGORITHMS FOR SOLUTION OF CHEMISTRY EQUATIONS

2.8.1. *Improved Algorithms for Solution of Stiff Atmospheric Chemistry Equations*

The large computational requirements in the study of chemically perturbed environments arise from the complexity of the chemistry of the atmosphere. Integration of the chemistry rate equations typically consumes as much as 90 percent of the total CPU time! This computational burden increases dramatically with the inclusion of aerosols. Obviously, more efficient integration schemes for the chemistry solvers would result in immediate benefits through the reduction of CPU time needed for each simulation. As more and more chemical species and reactions are added to the chemical scheme for valid scientific reasons, the need for faster yet more accurate chemical integrators becomes even more critical.

It is well known that the chemistry rate equations comprise a system of stiff ordinary differential equations (ODE). This fact precludes the use of explicit integration schemes. Indeed, theoretical considerations and numerical practice show that when an explicit integrator is applied to stiff ODEs, the code will choose prohibitively small integration steps in order to preserve stability. On the other hand, implicit integrators have infinite stability regions, and consequently, the step-size selection is done by monitoring the local truncation errors to meet the accuracy requirements set by the user. For integration of the chemistry rate equations, implicit integrators are likely to work with relatively large step-sizes when the accuracy requirements are not too stringent.

However, at each integration step, a nonlinear system of equations has to be solved. This involves the repeated evaluation of Jacobians and the solution of linear algebraic systems of dimension n, the number of species considered in the model. General stiff ODE solvers do not take advantage of the sparsity pattern of the Jacobian, and the number of arithmetic operations required for the numerical solution of the corresponding linear system is proportional to n^3. This is one of the reasons why general stiff ODE solvers are not very efficient for integration of the chemistry rate equations with a moderate to large number of species. A comparison of the exactness and time efficiency of different integrators can be found in the paper of [73].

More efficient chemistry integration algorithms for atmospheric chemistry have been obtained by carefully exploiting the particular properties of the model. One of the commonly used methods is the QSSA method from [39]. There are many other methods in use in atmospheric chemistry models. For example, S. Sillman in [74] developed an integration scheme based upon the analysis of sources and sinks of odd-hydrogen radicals in the troposphere. The basis of the method is the implicit Euler formula. Observing that nearly all tropospheric reactions include OH, HO_2, NO, NO_2 or NO_3 as a reactant and tropospheric reactions form sequential chains with a few recursive relationships, the author reorders and decomposes the vector of species so that the resulting Jacobian (used in the numerical solution of the implicit Euler equations) is nearly diagonal, thus enabling an elegant "decoupling" of the system of equations. Each decoupled part is treated differently. A fully implicit scheme is used for short-lived species (no matrix inversion needed here, because of the triangular upper left corner of the Jacobian) while a semi-implicit method is sufficient for long-lived species. This greatly helps the numerical computations. However, the scheme is difficult to generalize to solve more general problems.

Recently, Hertel et al. [38] proposed an algorithm based on the implicit Euler method. The algorithm uses linear operators only and preserves the mass. The nonlinear system is solved using functional iterations. The main idea is to speed up these iterations using explicit solutions for several groups of species. The method seems to work fine for very large step-sizes.

A particularly clear approach is taken from [33]. They divide the species into slow and fast ones, according to the size of their lifetimes; the slow species are estimated using an explicit Euler scheme; the fast ones are integrated with the implicit Euler scheme (and Newton-Raphson iterations for solving the nonlinear system); as a last step, the slow species are "corrected", reiterating the explicit Euler step. The hybrid technique allows a significant increases of the time step.

A fancy projection/forward differencing method was proposed in [29]. The species are grouped together in families; the distribution of the constituents inside a family is recalculated before each integration step using an implicit relation and solving the corresponding nonlinear system (this "projection" can be viewed as a "predictor"); then the integration is carried out for families using a significantly improved time step.

QSSA type methods were also explored in [61]. [46] investigates the extension of QSSA to a second order consistent scheme. It is tested against the "two step method", which is the second order BDF plus Gauss-Seidel iterations for solving the nonlinear system (according to the author, these iterations perform similarly to the modified Newton method, but with less overhead). The two step method seems to allow very large step-sizes.

A different approach is taken in [44]. The 3-D calculations are vectorized around the grid-cell dimension (very interesting idea) and advantage is taken of the sparse structure of Jacobians and a specific reordering of species (that makes Jacobians close to lower triangular form).

Although the above mentioned methods are more effective than general stiff ODE solvers, they are generally based on specific properties of the particular chemistry model rather than on general mathematical principles, and therefore, they are not easily applied directly to new situations. Furthermore, the main concern has been computational speed, and very little attention has been paid to formal error analysis.

We have looked in detail at the widely used QSSA method and have demonstrated that by combining a solid understanding at the mathematical level of the structure of this efficient ad hoc chemistry integration algorithm, with recent advances in the theory and practice of numerical methods for stiff ODE's, significant improvements in the efficiency and general applicability of this method are achieved in [45]. We have shown that the plain QSSA is an order one method, and that by using evaluations at one more point we can build an order two method. It is also shown that the global error of the plain QSSA has an asymptotic expansion in integer powers of the stepsize h and therefore extrapolation methods of arbitrarily high order can be constructed in the nonstiff case. In particular we obtained an order two method that uses two function evaluations per step which we call the Extrapolated QSSA Method. We also constructed a nontrivial modification of the well known GBS extrapolation algorithm based on an appropriate QSSA approximation whose global error has an assymptotic expansion in even powers of h (the proof of this fact is quite involved). In particular we obtained an order two method that uses three function evaluations per step which we call (for good reason) the Symmetric QSSA Method. In the stiff case the extrapolated methods are no longer so effective. However, we proved that the Extrapolated QSSA method and the Symmetric QSSA Method have a much smaller error constant which explains their superior performance.

Numerical experiments show that these techniques perform well, compared to the plain or iterated QSSA (especially when precision requirement is 10% or higher). In three dimensional atmospheric models, the optimal value for the relative error in the solution of chemistry kinetic equations is considered to be approximately 1%. More precision is redundant, due to inexactities in the transport scheme, and less precision can have an impredictible effect on overall accuracy, as it is hard to asses the error propagation through the transport scheme in an operator splitting algorithm. For this level of accuracy (two significant digits) the Extrapolated QSSA and the Symmetric QSSA Methods seem to be more efficient than Clasical QSSA, CHEMEQ % [79], and VODE [6] algorithms.

To test the properties of the new numerical methods we have chosen the chemical mechanism that is presently used in the STEM-II regional-scale transport/ chemistry/ removal model ([9]). This mechanism consists of 86 chemical species and 178 gas phase reactions. The mechanism, based on [53] and [2] is representative, of those presently being used in the study of chemically perturbed environments. Fig. 3 presents a work-precision comparison among different methods. The chemical mechanism was run for 5 days. The rate constants and initial conditions follow the IPCC scenario 3 ("Bio"). At the beginning of each hour, an injection of 0.01 $ppb/hour$ NO and NO_2 and 0.1 $ppb/hour$ of isoprene is carried out, to simulate an operator-splitting environment. The rate constants are updated at the beginning of each hour to their half interval values.

2.8.2. Efficient Treatment of Sparsity
A different approach to solving stiff ODE's arising from atmospheric chemistry is to use implicit integrators. Implicit integrators are very useful for solving stiff systems of ODE's arising from atmospheric chemistry kinetics. However the performance of implicit methods depends critically on the implementation of linear algebra (which appears in connection with Newton steps) and on their restart overhead (when used in an operator splitting environment). We have begun to explore these ideas [69]. We believe that it is possible to exploit sparsity in the Jacobian systematically and hence to develop more efficient implementations of fully implicit methods for solving atmospheric chemistry problems.

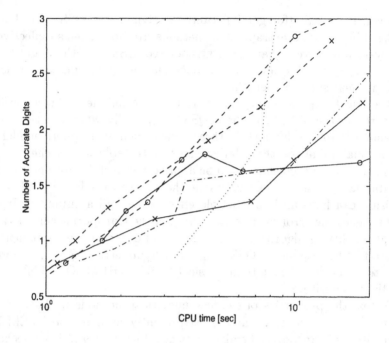

Figure 3. Work-precision diagram. QSSA (solid), VODE(dotted), Extrapolated QSSA (dashed with x), Symmetric QSSA (dashed with o) and CHEMEQ (dash-dots)

1. We evaluated reordering techniques that lead to minimal fill-in during LU decomposition. Our preliminary results indicate that diagonal Markowitz criterion performs very well (best, among the strategies considered so far).

2. We tested various off-the-shelf as well as tailored linear system solvers. We found that a simple Doolittle code, without pivoting, which relies on the off-line analysis of the sparsity structure performed by the symbolic preprocessor KPP, was the fastest method.

3. Finally we explored the effect of using this information to improve the efficiency of some state-of-the-art stiff ODE solvers. We chose two test problems consisting of comprehensive chemical mechanisms used to study stratospheric and tropospheric chemistry mechanism discussed previously. Information about the speed-ups obtained with sparse versus full linear algebra are presented in Table V.

In the usual practice, atmospheric modellers sometimes employ Gear type methods (see [44]). Our results indicate that when linear algebra is

Table V. Average speed-ups obtained.

CODE	STRATO	TROPO
VODE	2.48	4.63
SDIRK4	2.52	4.48
RODAS	2.01	3.50

treated efficiently, implicit methods from different families, like Singly-Diagonal-Implicit-Runge-Kutta or Rosenbrock, [69] can be competitive.

Fig. 4 compares results for three sparse codes with the widely used QSSA and CHEMEQ. A restart is carried out every hour or every 15 minutes (Fig. 4). This restart corresponds to an equal step-size of the transport scheme, in an operator splitting code. Since off-the-shelf integrators, endowed with the above sparse techniques, are quite competitive, it is clear that further work needs to be done for obtaining other implicit algorithms, that better suit the application. Considering the modest accuracy requirements, we expect that low order SDIRK methods and low order generalized linear methods are well suited for 3-dimensional air quality models, when sparsity is exploited to reduce the linear algebra overhead.

3. Sensitivity Analysis

Comprehensive sensitivity analysis of air pollution models remains the exception rather than the rule. Most sensitivity analysis applied to air pollution modeling studies has been focused on the calculation of the local sensitivities (first order derivatives of output variables with respect to model parameters) in box model studies of gas phase chemistry [15]. The most common form of sensitivity studies with comprehensive atmospheric chemistry/ transport models has been the so-called "brute force" method, i.e., a number of input parameters are selected to be varied and the simulation results are then compared. This method becomes less viable as the model becomes more comprehensive. A variety of alternative techniques are available including Green's function analysis [15], adjoint models and several variations of the direct decoupled methods [28].

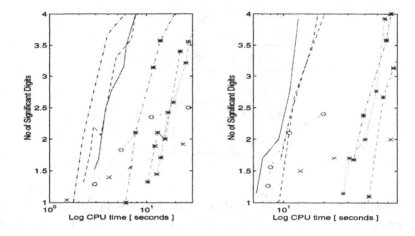

Figure 4. Work-precision diagram with a restart carried every hour (left) and every 15 minutes (right). Sparse VODE(solid), Sparse Rodas (dash - dots), Sparse Sdirk4 (dashed), QSSA (dots with "x") and CHEMEQ (dots with "o")

A recently developed technique for sensitivity study is automatic differentiation technology. Automatic differentiation is implemented by pre-compilers that analyze the code for evaluating a function of several variables and add instructions needed to compute the required derivatives. The resulting expanded code can simultaneously and efficiently evaluate derivatives and function values. This approach is superior to finite difference approximation of the derivatives because the numerical values of the computed derivatives are much more accurate and the computational effort is significantly lower [5], [4].

A promising new implementation developed at Argonne National Laboratory and Rice University over the last couple of years and which has recently been awarded the 1995 Wilkinson Prize for Numerical Software is the package ADIFOR (Automatic Differentiation in FOR-TRAN, [5]). It adopts a hybrid approach to computing derivatives that is generally based on the forward mode, but uses the reverse mode to compute the derivatives of assignment statements containing complicated expressions.

In [67] we analyzed forward mode Automatic Differentiation when applied to chemistry ODE solvers. This analysis shows that applying forward mode Automatic Differentiation is equivalent to using a direct method, i.e., to integrate forward in time the variational equations. This equivalence being established, a form of the direct decoupled method, which uses dedicated integrators, is considered.

We are presently extending the above described algorithms to three dimensional models, especially to computing the sensitivities of concentrations with respect to the intensities and locations of pollutant emission sources and to explore more fully the capabilities of ADIFOR.

3.1. PRELIMINARIES

If $c_i(t)$ is the concentration of the ith species, the kinetics of a chemical system is described as an initial value problem:

$$\frac{dc_i(t)}{dt} = f(t, c_1, ..., c_n, \beta_1, ..., \beta_m) , \tag{10}$$
$$c_i(t_0) = c_i^0 , \qquad i = 1 \cdots n$$

where β_j, $j = 1, ..., m$ are the parameters of the system (for example, reaction rate constants, etc).

The system (10) can be rewritten in matrix production - destruction form as:

$$\frac{dc(t)}{dt} = P(c(t)) - D(c(t)) \cdot c(t) \tag{11}$$

where $P \in \Re^n$, $D \in \Re^{n \times n}$, $D = diag(D_i)$ are the production and the destruction terms, respectively.

The *local sensitivity coefficients* are defined as:

$$s_{i,j}(t) = \frac{\partial c_i}{\partial \alpha_j} \tag{12}$$

where α_j represent either the initial values c_j^0 or the parameter β_j. The term *local* refers to the fact that these sensitivities describe the system around a given set of values for the parameters α. The system being considered to respond linearly for small perturbations, $s_{i,j}$ measures the ratio between the effect (*absolute* variation of the output Δc_i) and the cause (*absolute* variation of the input $\Delta \alpha_j$).

3.2. OVERVIEW OF SOME COMPUTATIONAL METHODS FOR SENSITIVITY ANALYSIS

There are many ways to compute sensitivities. Here we employ three different methods to compute the local sensitivity coefficients[5]. Conceptually, all three are equivalent, in the sense that a small perturbation

---

[5] Other techniques for sensitivity analysis, besides those described here, are available as well; one could mention *Green's function method* and *adjoint models*. Their description is beyond the scope of this paper.

of *a certain input* is propagated forward through the system, and the corresponding deviation of *all outputs*is estimated. Thus, all the methods described below may be called forward propagation methods. They are effective when the sensitivity of all (or many) outputs with respect to one (or few) entries are desired.

3.2.1. *"Brute-Force" Approach*

Equation (10) is first solved for parameters $\alpha_1, ..., \tilde{\alpha}_j, ..., \alpha_m$ then for $\alpha_1, ..., \bar{\alpha}_j, ..., \alpha_m$, and the obtained outputs are $\tilde{c}(t)$ and $\bar{c}(t)$, respectively using one-sided and central difference approach.

The "Brute-Force" approach requires only one (for one-sided differences) or two (for central differences) extra function evaluations for each independent variable with respect to which sensitivities are desired. The main drawback is that the accuracy of the method is hard to analyze. The smaller the perturbation ϵ, the lower the truncation error resulting from the omission of higher order terms (see the expansion of finite difference formulas in Taylor series), but the higher the loss-of-significance errors, resulting from subtracting two almost equal numbers. At the very best, the brute force approach results in a sensitivity approximation that has half the significant digits of f.

3.2.2. *Variational Equations*

By differentiating (11) with respect to the vector of parameters we obtain the variational equations:

$$\frac{d}{dt}\nabla c_i(t) \doteq \nabla P^i(c) - \nabla D^i(c) \cdot c_i - D^i \cdot \nabla c_i, \quad i = 1, ..., n \qquad (13)$$

The fact that the sensitivities satisfy (13) can be proved rigorously, see for example [1]. The notation ∇A stands for the sensitivity coefficients vector $\frac{\partial A}{\partial \alpha}$, and "$\doteq$" represents a vector (element-by-element) assignment.

To obtain $\nabla c_i(t)$, one has to numerically integrate the large system obtained by appending together (10) and (13). This method is usually referred as *the direct approach*. The initial values $\nabla c_i(0)$ must be set properly [6]. There are two main drawbacks of this approach:

- The generation of the variational equations requires significant extra effort;

- The integration of the large appended system may be very time consuming.

---

[6] If $\nabla x = \frac{\partial x}{\partial c_i}$ then $\nabla c_i(0) = 1$, otherwise $\nabla c_i(0) = 0$.

3.2.3. *Automatic Differentiation*

Automatic differentiation[7] techniques are based on the fact that any function (regardless of its complexity) is executed on a computer as a well-determined sequence of elementary operations like additions, multiplications and calls to elementary (intrinsic) functions such as sin, cos, etc.

By repeatidly applying the chain rule:

$$\frac{\partial}{\partial t} f(g(t))|_{t=t_0} = \left(\frac{\partial f(s)}{\partial s}|_{s=g(t_0)} \right) \cdot \left(\frac{\partial g(t)}{\partial t}|_{t=t_0} \right) \tag{14}$$

to the composition of these elementary operations one can compute, completely automatically, derivatives of f that are correct up to machine precision.

According to how the chain rule is used to propagate derivatives through the computation, one can distinguish two approaches to AD: the "forward" and the reverse "modes" (see [4], [5] for a detailed discussion).

- **The Forward Mode.** Is similar to the way in which the chain rule of differential calculus is usually taught. At each computational stage derivatives of the intermediate variables with respect to input variables are computed. These derivatives are propagated forward through the computational stages. From now on we will refer to forward automatic differentiation as FAD.

- **The Reverse Mode.** Adjoint quantities - the derivatives of the final result with respect to intermediate variables - are computed at each step. To propagate adjoints, one has to be able to reverse the flow of a program, and remember or recompute any intermediate value that nonlinearly impacts the final result.

3.3. SENSITIVITY CALCULATION VIA FAD-GENERATED VARIATIONAL EQUATIONS

3.3.1. *General Setting*

As shown by our previous considerations, automatic differentiation implicitly generates the linearized equations (13). To take advantage of that without facing any algorithmic problems, we propose the following hybrid approach: generate (13) via automatic differentiation, then solve the variational system using an integrator of choice. This approach should work better than the "blind" automatic differentiation, which is

---

[7] Reffered throughout the paper as AD.

in principle equivalent to the direct approach, with a fixed method for integrating the system; in the hybrid approach we have the extra degree of freedom of choosing the integration method and of doing different optimizations and easier than the usual way of implementing the direct approach where the variational equations are derived either by hand or by using symbolic manipulation.

To be more precise, suppose that the chemistry kinetics equations are described by the subroutine:

subroutine compute $(in : \alpha, \beta, c_0; out : cdot)$

where α, β are some parameters, c is the vector of concentrations and $cdot = \frac{dc}{dt}$ at given input arguments.

Forward automatic differentiation will generate:

subroutine g_compute $(in : \alpha, \beta, \nabla\beta, c_0, \nabla c_0;$
$out : cdot, \nabla cdot)$

Under the assumptions that sensitivities obey (13) we have that:

$$\nabla cdot = \nabla \frac{dc}{dt} = \frac{d}{dt}\nabla c. \tag{15}$$

Hence the subroutine *g_compute* describes completely (13). All that we have to do from now on is to properly set the initial conditions and to employ a standard numerical integrator.

3.4. APPLICATION OF FAD TO A COMPREHENSIVE CHEMICAL MECHANISM

The chemical mechanism used in this study is that presently used in the STEM-II regional scale transport/chemistry/removal model (see [9]). This mechanism consists of 86 chemical species and 178 gas phase reactions. The mechanism, based on the work of Lurmann et al. ([53]) and Atkinson et al. ([2]) is representative of those presently being used in the study of chemically perturbed environments. The mechanism represents the major features of the photochemical oxidant cycle in the troposphere and can be used to study the chemistry of both highly polluted (e.g., near urban centers) and remote environments. The photochemical oxydant cycle is driven by solar energy and involves nitrogen oxides, reactive hydrocarbons, sulphur oxides and water vapour. The chemistry also involves naturally occuring species as well as those produced by anthropogenic activities. Many of the chemical reaction rate coefficients vary with the intensity of solar radiation (photolysis rates), and thus follow a strong diurnal cycle. Others vary with temperature.

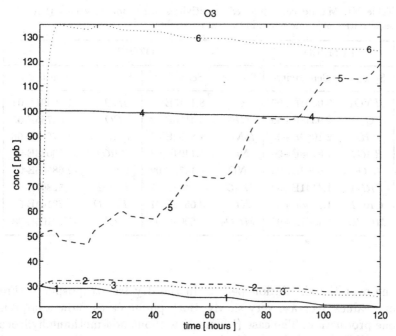

Figure 5. The variations of ozone under different IPCC scenarios (see Table VI for a detailed description).

To test the robustness of the above numerical algorithms, we have employed six different scenarios[8]. These conditions represent various chemical environments ranging from: low NO_x oceanic boundary layer regions (Marine); high NO_x continental boundary layer regions without (Land) and with isoprene (Bio); dry upper tropospheric regions (Free); biomass burning plumes without (Plume 1) and with (Plume 2) reactive hydrocarbon species. Further details are presented in Chapter 7 of the current WMO Ozone Assesment (see [30]). ADIFOR was used to calculate sensitivities of ozone with respect to initial conditions and reaction rate parameters. In the simulations of these cases the QSSA method with a fixed stepsize of 10 seconds was used. This algorithm is suited for direct automatic differentiation, and is easy to implement when solving (13). Its use leads to results having 1-2 significant digits. The calculated ozone concentrations for the five cases are presented in Fig. 5. In the marine boundary layer case, ozone is continuously destroyed throughout the 5 day period. The land and bio conditions show initially ozone production, followed by a net slight destruction of

---

[8] These scenarios follow the IPCC (Intergovernmental Panel on Climate Change) photochemistry intercomparison.

Table VI. Marine case, lumped sensitivities w.r.t. initial values (left) and sensitivities of ozone w.r.t. initial values (right).

LUMPED		OZONE			
Species	Sensitivity	Species	Sensit > 0	Species	Sensit < 0
HNO_3	3.0591E+00	O_3	8.1813E-01	H_2O_2	-1.12491E-02
O_3	2.9348E+00	CH_4	5.4110E-02	CO	-1.04765E-02
CH_4	2.1060E+00	HNO_3	4.8292E-02	$PRN1$	-1.17844E-06
$RAO2$	1.8158E+00	NO_2	1.1494E-02	$MRO2$	-3.33028E-07
R_3O_2	1.6063E+00	NO	1.1475E-02	$VRO2$	-2.68546E-07
$PRN1$	1.4141E+00	MAO_3	3.6442E-07	$RAO2$	-1.52807E-07
$CRO2$	1.2698E+00	$KO2$	3.0516E-07	$MVKO$	-8.79144E-08
$MCRG$	1.1467E+00	MCO_3	2.6008E-07	OH	-7.02106E-08

ozone over the simulation period. In the dry free troposphere (Free) ozone values decrease very slowly. Both plume cases show a large net ozone production. The case (Plume-1) without non-methane hydrocarbons (NMHC) shows a much slower net ozone production rate, and a distinct diurnal behavior. The Plume-2 case (with NMHC) shows a very rapid increase in ozone, followed by a period of slow ozone destruction.

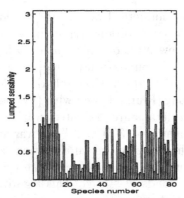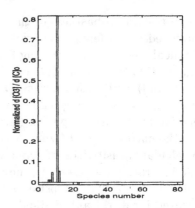

Figure 6. Marine case sensitivities

The calculated local sensitivities of ozone with respect to the initial conditions of each species for the Marine case are shown in Fig. 6. Plotted are the normalized sensitivities at 120 hours of simulation. Also shown are the 8 largest (+) sensitivities (indicating ozone production) and (-) sensitivities. Under these conditions ozone concentrations are most sensitive to the initial concentration of ozone (as expected since this case has a net destruction of ozone). Ozone levels increase with increases in CH_4, $NO_x = NO + NO_2$ and HNO_3, species which both lead to the production of ozone and also help to modulate the HO_2 concentrations which is the principal lose mechanism for ozone under these conditions. note also that HNO_3 is the principal source of NO_x in this case since its initial condition is an order of magnitude higher than NO_x. The largest negative sensitivity is that with respect to H_2O_2, which is the dominant source of HO_2 radicals.

Also shown are the lumped sensitivities. Lumped sensitivities can help describe the overall effect of a given perturbation. The lumped sensitivity of the system with respect to parameter α_j is defined as:

$$L(\alpha_j) = \sqrt{\sum_{i=1}^{N} \left(\frac{\partial c_i(t)}{\partial(\alpha_j)} \right)^2} \qquad (16)$$

where c_i, $i = 1, ..., N$ are the concentrations of the component species. Since $L(\alpha_j), j = 1, ..., m$ are functions of time, and since we are interested in the global effect of α_j over the system, we employ the mean values of the lumped sensitivity coefficients over the selected time horizon:

$$\bar{L}(\alpha_j) = \frac{1}{t_2 - t_1} \cdot \int_{t_1}^{t_2} L(\alpha_j) \cdot dt \qquad (17)$$

If the mean sensitivities:

$$S_j = \frac{\bar{L}(\alpha_j)}{N}$$

are far less then one, the system is considered stable with respect to the initial conditions (see [15]). A powerful aspect of the use of ADIFOR is that it is readily applied to combined transport/chemistry problems.

3.5. AEROSOL CONSIDERATIONS

As discussed, the photochemical oxidant cycle is a complex system, involving interactions among ozone, NO_x ($NO + NO_2$), HO_x ($OH + HO_2$), various HC (hydrocarbons), solar radiation, and aerosols. While significant efforts have been directed towards understanding the photochemical ozone formation and the nonlinear interactions between NO_x and HC, the impact of aerosol on tropospheric ozone production has received less attention because their impacts were thought to be small [EPA, 1991; SAI, 1995]. However, aerosols perturb the radiation balance, and thus affect photodissociation reactions. For example, NO_2 photolysis is the only primary source for tropospheric ozone production. Thus aerosol effects on NO_2 photodissociation could lead to significant impact on ozone levels. This effect has recently been studied and the presence of aerosol was found to increase or decrease the ground level ozone concentration by more than 20ppbv, depending on the single scattering albedo of particle [Dickerson, et al. 1997 [26]]. The significant role of aerosol in the photochemical oxidant cycle has not been fully exploited yet due to the large number of parameters and complicated interactions among aerosol, radiation, ozone and NO_x. ADIFOR is a powerful tool for sensitivity analysis, and can help in studying the role of aerosol-radiation interactions in the photochemical oxidant cycle.

To demonstrate this ADIFOR is applied to a coupled chemistry/ transport/ radiative transfer model, and is based on the STEM-III model (Crist, 1994 [17]).

3.6. ADIFOR APPLICATION ON STEM

In order to study the role of aerosol-radiation interactions in the photochemical oxidant cycle, ADIFOR is applied on a 1D STEM model. While the STEM 1D model simulates the tropospheric chemistry from the Earth's surface up to 8km altitude with 400km resolution, the radiative transfer model estimates perturbation of solar radiation from the Earth's surface up to 60km altitude. The radiation wavelength covers from 203nm to 435nm within 80 intervals. The modeling site studied is located at Bridgeport Connecticut, a typical urban region, 73.3 degrees W in longitude, 41.5 degrees N in latitude. The model is driven by local meteorological data. Relative humidity (RH) is at 70% for a June period. Typical aerosol mass concentrations in boundary layer vary from 10 to $70\mu g/m^3$ on the East Coast of US ([57]). For the base simulation, urban aerosol optical properties are used with concentration of $50\mu g/m^3$ within the boundary layer (0~2km altitude). The physical and optical properties of urban aerosols are obtained in [21]. At λ=550nm, aerosol extinction coefficient $\sigma_e x = 6.985E^{-6}/m$, single scattering albedo $\omega = 0.9383$, asymmetric factor $g = 0.647$. Simulations start at 7:00 pm local time and continually run with a 15 minutes interval for 3 days.

The top graph in Fig. 7 shows the calculated diurnal cycle of ground level ozone, NO_x, and NMHC for these conditions. As the sun rises, photodissociation begins with NO_2 photolysis providing atomic oxygen for ozone formation. The concentration of ozone rapidly builds up, while NO_x and NMHC are consumed and their concentrations decrease. The ozone curve reaches its peak and NO_x and NMHC curves reach their minimum in the early afternoon. As the sun sets, the ozone production decreases due to the inactive photodissociation, and NO_x and NMHC accumulate due to the continuous NO_x and NMHC emissions. The input temperature and eddy diffusion show a diurnal variation with high value at day time and low value at night, which cause chemical reactions and vertical transport more active in day time than at night. These facts can also contribute to the diurnal cycling of ozone, NO_x and NMHC.

The first-order sensitivities of ozone concentration to the independent variables TOMS, NMHC and NO_x emissions are also shown in Fig. 7. These sensitivities also show a diurnal cycle with peaks at sun rise. TOMS and NMHC emissions have positive impacts on ozone, but NO_x emissions have a negative influence on ozone concentration. Comparing the results from aerosol-free condition and with $50\mu g/m^3$ urban aerosol loading in the boundary layer, one finds a significant reduction of ozone (up to 31% at local noon after 3 days) when aerosols are present. Also, the presence of aerosol in the air enhances the impact of NMHC and NO_x on ozone, and weakens the influence of TOMS on ozone concentration.

4. Kinetic PreProcessor

The issues raised above regarding aerosol equilibrium, solution of stiff ODE's, sensitivity analysis, place great demands on the software. We have been developing software tools to assist model development and testing. The KPP preprocessor is shawn in figure 8. This tool facilitates change chemical mechanisms, evaluating ODE solvers, assisting sensitivity analysis, etc. Further details can be found in [22] and at http://www.cgrer.uiowa.edu/people in vdamian.

5. Closing Comments

Most of the recent attention regarding tropospheric aerosols has focused on their radiative properties. Secondary aerosols formed via the photochemical oxidant cycle play a significant role in radiative transfer by absorption and scattering of solar and terrestrial radiation, and by changing the optical properties of clouds through modification of the distribution of cloud condensation nuclei (CCN) ([14]). Current estimates suggest that the climate forcing due to aerosols linked to fossil fuel combustion and biomass burning largely offset that due to greenhouse gases in vast portions of the tropics and industrialized areas ([48]). However, large uncertainties remain in assessing the role

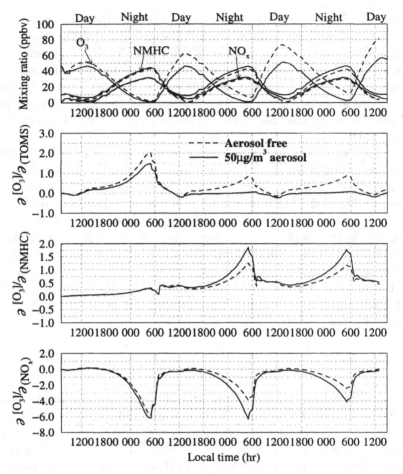

Figure 7. Diurnal Cycle of O_3, NO_x, and NMHC, and Sensitivity of O_3 concentration to TOMS, Emissions of NMHC, and NO_x.

of aerosols in climate and atmospheric dynamics due to the lack of detailed information on size distribution, chemical composition, surface properties, source strengths, and atmospheric transport and removal processes. Secondary aerosols also have a major impact on ambient visibility ([56]), and have significant adverse impacts on human health ([27]).

While the importance of inorganic aerosols (nitrate-ammonium-sulfate mixtures) has been widely appreciated and studied, there is growing awareness that organic aerosols are also important. Secondary organic aerosols (SOA) are a major component of urban environments ([59], [66]), and have also been found to represent an appreciable component

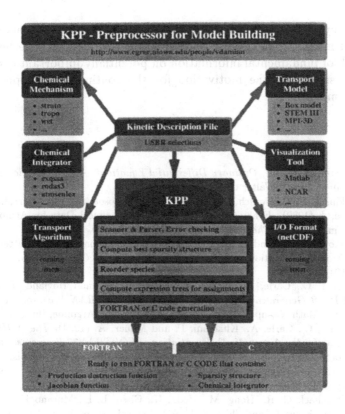

Figure 8. Schematic of the KPP Software Tool for Model Building.

of the fine aerosol mass (comparable to that for sulfate) at the Grand Canyon ([56]), as well as in the marine atmosphere ([55]). SOA are produced as a by-product of the oxidation of many organic molecules, including aromatics ([59]), monoterpenes ([60]), and carbonyls ([34]). These reactions yield semi-volatile compounds, which along with primary semi-volatile organics, partition themselves between the gas and particle phases. Furthermore, the presence of polar organic species on the aerosol surface can make the aerosol more hygroscopic ([70]), and thus increase the light scattering by particles. These organic aerosols may represent an important source of CCN ([58]).

Since aerosols and ozone share common chemical precursors and emissions sources, and are intricately linked through the photochemical oxidant cycle, it is important that they be studied in a simultaneous and coupled manner. The need to analyze ozone and aerosols together, and the lack of fundamental information on potentially important chemical processes, provide the motivation for the continued development of aerosol models.

References

1. Amann, H. (1990), *Ordinary Differential Equations: An Introduction to Nonlinear Analysis*, Walter de Gruyter, Berlin.

2. Atkinson, R. D., Baulch, D. L., Cox, R. A., Hampson, R. F. Jr., Kerr,J. A. and Troe, J. (1989),. Evaluated Kinetic and Photochemical Data for Atmospheric Chemistry, *J. of Chemical Kinetics*, **21**, pp. 115–190.

3. Binkowski, F. S. and Shankar, U. (1995), The Regional Particle Matter Model 1. Model Description and Preliminary Results, *J. Geophys. Res.*, **100**, pp. 26191–26209.

4. Bischof, C., Carle, A., Corliss, G., Griewank, A. and Hovland, P. (1992), *ADIFOR Generating Derivative Codes from FORTRAN Programs*, Technical report, Math. Comp. Sci. Div., Argonne Nat. Lab., Argonne, Illinois.

5. Bischof, C., Carle, A., Khademi, P. and Mauer, A. (1994), *The ADIFOR2.0 System for the Automatic Differentiation of FORTRAN77 Programs*, Technical report, Math. Comp. Sci. Div., Argonne Nat. Lab., Argonne, Illinois.

6. Brown, P. N., Byrne, G. D. and Hindmarsh, A. C. (1989), VODE: A Variable Step ODE Solver, *SIAM J. Sci. Stat. Comput.*, **10**, pp. 1038–1051.

7. Carmichael, G. R., Hong, M., Ueda, H., Chen, L. L., Murano K., Park, J., Lee, H., Kang, C. and Shim, S. (1997), Aerosol Composition at Cheju Island, Korea, *J. Geophys. Res.*, **102**, pp. 6047–6061.

8. Carmichael, G. R. and Peters, L. K. (1984), An Eulerian Transport/ Transformation/ Removal Model for SO_2 and Sulfate - I. Model Development, *Atmos. Environ.*, **18**, pp. 937–951.

9. Carmichael, G. R., Peters, L. K. and Kitada, T. (1986), A Second Generation Model for Regional-Scale Transport/Chemistry/Deposition, *Atmos. Environ.*, **20**, pp. 173–188.

10. Carmichael, G. R., Peters, L. K. and Saylor, R. D. (1991), The STEM-II Regional Scale Acid Deposition and Photochemical Oxidant Model - I. An Overview of Model Development and Applications, *Atmos. Environ.*, **25**, pp. 2077–2090.

11. Carmichael, G. R., Sandu, A., Potra, F. A., Damian, V. and Damian, M. (1996), The Current State and the Future Directions in Air Quality Modeling, *SAMS*, **25**, pp. 75–105.

12. Carmichael, G. R., Uno, I., Phadnis, M. J., Zhang, Y. and Sunwoo, Y. (1997), Tropospheric Ozone Production and Transport in the Springtime in East Asia, *J. Geophys. Res.*, submitted.

13. Chang, T. S., Carmichael, G. R., Kurita, H. and Ueda, H. (1989), The Transport and Formation of Photochemical Oxidants in Central Japan, *Atmos. Environ.*, **23**, 363–393.

14. Charlson, R. J., Schwartz, S. E., Hales, J. M., Cess, R. D., Coakley, J. A. Jr., Hansen, J. M. and D.J. Hofmann, D. J. (1992), Climate Forcing by Anthropogenic Aerosols, *Science*, pp. 423–430.

15. Cho, Y. S. (1986), Ph.D. Thesis, Univ. Iowa, Iowa City, Iowa.

16. Covert, D. S. and Heintzenberg, J. (1984) Measurement of the Degree of Internal/External Mixing of Hygroscopic Compounds and Soot in Atmospheric Aerosols, *The Science of the Total Environ.*, **36**, pp. 347–352.

17. Crist, K. (1994) *Comprehensive Regional Scale Air Pollution/Radiative Modeling*, PhD. Thesis. Dept. of Chemical Engineering, Univ. Iowa, Iowa City, Iowa.

18. Dabdub, D. and Seinfeld, J. H. (1994) Air Quality Modeling on Massively Parallel Computers, *Atmos. Environ.*, **28**, pp. 1679–1689.

19. Dabdub, D. and Seinfeld, J. H. (1994) Numerical Advective Schemes Used in Air Quality Models - Sequential and Parallel Implementation, *Atmos. Environ.*, **28**, pp. 3369–3385.

20. Dabdub, D. and Seinfeld, J. H. (1995) Extrapolation Techniques Used in the Solution of Stiff ODEs Associated with Chemical Kinetics of Air Quality Models. *Atmos. Environ.*, **29**, pp. 403–410.

21. d'Almeida, A. G., Koepke, P. and Shettle, E. P. (1991), *Atmospheric Aerosols Global Climatology and Radiative Characteristics*. A. Deepak Publishing, Hampton, Virginia, USA.

22. Damian, V. (1998), *Tools for Air Quality Modeling*, PhD. Thesis, Dept. of Comp. Sci., Univ. Iowa, Iowa City.

23. Demore, W. B., Howard, C. J., Golden, D. M., Kolb, C. E., Hampson, R. F., Kurylo, M. J. and Molina M. J. (1992), *Chemical Kinetics and Photochemical Data for Use in Stratospheric Modeling*, JPL Publication 92-20.

24. Dentener, F. J., Carmichael, G. R., Zhang, Y., Lelieveled, J. and Crutzen, P. J. (1996), Role of Mineral Aerosol as a Reactive Surface in the Global Troposphere, *J. Geophys. Res.*, **101**, pp. 22869–228896.

25. Dentener, R. and Crutzen, P. (1993), Reaction of N_2O_5 on Tropospheric Aerosols: Impact of the Global Distributions of NO_x, O_3 and OH, *J. Geophys. Res.*, **98**, pp. 7149–7163.

26. Dickerson, R. R., Kondragunta, S., Stenchikov, G., Civerolo, K. L., Doddridge, B. G. and Holben, B. N. (1997), The Impact of Aerosols on Solar Ultraviolet Radiation and Photochemical Smog, *Science*, **278**, pp. 827–830.

27. Dockery, D., Pope, A., Xu, X., Spengler, J., Ware, J., Fay, M., Ferris, B. and Speizer, F. (1993), An Association between Air Pollution and Mortality in Six U.S. Cities, *The New England J. of Medicine*, **329**, pp. 1733–1759.

28. Dunker, A. M. (1984), The Decoupled Direct Method for Calculating Sensitivity Coefficients in Chemical Kinetics, *J. Chemical Physics*, **81**, pp. 2385.

29. Elliot, S., Turco, R. P. and Jacobson, M. Z. (1993), Tests on Combined Projection/Forward Differencing Integration for Stiff Photochemical Family Systems at Long Time Step, *Comp.& Chem.*, **17**, pp. 91–102.

30. Prather, M. et. al. (1995), *Intercomparison of Tropospheric Chemistry/Transport Models*, Scientific assesment of ozone depletion, WMO, Jeneva.

31. Fuchs, N. A. (1964), *The Mechanics of Aerosols*. Pergamon Press.

32. Fuchs, N. A. and Sutugin, A. G. (1970), *Highly Dispersed Aerosols*. Ann Arbor Science, 1970.

33. Gong, W. and Cho, H. R. (1993), A Numerical Scheme for the Integration of the Gas Phase Chemical Rate Equations in 3D Atmospheric Models, *Atmos. Environ.*, **27A**, pp. 2147–2160.

134

34. Grosjean, E., Grosjean, D., Frazer, M. and Cass, G. (1996), Air Quality Model Evaluation Data for Organics. 2. C1 - C14 Carbonyls in Los Angeles Air, *Environ. Sci. Technol.*, **30**, pp. 2687–2703.

35. Hanel H. (1976), The Properties of Atmospheric Aerosol Particles as a Function of the Relative Humidity at Thermodynamic Equilibrium with the Surrounding Moist Air, *Advance Geophysics*, **19**, pp. 73–188.

36. Harrison, R. M. and Pio, C. A. (1983), Size-Differentiated Composition of Inorganic Atmospheric Aerosols of Both Marine and Polluted Continental Origin, *Atmos. Environ.*, **17**, pp. 1733–1738.

37. Heikes, B. and Thompson, A. M. (1983), Effects of Heterogeneous Processes on NO_3, $HONO$, and HNO_3 Chemistry in the Troposphere, *J. Geophys. Res.*, **88**, pp. 10883–10895.

38. Hertel, O., Berkowicz, R., Christensen, J. and Hov, Ø. (1993), Test of Two Numerical Schemes for Use in Atmospheric Transport-Chemistry Models, *Atmos. Environ.*, **27A**, pp. 2591–2611.

39. Hesstvedt, E., Hov, Ø. and Isaacsen, I. (1978), Quasi-Steady-State-Approximation in Air Pollution Modelling: Comparison of Two Numerical Schemes for Oxidant Prediction, *Int. J. Chem. Kinet.*, **10**, pp. 971–994.

40. Hitchcock, D. R., Spiller, L. L. and Wilson W. E. (1980), Sulfuric Acid Aerosols and HCl Release in Coastal Atmosphere: Evidence of Rapid Formation of Sulfuric Acid Particulates. *Atmos. Environ.*, **14**, pp. 165–182.

41. Jacob, D. J., Logan, J. A., Gardner, G. M., Spivakovsky, C. M., Yevich, R. M., Wofsy, S. C., S. Sillman, and M.J. Prather, M. J. (1993), Factors Regulating Ozone over the United States and its Export to the Global Atmosphere. *J. Geophys. Res.*, **98**, pp. 14817–14826.

42. Jacobson, M. Z. (1997), Development and Application of a New Air Pollution Modeling System - II. Aerosol Module Structure and Design. *Atmos. Environ*, **31**, pp. 131–144.

43. Jacobson, M. Z. (1997), Development and Application of a New Air Pollution Modeling System - III. Aerosol-Phase Simulations, *Atmos. Environ*, **31**, pp. 587–608.

44. Jacobson, M. Z. and Turco, R. P. (1994), SMVGEAR: A Sparse-Matrix, Vectorized Gear Code for Atmospheric Models, *Atmos. Environ.*, **17**, pp. 273–284.

45. Jay, L. O., Sandu, A., Potra, F. A. and Carmichael, G. R. (1995), *Improved QSSA Methods for Atmospheric Chemistry Integration*, Technical report 67, Univ.Iowa, Dept. Math., Iowa City, Iowa.

46. Verwer J. G. (1994), *Explicit Methods fo Stiff ODEs from Atmospheric Chemistry*, Preprint NM-R94, EMEP MSC-W Norwegian Meteorological Institute, Oslo, Norway.

47. Kaneyasu, N., Ohta, S. and Murano, N. (1995), Seasonal Variation in the Chemical Composition of Atmospheric Aerosols and Gaseous Species in Sapporo, Japan, *Atmos. Environ.*, **13**, pp. 1559–1568.

48. Kiehl, J. and Briegleb, B. (1993), The Relative Roles of Sulfate Aerosols and Green House Gases in Climate Forcing, *Science*, **260**, pp. 311–314.

49. Kim, Y. P., Seinfeld, J. H. and Saxena P. (1993), Atmospheric Gas-Aerosol Equilibrium I. Thermodynamic Model, *Aerosol Sci. Technol.*, **19**, pp. 157–181.

50. Kotamarthi, V. R. and Carmichael, G. R. (1990), The Long Range Transport of Pollutants in the Pacific Rim Region, *Atmos. Environ.*, **24**, pp. 1521–1534.

51. Kotamarthi, V. R. and Carmichael, G. R. (1993), A Modeling Study of the Long-Range Transport of Kosa Using Particle Trajectory Methods, *Tellus*, **45**, pp. 426–441.

52. Luria, M. and Sievering, H. (1991), Heterogeneous and Homogeneous Oxidation of SO_2 in the Remote Maritime Atmosphere, *Atmos. Environ*, **25**, pp. 1489–1496.

53. Lurmann, F. W., Loyd, A. C. and Atkinson, R. (1986), A Chemical Mechanism for Use in Long-Range Transport/Acid Deposition Computer Modeling, *J. Geophys. Res.*, **91**, pp. 10905–10936.

54. Lurmann, F. W., Wexler, A. S., Candis, S. N., Musarra, S., Kumar, N. and Seinfeld, J. H. (1997), Modeling Urban and Regional Aerosols: II. Application to California's South Coast Air Basin, *Atmos. Environ.*, submitted.

55. Matsumoto, K., Tanaka, H., Nagao, I. and Ishizaka, Y. (1997), Contribution of Particulate Sulfate and Organic Carbon to Cloud Condensation Nuclei in the Marine Atmosphere, *Geophy. Res. Lett.*, **24**, pp. 655–658.

56. Mazurek, M., Masonjones, M., Masonjones, H., Salmon, L., Cass, G., Hallock, K. and Leach, M. (1997), Visibility-Reducing Organic Aerosols in the Vicinity of Grand Canyon National Park: Properties Observed by High Resolution Gas Chromatography, *J. Geophys. Res.*, **102**, pp. 3779–3793.

57. Novakov, T., Hegg, D. and Hobbs, P. (1997), Airborne Measurements of Carbonaceous Aerosols on the East Coast of the United States, *J. Geophys. Res.*, **102**, pp. 30023–30030.

58. Novakov, T. and Penner, P. (1993), Large Contribution of Organic Aerosols to Cloud-Condensation Nuclei Concentrations, *Nature*, **365**, pp. 823–826.

59. Odum, J., Jungkamp, T., Griffin, R., Flagan, R. and Seinfeld (1997), The Atmospheric Aerosol Forming Potential of Whole Gasoline Vapor, *Science*, **276**, pp. 96–99.

60. Pandis, S., Harley, H., Cass, G. and Seinfeld., J. (1992), Secondary Organic Aerosol Formation and Transport, *Atmos. Environ.*, **26**, pp. 2269–2282.

61. Pandis, S. N. and Seinfeld, J. H. (1989), Sensitivity Analysis of a Chemical Mechanism for Aqueous-Phase Atmospheric Chemistry, *J. Geophys. Res.*, **94**, pp. 1105-1126, 1989.

62. Pankow, J. (1994), An Absorption Model of Gas/Particle Partitioning of Organic Compounds in the Atmosphere, *Atmos. Environ.*, **28**, pp. 185–188.

63. Peters, L. K. and Easter, D. (1994), Binary Homogeneous Nucleation as a Mechanism for New Particle Formation in the Atmosphere, *Monthly Update*, **3**, pp. 1–5.

64. Pilinis, C. and Seinfeld, J. H. (1987), Continuous Development of a General Equilibrium Model for Inorganic Multi-Component Atmospheric Aerosols. *Atmos. Environ.*, **21**, pp. 2453–2466.

65. Pilinis, C. and Seinfeld, J. H. (1988), Development and Evaluation of an Eulerian Photochemical Gas-Aerosol Model, *Atmos. Environ.*, **22**, pp. 1985–2001.

66. Rogge, W. F., Mazurek, M. A., Hildemann, L. M., Cass, G. R. and Simoneit, B. R. T. (1993), Quantification of Urban Organic Aerosols at a Molecular Level: Identification, Abundance and Seasonal Variation, *Atmos. Environ*, **27**, pp. 1309–1330.

67. Sandu, A., Carmichael, G. R. and Potra, F. A. (1995), *Sensitivity Analysis for Atmospheric Chemistry Models via Automatic Differentiation*, Technical report 73, Univ.Iowa, Dept. Math., Iowa City, Iowa.

68. Sandu, A., Potra, F. A., Carmichael, G. R. and Damian, V. (1996), Efficient Implementation of Fully Implicit Methods for Atmospheric Chemical Kinetics, *J. Comput. Phys.*, **129**, pp. 101–110.

69. Sandu, A., Potra, F. A., Damian, V. and Carmichael, G. R. (1995), *Efficient Implementation of Fully Implicit Methods for Atmospheric Chemistry*, Technical report 79, Univ.Iowa, Dept. Math., Iowa City, Iowa.

70. Saxena, P., Hildemann, L., McMurry, P. Seinfeld, J. (1995), Organics Alter Hygroscopic Behavior of Atmospheric Particles, *J. Geophys. Res.*, **100**, pp. 18755–18770.

71. Saylor, R. D. (1997), An Estimation of the Potential Significance of Heterogeneous Loss to Aerosols as an Additional Sink for Hydroperoxy Radicals in the Troposphere, *Atmos. Environ.*, **31**, pp. 3653–3658.

72. Seinfeld, J. H. (1986), *Atmospheric Chemistry and Physics of Air Pollution*, John Wiley and Sons, 1986.

73. Shyan-Shu Shieh, D, Chang, Y. and Carmichael, G. R. (1988), The Evaluation of Numerical Techniques for Solution of Stiff ODE Arising from Chemical Kinetic Problems, *Environ. Software*, **3**.

74. Sillman, S. (1991), A Numerical Solution for the Equations of Tropospheric Chemistry Based on an Analysis of Sources and Sinks of Odd Hydrogen, *J. Geophys. Res.*, **96**, pp. 20735–20744.

75. Song, C. H. and Carmichael, G. R. (1998), Gas-to-Particle Conversion and Thermodynamic Distribution of Sulfuric Acid and Volatile Inorganic Species in the Marine Boundary Layer, *Atmos. Environ.*, submitted.

76. Waggoner, A. P., Weiss, R. E., Ahlquist, N., Voert, P., Will, S. and Charson, R. J. (1981), *Atmos. Environ.*, **15**, pp. 1891.

77. Wexler, A. S., Lurmann F. W. and J.H. Seinfeld, J. H. (1994), Modeling Urban and Regional Aerosols - I. Model Developmens, *Atmos. Environ.*, bf 28, pp. 531–546.

78. Xiao, H., Carmichael, G. R., Durchenwald, J. N., i Thornton, D. and Brady, A. (1997), Long Range Transport of SO_x and Dust in East Asia During the PEM West - B Experiment, *J. Geophys. Res.*, **102**, pp. 28589–28612.

79. Young, T. R. and Boris, J. P. (1997), A Numerical Technique for Solving Stiff ODE Associated with the Chemical Kinetics of Reactive Flow Problems, *J. of Physical Chemistry*, **81**, pp. 2424–2427.

80. Zhang, Y. (1994), *The Chemical Role of Mineral Aerosols in the Troposphere in East Asia*, PhD. Thesis, Univ. Iowa, Iowa City, Iowa.

81. Zhang, Y., Sunwoo, Y., Kothamarthi, V. and Carmichael, G. R. (1994), Photochemical Oxidant Processes in the Presence of Dust: An Evaluation of the Impact of Dust on Particulate Nitrate and Ozone Formation, *J. Appl. Meteorol.*, **33**, pp. 813–824.

ETEX: A EUROPEAN TRACER EXPERIMENT
OBSERVATIONS, DISPERSION MODELLING AND EMERGENCY RESPONSE

H. VAN DOP
Institute for Marine and Atmospheric Research, Utrecht University, p.o. box 80005, 3508 TA Utrecht, The Netherlands

G. GRAZIANI
CEC, Joint Research Centre, Environment Institute, 21020 Ispra (Va), Italy

W. KLUG
Mittermayerweg 21, 64289-Darmstadt, Germany

1. Introduction

In 1986 the Chernobyl accident served as a rude reminder that relatively little effort had been devoted to atmospheric long range transport models. Deficiencies covered both our ability to understand and represent dispersion processes on the spatial and temporal scales involved and our capacity to acquire and treat the necessarily large amounts of data. Appreciable progress has been achieved on both fronts especially in data communications and computing power, but a further barrier exists in the form of the relatively sparse experimental data against which model results can be compared. Thus validation work post-Chernobyl has frequently relied on data resulting from that accident. Such data, however, was heterogeneous [25] often lacking in quality control and, being scattered over many institutes, usually only selectively available to individual modellers [2, 29, 22, 1, 13, 23, 20, 11, 15, 12].

The European Tracer Experiment (ETEX) was initiated in 1992, cosponsored by the European Commission (EC), the International Atomic Energy Agency (IAEA) and the World Meteorological Organization (WMO). It involved controlled tracer releases, systematic environmental monitoring at up to 2000 km, the collection of model predictions in simulated emergency response conditions and the evaluation of these predictions.

This paper contains a brief summary of the preparations for the field campaign and the experiment and contains the main findings of the dispersion modelling results and emergency response analysis including statistical aspects of comparison between model output and observation, and emergency response modelling.

Z. Zlatev et al. (eds.), Large-Scale Computations in Air Pollution Modelling, 137–150.
© 1999 *Kluwer Academic Publishers. Printed in the Netherlands.*

2. Preparatory work

The preparations for ETEX started in 1992, well in advance of the experiment, in order to select a suitable tracer and to co-ordinate simultaneously and on short notice the release activities, the start of sampling at the stations and the aircraft operations as well as communications. All these procedures had to be tested out and checked in the period well in advance of the experiment.

Perfluorocarbon compounds are suitable tracer substances for experiments on transport over long distances [7, 4]. They are non-toxic, non-depositing, non-water-soluble, inert and environmentally safe. At ambient pressure and temperature they are odourless and clear liquids which are released into the atmosphere by spraying the liquid into a hot air stream, causing it to evaporate.

A total of 168 stations, all part of the synoptic network of national meteorological services in 17 countries were equipped with sequential air samplers. The air samples were collected in metal tubes filled with absorbing material. They were distributed prior to the experiment and sent back to the Joint Research Centre (JRC) in Ispra(It) afterwards for chemical analysis by thermally desorbing the metal tubes and using gas chromatography with electron capture detection. Two studies were carried out to determine the ambient levels of the PFC tracers in Europe. The first took place in early 1994 and the second in October 1994 just before the experiment [21]. At all ETEX stations adsorption tubes were exposed for 14 days and their contents analysed at Brookhaven National Laboratory (BNL), USA to determine PFC levels. For quality control reasons, at a couple of measuring sites duplicate sampling was performed. A high number of field and laboratory blanks were distributed. Also the PFC standards in use at BNL and JRC were compared [19].

3. The experiment

To conduct a tracer experiment in Europe where meteorological conditions with westerly-south-westerly air flow prevail, a release location in the western part of France was selected. During the first release on 23 October 1994, starting at 16:00 UTC, perfluoromethylcyclohexane (PMCH) was used. In the second release on 14 November 1994 at 15:00 UTC perfluoromethylcyclopentane (PMCP), was used to avoid cross contamination. Both releases lasted 12 hours. A total of 340 kg PMCH and 490 kg PMCP was emitted, respectively, corresponding with average release rates of 7.95 and 11.56 g/s.

During the first release a rather strong west to south-westerly flow was advecting the tracer in the direction of the sampling network. A day later unstable flow was still observed over the release site and the advection area. The winds decreased after the end of the release.

During the second release, strong south-westerly winds were blowing at the release site, but decreased and veered to west, after the cold front passed in the early morning hours [10]. During both releases dispersion characteristics of the atmospheric boundary layer near and at the release site were monitored with constant level balloons floating at different heights in the atmosphere [14, 28].

Each station took 24 samples with a sampling time of three hours per sample, in total covering a period of 72 h. The sampling was initiated just before the expected arrival time of the tracer plume at each station. The last measurements were taken 96 hours after the start of the release in the easternmost countries. During each release, over 5000 air samples were taken. The laborious chemical analysis and quality control took approximately two years [18, 17]. During the first release tracer was also measured by aircraft at four different altitudes enabling in principle a 3-dimensional analysis of the dispersing plume.

4. Dispersion modelling results

Twenty four institutions (table 1) took part in the real-time forecasting of the cloud evolution, with 28 long-range dispersion models, using the meteorological forecast data from various sources and later from the ECMWF.

Surface concentration evolution at the locations where the tracer was sampled was simulated. The results of these calculations were compared for both experiments with the measurements. A short overview of the basic principles used in dispersion modelling is given below.

Modellers were informed in advance that the two releases would occur within the time window 15 October to 15 December 1994, and that they would be called to react in real time. They were, however, notified of the exact time and location only after the start of the release. The procedures adopted required the modellers to transmit their results initially by fax to the evaluation team at the JRC as soon as possible, and to send subsequently the same information in digital form for statistical processing.

The quantitative evaluation of the predicted concentrations led to the conclusion that there is a large number of models which give satisfactory results for the first release [3]. The fair agreement between model prediction and data for the first release does not seem to be

Table I. ETEX participants

COUNTRY	INSTITUTE acronym
Belgium	KMI
Bulgaria	BMI
Canada	CMC
Czech R.	CHI
Denmark	DMI
Finland	FMI
France	METEOFRANCE
France	IPSN
Germany	DWD
Israel	IIBR
Italy	ANPA
Japan	JAERI
Netherlands	RIVM
Netherlands	KNMI
Norway	NMI
Rumania	RMI
Russia	TYPHOON
Slovak R.	SHMU
Sweden	SMHI
Switzerland	SMI
United Kingdom	MetOffice
United States	NOAA
United States	ARAC
United States	SRS

strictly dependent on the type and complexity of the model used, although, not surprisingly, coarser meteorological input fields produce concentration values that are less accurate than those produced using finer scale meteorological models.

For the majority of models the differences between using the forecast data and analysed meteorological data are not significant (see table 2). This may indicate that the current meteorological (mesoscale) models could be considered adequate for emergency response purposes, at least in this selected simple meteorological situation.

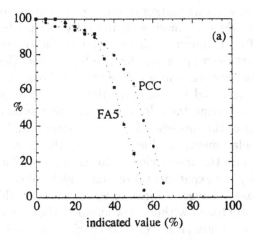

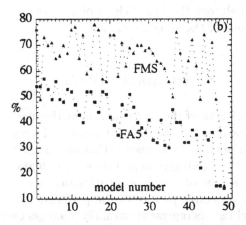

Figure 1. Model performance as derived from a statistical comparison between 72 3-hourly observed and predicted concentrations at 168 stations for the first release. The data used here result from a post factum analysis, using observed meteorological fields and are generally superior to the real time results. In total 49 models were included in this analysis. (a) Along the vertical axis the percentage of models is shown for which the statistical parameter PCC and FA5 exceeds the indicated value along the horizontal axis. PCC denotes the (linear) correlation coefficient on the logarithmic vallues. FA5 denotes the percentage of predictions which are within a factor of 5 from the observed values. (b) Along the horizontal axis the models are ranked from high to low according to their value of the correlation coefficient (PCC). Along the vertical axis the statistical parameters FA5 and FMS are shown. The FMS is the figure of merit in space for the time integrated concentration. If A_m and A_o respectively, denote the relative areas where the model and observed dose exceeds a given value, the dose FMS is defined as the ratio between intersection and union of these areas($(A_m \cap A_o)/(A_m \cup A_o)$). The best models shown here have a PCC> 0.6, an FA5> 0.5 and an FMS > 0.7.

The second release showed larger discrepancies between observations and model results where all models significantly overpredicted surface concentrations after 12 hours from the start of the release. No clear explanation has yet been given for this result. Further, the evaluation of model performances was more difficult, since both the number of sites hit by the tracer and the concentration levels were reduced, due to the strong winds during the release period. The results obtained by the large majority of the models indicate consistently an initial north-easterly cloud displacement, bending more towards the east with time. After about 24 hours the tracer cloud had left France. This evolution is not confirmed by the experimental results, which indicated that only a part of the release behaved as foreseen by the models, while a part of the material was still found within a few hundred kilometres east of the release site 24 to 48 hours later. No final conclusions could be derived yet for the second release. Possible stratification of the released air-tracer mixture is being investigated, together with other local effects which might have altered the near-field evolution of the plume in the particular conditions of the second release.

5. Statistical analysis

For the whole set of data of the first experiment the model results were compared with measured ground concentrations. The reason why more than one statistical measure was used is that any single test only offers a partial view of the (dis)agreement between simulation and reality. By presenting the results of a number of statistical tests we get a more complete and still objective picture of the performance of the various simulations. Model results represent normally averages over a few hours and are therefore compatible with the sampling time of three hours. However, a problem which is encountered in the comparison of model results with data is that the former represent spatial 3-dimensional grid averages with typical horizontal dimensions of 40-200 km, whereas the measurements are virtually point measurements. For the statistical comparison we have assumed observed data are spatially smooth (the typical distance over which the concentration field significantly changes is of the order of or greater than the horizontal grid-cell dimension). It should also be noted that the spatial interpolation of the model or measurement data in itself will effectively result in some smoothing of the data, thus changing the concentration levels. In particular in an early stage of development, where the (horizontal) grid-cell dimension and the sampling site interdistances greatly exceed the dimensions of the plume, this may greatly influence the dispersion pattern. These are all potential sources of error. Model performance is summarized in table 2 by a number of statistical measures explained below.

The bias, is defined as the difference between the mean values of the computed and the observed data and can be used to examine a model's tendency for over- or under-prediction. The Normalised Mean Square Error (NMSE) estimates the relative deviation between prediction and measurement. PCC denotes the (linear) correlation coefficient ranging between -1 and +1. FA2 and FA5 denote the percentage of predictions which are within a factor of 2 and 5, respectively from the observations. FOEX is also a measure for the bias. It equals the percentage of points above/below the $y = x$ line. It ranges from -50 (all below) to +50 (all above). In figure 1 some statistical results of this evaluation are summarized. A more comprehensive review is given in [9].

6. Mathematical concepts in dispersion modelling

Turbulence affects the tracer at all spatial scales, from the orifice through which it is released (0.02 m) to the continental scale (2000 km), covering eight orders of magnitude. In addition, the tracer moves initially close to the surface in a layer with normally a high wind shear and varying stability due to the thermal stratification affecting both the horizontal and vertical displacement. A slight dislocation of the cloud in an early stage may has a tremendous effect on its future evolution. Either distributing the tracer material through the whole boundary layer at the time of release or mixing the tracer through the boundary layer rapidly, resulted in more accurate forecasts of the position of the tracer cloud. For later stages it became evident that deformation and transport by the mean wind field dominated plume spread rather than small scale atmospheric diffusion.

These conditions are very demanding for an accurate description of the dispersion process: Not only a detailed knowledge is required of the turbulent flow field (the meteorological input data), but also model resolution, at least close to the source, should be sufficiently high.

The dispersion of passive (not interacting with the dynamics of the fluid in which it disperses) tracer is generally described by a (tracer) mass-conservation equation which can be formulated as

$$\frac{\partial \rho c}{\partial t} + \nabla(\rho u c) = 0, \qquad (1)$$

where ρ denotes the atmospheric density, u, the three dimensional velocity field and c the tracer mixing ratio field. Using the fluid mass continuity equation this can be transformed to

$$\frac{dc}{dt} = 0, \qquad (2)$$

where the operator $d/dt \equiv \partial/\partial t + u.\nabla$, expressing that the mixing ratio is conserved along a fluid trajectory. Molecular diffusion is neglected. Decomposing the variables in mean and fluctuating quantities and averaging, the resulting equation is

$$\frac{\partial C}{\partial t} + U.\nabla C = \frac{1}{\rho_0}\nabla.(\rho_0 K \nabla C), \qquad (3)$$

where upper case symbols denote mean quantities and ρ_0 expresses the atmospheric density which is assumed to be a function of height only. The term at the right-hand-side is an approximation for the turbulent transport induced by the fluctuations. It contains the eddy-diffusivity tensor K. It is a dominant term certainly in the vertical direction and also in low horizontal mean-wind conditions. In order to solve the last equation U and K are required as input data. These data are normally obtained from weather prediction models.

An alternative for the above, Eulerian description is the so-called Lagrangian description of dispersion where the position of individual fluid particles is evaluated from

$$X(t) - X(t_0) = \int_{t_0}^{t} U(t')dt'. \qquad (4)$$

Here, X denotes the fluid particle position and U its velocity, which also may consist of a mean and a fluctuating component. Any initial concentration field at time t_0 can now be mapped onto a new field at arbitrary time t, through integration of the above equation which displaces fluid particles (with conserved mixing ratios). The Eulerian and Lagrangian description are mathematically equivalent, though the absence of numerical diffusion in the latter method offers some advantages certainly in the case of a single point source.

7. Data assimilation

Data assimilation can be beneficial in improving and assessing source characteristics and model results. Robertson and Langner [26] use a variational method to reconstruct the source characteristics. First a cost function based on the 4D RMS difference between observed and predicted field is defined. Then, by making iterative runs forward and (adjoint) backward in time, the cost function is minimized. Thus the source term was adapted in order to get a best fit with the time sequence of the observed concentration fields. The result is illustrated in figure 2. A fair agreement with the real source is found though the same exercise repeated for the second release was less conclusive. The use of this technique, can certainly improve the forecast by making use of the additional information that a real-time-measured concentration field provides.

Furthermore, it was pointed out that the source term is an essential quantity in order to perform a dispersion forecast that it is usually unknown during an emergency and that this technique can probably be helpful in this sense. The ETEX database can be used to test and improve data assimilation algorithms and has much potential to elucidate this issue.

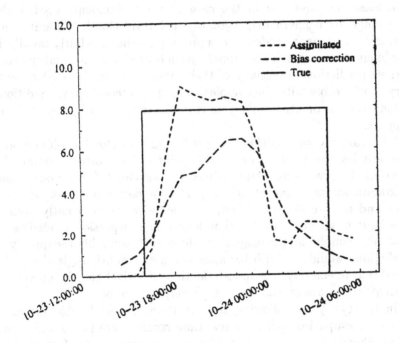

Figure 2. The true and assimilated temporal variation of the emission rate; solid line: true emission; dashed: full assimilation; long dashes:simplified approach. (Figure reproduced by courtesy of Robertson and Langner)

8. Discussion

ETEX demonstrated the capability of conducting a continental scale tracer experiment across Europe using the perfluorocarbon tracer technique and assembled a unique experimental database including tracer concentrations and meteorological data. It successfully established a communication network through which meteorological services and other institutes disseminated on demand in real-time their forecasts of concentrations of material released into the atmosphere. It further created widespread interest and resulted in considerable dispersion model improvement and development (see e.g.,[24, 26]), as well as in reinforcement of communication and collaboration between national institutes and international organizations. As a consequence, a large number of institutes can (and will in the case of a real accident) predict the long-range atmospheric dispersion of a pollutant cloud. While it is encouraging that many modellers can produce predictions fairly rapidly, it is evident that a much better appreciation is needed of the quality of the various predictions, especially of their inherent limitations which may vary considerably with the complexity of the meteorological conditions encountered in practice; this is an essential input for effective decision support.

The rapidity of modellers in predicting the cloud evolution and transmitting the results to a central point (within 3 to 6 hours), the precision in predicting arrival times of the cloud (3-6 hours) and maximum surface concentrations (within a factor of 3) at selected, close and remote locations should be viewed as the currently achievable limit of accuracy in real-time long-range dispersion modelling in relatively 'simple' meteorological conditions. Presumably the quality of predictions would diminish for more complex conditions. It should be noted that in emergency response the time of arrival and the maximum concentrations are crucial issues for emergency response.

In the two years following the experiment, considerable improvements in comparison with the real-time results were found which can be attributed to improvement of the dispersion model formulation, the use of more detailed meteorological data or to a tuning of model parameters. Nevertheless, a considerable disparity in the concentration values as presented by the different models persisted.

Major uncertainties in the position and intensity of a dispersing cloud will remain to exist also in the future since they will always include the basic uncertainty of a meteorological prediction. Deficiencies in the (diagnostic) description of a dispersing cloud using observed meteorological data should be added to this uncertainty.

The experiment contained a number of artificialities compared with what would be experienced in nuclear accident (e.g., no deposition, no monitoring data in real time as material disperses). The availability of such data would enhance the quality of the dispersion forecasts compared with the ETEX experiment. Few of the current models, however, have a capability to utilise monitoring data in real time to enhance predictions and this is an area for future developments [26].

Table II. ETEX first experiment. The first set of statistical measures is obtained using forecast meteorological data, the second is based on analysed meteorological data . FA2,FA5, FOEX(%), NMSE, bias (ngm^{-3}) and PCC (for explanation see the text).

Model No.	forecast						analysed					
	FA2	FA5	FOEX	NMSE	Bias	PCC	FA2	FA5	FOEX	NMSE	Bias	PCC
1	28	28	-33	101.3	-0.27	-0.01	26	27	-29	80.6	-0.27	0.00
2	26	28	-13	61.0	1.93	0.14	28	29	-23	158.4	1.60	0.18
3	20	25	-7	12.5	0.06	0 19	19	25	-7	21.0	0.08	0 09
4	28	40	1	6.9	0.17	0.49	26	39	5	12.7	0.16	0.38
5	25	28	-18	14.9	-0.13	0.06	25	29	-11	25.6	0.12	0.08
6	22	34	9	29.7	0.09	0.36	23	40	27	133.7	0.44	0.59
7	29	29	-32	75.2	-0.03	0.16	29	30	-29	1281.0	0.59	0.29
8	29	30	-29	30.3	-0.06	0.13	28	29	-28	37.7	-0.06	0.08
9	13	25	12	41.8	0.43	0.21	15	27	16	71.3	1.03	0.44
10	14	29	12	8.6	0.05	0.22	13	26	25	35.6	1.36	0.36
11	19	33	30	22.4	1.18	0.56	19	35	31	49.8	1.58	0.51
12	19	36	22	10.7	0.41	0.63	15	32	15	15.5	0.48	0.50
13	30	35	-23	49.5	-0.12	0.22	26	33	-21	114.9	0.08	0.32
14	28	32	-16	325.8	1.05	0.29	30	36	-17	134.8	0.35	0.44
15	27	39	3	7.5	-0.11	0.32	30	45	4	19.6	-0.06	0.43
16	29	32	-20	204.3	1.03	0.26	28	31	-19	315.5	1.10	0.18
17	27	29	-23	46.8	0.26	0.05	27	30	-22	59.7	0.27	0.00
18	34	44	-2	7.5	0.20	0.50	34	44	-2	13.6	0.24	0.50
19	18	36	28	3.5	0.25	0.40	18	36	26	4.4	0.17	0.34
20	28	31	-17	20.2	019	0.16	32	36	-17	32.0	0.44	0.30
21	27	29	-21	79.0	0.24	0.05	26	28	-22	4550.0	6.46	0.00
22	28	34	-4	55.8	0.65	0.30	28	34	-7	64.6	0.62	0.09
23	24	36	3	83.0	0.57	0.48	26	39	14	83.4	0.64	0.51
24	29	33	-9	16.5	0.22	0.17	26	30	-11	16.8	0.06	0.10
25	31	43	0	21.8	0.49	0.42	35	47	4	31.5	0.35	0.58
26	32	42	-10	1389.4	1.50	0.44	33	42	-12	2085.0	1.97	0.41
27	27	32	-13	7867.2	7.43	0.19	34	43	-5	9965.0	12.20	0.43
28	25	26	-26	35.3	-0.24	-0.08	26	31	-16	22.9	-0.08	0.28

9. Conclusions

1. Based on the experiments no preference was found for the mathematical description (Lagrangian vs. Eulerian models) with regard to its performance. Some Eulerian model results improved by initiating the dispersion with a Lagrangian particle dispersion model in the area near the source.

2. Present meteorological data have a typical horizontal resolution of 100 km and a vertical resolution of a few tens of metres. This is clearly insufficient to describe dispersion in the souce area. A solution for this problem may be to couple models which give better description of the small scale dynamics and dispersion to long-range tranport models. One of the findings of this study is that especially higher resolution meteorological data improve the model performance.

3. A standard approach in turbulent dispersion modelling is that all motion that is not resolved by the mean wind field is treated as (turbulent) diffusion (similar to molecular diffusion), where transport of tracer is taken proportional to the mean tracer gradient. In fact, in the atmospheric boundary-layer, which is a highly turbulent layer extending upwards to 1-2 km, turbulent diffusion is the major transporting mechanism. Noting that in the same layer horizontal winds may change strongly with height (both direction and magnitude), it is obvious that the intensity of vertical transport shall have a significant effect on both vertical and horizontal distribution.

4. In situations where (convective) clouds develop, vertical motion is strongly enhanced in the atmosphere and can extend up to the tropopause level (10 km). Since these processes, referred to as deep convection, are not resolved by weather prediction and general circulation models they are separately formulated in these models in order to get a better representation of cloud cover and precipitation. Deep convection is not (yet) generally present in the dispersion algorithms of the dispersion models currently used. Evidence from the second experiment indicates that this may be a serious defect, which may have contributed to the large overprediction of surface concentrations.

5. Increasing the resolution of meteorological data improves model performance (see e.g., [27, 16]).

6. Successful application of theories of atmospheric boundary layer structure and turbulence in an arbitrary environment in varying and complex conditions is still far from perfect: it appeared that simplified dispersion schemes, using e.g. constant boundary-layer height, still rank among the ones with the best performance [5, 6].

7. Dispersion predictions using observed meteorological fields differ from those using forecast meteorological fields. These differences are,

however, not significant in view of the generally large uncertainties in model predictions.

8. The ETEX database can be used to test and improve data assimilation algorithms and has much potential to elucidate this issue.

9. Since the accident in Chernobyl, slightly more than a decade ago, we have made enormous progress in communication- and computer-technology, so that we are now capable of providing swiftly information and predictions. Major advancements were made, due to increasing computer power (so that higher resolution models could be used) and to a lesser extent due to improved insights in long-range dispersion processes.

References

1. Albergel, A., Martin, D., Strauss, B. and Gros, J. M. (1988), The Chernobyl accident: modelling of dispersion over Europe of the radioactive plume and comparison with activity measurements, *Atmos. Environ.* **22**, 2431-2444.

2. Apsimon, H.M. and Wilson, J.J.N. (1987), Modelling atmospheric dispersal of the Chernobyl release across Europe, *Boundary Layer Meteorology* **41**, 123-133.

3. Archer, J., Girardi, F., Graziani, G., Klug, W., Mosca, S. and Nodop, K. (1996), The European Long range Tracer Experiment (ETEX). Preliminary Evaluation of Model Intercomparison Exercise. *Procs. of 21st International Technical Meeting on Air Pollution Modelling and its Application, Baltimore, USA, November, 6-10, 1995*, Gryning and Schiermeier (eds.), Plenum Press, New York.

4. Begley, P., Foulger, B. and Simmonds, P. (1988), Femtogram detection of perfluorocarbon tracers using capillary gas chromatography-electron-capture negative ion chemical ionisation mass spectrometry. *Journal of chromatography*, 445-453.

5. Brandt, J., Bastrup-Birk, A., Christensen, J. H., Mikkelsen, T., Thykier-Nielsen, S. and Zlatev, Z., (1998), Testing the importance of accurate meteorological input fields and parameterizations in atmospheric transport modelling, using DREAM - validation against ETEX-1, *Atmos. Environ.*, **32**, pp. 4167-4186.

6. Desiato, F., Anfossi, D., Castelli, S. T., Ferrero, E. and Tinarelli, G. (1998), The role of wind field, mixing height and horizontal diffusivity investigated through two Lagrangian models. *Atmos. Environ.*, **32**, pp. 4157-4165.

7. Dietz, R.N. (1986), Perfluorocarbon Tracer technology . in *Regional and long-range transport of air pollution* (S. Sandroni, ed.), Elsevier Science Publishers.

8. Eliassen, A. (1980), A review of long-range transport modeling. *Journal of Applied Meteorolgy* **19**, 231-240.

9. Graziani,G., W. Klug and S. Mosca (1998), *Real-time long-range dispersion model evaluation of the ETEX first release*, ISBN 92-828-3657-6. Office for Official Publications of the European Communities, Luxembourg.

10. Gryning, S.-E., Batchvarova, E., Schneiter, D., Bessemoulin, P. and Berger, H. (1998),, Meteorological conditions at the release site during the two tracer experiments, *Atmosr. Environ.*, **32**, pp. 4123-4137.

11. Hass, H., Memmesheimer, M., Geiss, H., Jakobs, H.J., Laube, M. and Ebel, A. (1990), Simulation of the Chernobyl radioactive cloud over Europe using the EURAD model. *Atmos. Environ.* **24A**, 673-792.

12. Ishikawa, H. (1995), Evaluation of the effect of horizontal diffusion on the long-range atmospheric transport simulation with Chernobyl data, *J. Appl. Meteor.* **34**, 1653-1665.

13. Kimura, F., and Yoshikawa, T. (1988), Numerical simulation of global dispersion of radioactive pollutants from the accident at the Chernobyl nuclear power plant. *J. Met. Soc. Japan* **66**, 489-495.

14. Koffi(1998), *Evaluation of Long Range Atmospheric Models using Environmental Radioactivity*, Elsevier Science Publishers, Barking, England.

15. Maryon, R.H. , Smith, J.B., Conway, B.J. and Goddard, D.M. (1991), The UK nuclear accident model. *Prog. Nucl. Energy* **26**, 85-104.

16. Nasstrom, J. S. and Pace, J. C. (1998), Evaluation of the effect of meteorological data resolution on Lagrangian particle dispersion simulations using the ETEX experiment, *Atmos. Environ.*, **32**, pp. 4187-4194.

17. Nodop, R. Connolly, and Girardi, F. (1997), European Tracer Experiment. Experimental Results and Data Base. In: *Proc. 6th Topical Meeting on Emergency Preparedness and Response*, San Francisco, CA, April 22-25, 1997.

18. Nodop, R. Connolly, F. Girardi (1997), The European Tracer Experiment. Experimental Results and Database. In: *Proc. 22nd NATO/CCMS Intern. Technical Meeting on Air Pollution Modelling and its Application*, Clermont-Ferrand, France 2-6 June 1997.

19. Nodop, K. Comolly, R. and Girardi, F. (1998), The field campaigns of the European tracer experiment (ETEX): overview and results, *Atmosr. Environ.*, **32**, pp. 4095-4108.

20. Piedelivre, J.P., Musson-Genon, L.M. and Bompay, F. (1990), MEDIA - a E ulerian model of atmospheric dispersion: first validation on the Chernobyl release. *J. Appl. Meteor* **29**, 1205-1220.

21. Piringer, K. Baumann, H. Rtzer, J. Riesing and Nodop, K. (1997), Results on Perfluorocarbon Background Concentrations in Austria, *Atmos. Environ.* **31**, 515-527.

22. Pudykiewicz, J. (1988), Numerical simulation of the transport of radioactive cloud from the Chernobyl nuclear accident. *Tellus* **40B**, 241-259.

23. Pudykiewicz, J. (1989), Simulation of the Chernobyl dispersion with a 3-D hemispheric tracer transport model, *Tellus* **41B**, 391-412.

24. Pudykiewicz, J. A. and Koziol, A. S. (1998), An application of the theory of kinematics of mixing to the study of tropospheric dispersion. *Atmos. Environ.*, **32**, pp. 4227-4244

25. Raes, F., Graziani, G., Stanners, D. and Girardi, F. (1990), Radioactivity measurements in air over Europe after the Chernobyl accident. *Atmos. Environ.* **24A**, 909-916.

26. Robertson, L. and Langner, J. (1998), Source funftion estimate by means of varitional data assimilation applied to the ETEX-1 tracer experiment. *Atmosp. Environ.*, **32**, pp. 4219-4226.

27. Sørensen, J. H., Rasmussen, A., Ellermann, T. and Lyck, E. (1998), Mesoscale influence on long-range transport - evidence from ETEX modelling and observations, *Atmos. Environ.*, **32**, pp. 4207-4218.

28. Stohl, A. and Koffi, N. E. (1998), Evaluation of trajectories calculated from ECMWF data against constant volume balloon flights during ETEX, *Atmos. Environ.*, **32**, pp. 4151-4156.

29. Wheeler, D.A. (1988), Atmospheric dispersal and deposition of radioactive material from Chernobyl. *Atmosp. Environ.* **22**, 853-863.

PARALLEL 4D-VARIATIONAL DATA ASSIMILATION FOR AN EULERIAN CHEMISTRY TRANSPORT MODEL

H. ELBERN, H. SCHMIDT and A. EBEL
Institute for Geophysics and Meteorology, Project EURAD
University of Cologne
D–50931 Köln, Aachener Str. 201–209, Germany

1. Introduction

The problem of exploiting observations scattered in time for the analysis of the state of the atmosphere is considered in meteorology for initial value determination since the mid–eighties (e.g. [7]). While there is now a growing literature on that subject in meteorology and oceanography, applications in atmospheric chemistry data assimilation are still very rare.

The objective of the present paper is (i) to make a step beyond 0+1–dimensional variational assimilation and to demonstrate the feasibility of the four–dimensional variational data assimilation technique (4D-var) for comprehensive gas phase chemistry transport models, (ii) to describe the parallel implementation, and (iii) to give an account of first experiences with the analysis skill and encountered preconditioning problems.

2. Model description

The EURAD CTM2 is a comprehensive tropospheric Eulerian model operating on the mesoscale–α. A full description of the EURAD (European Air pollution Dispersion model) CTM2, which is an offspring of the Regional Acid Deposition Model RADM2 ([2]) may be found in [5]. In its standard configuration the model domain encompasses the area from the eastern North Atlantic to the Black Sea and from northern Africa to central Scandinavia. The model grid configuration is $33 \times 27 \times 3$ in west–east, south–north, and vertical direction, respectively. The chemistry transport model calculates the transport, diffusion, and gas phase transformation of about 60 chemical species with 158 reactions. For the gas phase chemistry a semi–implicit and quasi steady state approximation method (QSSA) is applied for the

Z. Zlatev et al. (eds.), Large-Scale Computations in Air Pollution Modelling, 151–160.
© 1999 *Kluwer Academic Publishers. Printed in the Netherlands.*

numerical solution of the stiff ordinary differential equation system as proposed by [6]. Horizontal and vertical transport is simulated by a fourth order Bott scheme, [1]. The processes are calculated sequentially by symmetric operator splitting, when stepping from t to $t + \Delta t$. The following sequence is implemented $c_i^{t+\Delta t} = T_h T_z D_z A D_z T_z T_h \circ c_i^t.$, where T and D denote transport and diffusion operators in horizontal (h) or vertical (z) direction, respectively, which applies a fixed dynamic time step Δt of 10 minutes. The chemistry time step Δt_c of the stiff ordinary differential equation solver is highly variable in time.

3. Variational data assimilation

The following cursory description of the variational calculus focusses at the practical requirements of implementation, namely the storage problem. A more comprehensive exposition of the 4D–var method may be found in Talagrand and Courtier [11] for meteorological modeling and in Elbern et al. [4]) for an atmospheric chemistry application. A first example of adjoint sensitivity calculation is provided in [8].

Data assimilation procedures seek to find an initial model state which assures an optimal compliance between observations and an ensuing model integration. An objective measure to quantify the difference between measurements and model state is conveniently defined by a distance function as follows:

$$J(\mathbf{x}(t)) = \frac{1}{2}(\mathbf{x}_b - \mathbf{x}(t_0))^T \mathbf{B}^{-1}(\mathbf{x}_b - \mathbf{x}(t_0)) +$$
$$\frac{1}{2}\int_{t_0}^{t_N} (\hat{\mathbf{x}}(t) - \mathbf{x}(t))^T \mathbf{O}^{-1}(\hat{\mathbf{x}}(t) - \mathbf{x}(t))dt \qquad (1)$$

where J is a scalar functional defined on the time interval $t_0 \leq t \leq t_N$ dependent on the vector valued state variable $\mathbf{x} \in \mathcal{H}$ with $\mathcal{H}$ denoting a Hilbert space. The first guess or background values $\mathbf{x}_b$ are defined at $t = t_0$, and $\mathbf{B}$ is the covariance matrix of the estimated background error. The observations are denoted $\hat{\mathbf{x}}$ and the observation and representativeness errors are included in the covariance matrix $\mathbf{O}$.

Let the differential equation of the model $\mathbf{M}$ be given by $\frac{d\mathbf{x}}{dt} = \mathbf{M}(\mathbf{x})$, where $\mathbf{M}$ acts as a generally nonlinear operator defining uniquely the state variable $\mathbf{x}(t)$ at time t, after an initial state $\mathbf{x}(t_0)$ is provided. The linear perturbation equation, giving the evolution of a small deviation $\delta\mathbf{x}(t)$ from a model state $\mathbf{x}(t)$ then reads $\frac{d\delta\mathbf{x}}{dt} = \mathbf{M}'\delta\mathbf{x}$, where $\mathbf{M}'$ is the tangent linear model of $\mathbf{M}$.

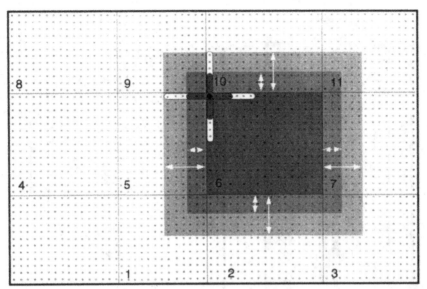

Figure 1. Horizontal domain decomposition for parallel computation with distributed memory. Three grid point single (short arrows, medium grey shading) and double (long arrows, medium + light grey) domain overlap as required by a fourth order upstream advection algorithm. The dark inner grid point cross indicates the grid points engaged in forward advection calculation of the central grid point, while the white outer cross denotes those grid points which require the adjoint value of the central grid point for the computation of their adjoint.

The evolution of an initial perturbation $\delta\mathbf{x}(t_0)$ at time t_0 to a perturbation state $\delta\mathbf{x}(t_n)$ at time t_n can be formally written by a sequence of stepwise integration operators $\tilde{\mathbf{R}}(t_i, t_{i-1})$ as

$$\delta\mathbf{x}(t_n) = \tilde{\mathbf{R}}(t_n, t_{n-1})\tilde{\mathbf{R}}(t_{n-1}, t_{n-2})\ldots\tilde{\mathbf{R}}(t_1, t_0)\delta\mathbf{x}(t_0) \qquad (2)$$

with $\tilde{\mathbf{R}}(t_i, t_{i-1})$ being a sufficiently accurate numerical operator for the stepwise calculation of (2).

Let λ and $\mathbf{M}'^{*}$ be the variables and model adjoint to $\delta\mathbf{x}$ and $\mathbf{M}'$, respectively. It can be shown that the adjoint differential equation then reads

$$-\frac{d\lambda(t)}{dt} - \mathbf{M}'^{*}\lambda(t) = \mathbf{O}^{-1}(\hat{\mathbf{x}}(t) - \mathbf{x}(t)). \qquad (3)$$

with the right hand side being the weighted observational forcing.

Finally, with $\tilde{\mathbf{S}}$ being the adjoint analog to $\tilde{\mathbf{R}}$, the gradient of the cost function $\mathcal{J}$ is given by

$$\nabla_{\mathbf{x}(t_0)}\mathcal{J} = \sum_{m=0}^{N} \tilde{\mathbf{S}}(t_0, t_1)\tilde{\mathbf{S}}(t_1, t_2)\ldots\tilde{\mathbf{S}}(t_{m-1}, t_m)\nabla_{\mathbf{x}}\mathcal{O}(t_m). \qquad (4)$$

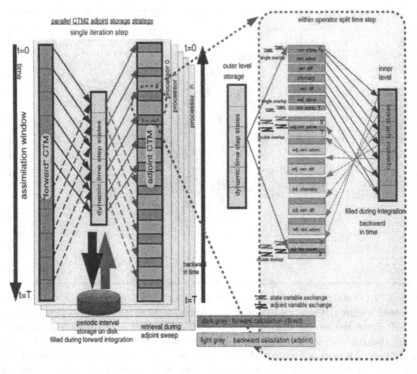

Figure 2. Storage and recalculation strategy for the parallel adjoint implementation. Left panel: Storage and retrieval for the full dynamic time step states of the entire assimilation window. Right panel: computation sequence, storage and retrival during a single dynamic time step for the intermediate operator split states. See text for further details.

The backward integration (4) may be regarded as a sequence of linear operators, linearized around time step states and all intermediate computational steps obtained during the forward sweep of the generally nonlinear forward model. For the proper construction of the linear adjoint operators these quantities must be readily available in reverse order.

The ensuing requirements of computer memory pose a principal problem in adjoint calculus of complex models. In atmospheric chemistry modeling this is the more the case as the number of constituent state variables are one order of magnitude higher than in the case of meteorological implementations.

An assessment of storage and calculation complexity of the EURAD–CTM can be based on the following estimates in terms of orders of magnitude: spatial dimension of the grid $\mathcal{O}(N_x \cdot N_y \cdot N_z) \approx 10^4 - 10^5$, number of constituents $\mathcal{O}(N_c) = 10^2$, dynamic time steps $\mathcal{O}(T_d) = 10^2$,

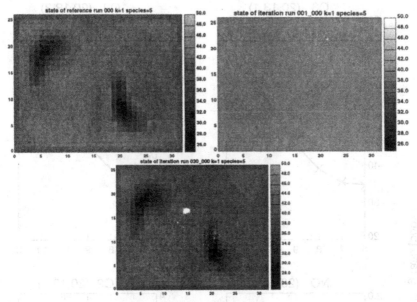

Figure 3. Identical twin experiments with preconditioning for ozone with respect to all other constituents. Displayed quantity is surface ozone. Upper left panel: reference state ('truth'), upper right panel: selected first guess for ozone, lower panel: analysis result after 30 iterations.

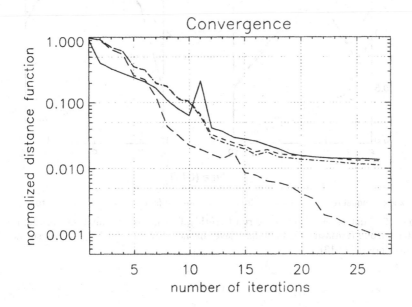

Figure 4. Convergence performance due to different prescaling factors: 1: solid; 10: long dashes; 100: dash–dots; 1000: dashes.

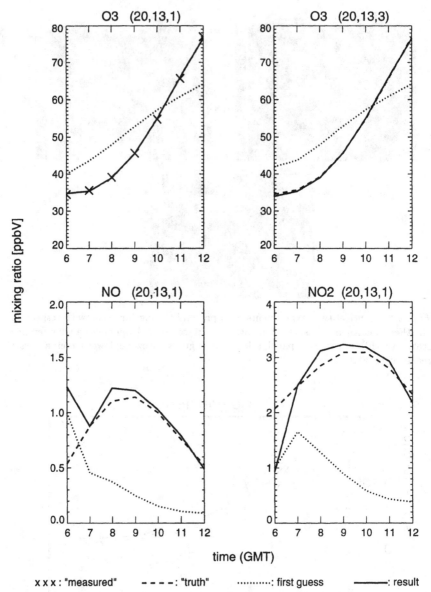

Figure 5. Time series of reference run ('truth'), first guess, and analysis (result) for surface ozone (mesasured), boundary layer ozone (230 m), and NO_x at the surface for grid point (20,13).

number of operator calls per dynamic time step $\mathcal{O}(M) = 10$, number of chemistry time steps within a dynamic time step $\mathcal{O}(T_c) = 10^2$, and number of intermediate results in the code $\mathcal{O}(L) = 10$. From runtime measurements it can be observed that the time of calculation of an adjoint piece of code T_{ad} takes about twice as much as the forward code T_{fw}. Table 1 comprises the storage and runtime estimates in terms of T_{fw} for three storage strategies:

Table I. Storage and recalculation strategies for nonlinear adjoint integration

storage strategy	forward runs per iteration	storage	$\mathcal{O}$	complexity $[T_{fw}]$
total storage	1	$N_x N_y N_z N_c M T_d T_c L$	$\approx 10^{12} - 10^{13}$	3
operator–wise	2	$N_x N_y N_z N_c M T_d$	$\approx 10^9 - 10^{10}$	4
dynamic stepwise	3	$N_x N_y N_z N_c (T_d + M)$	$\approx 10^8 - 10^9$	5

Due to storage limitations only the dynamic stepwise approach is feasible and adopted in this study. The computing time of about five forward integrations is required.

In the present implementation a Cray T3E with 128 MByte local storage per processor using a horizontal equal area partitioning as depicted in Fig. 1, is applied ([3]). On the basis of these computational and storage resources a single iteration step includes: first a forward model run with storage of all dynamic time steps $t_0, \ldots, t_T$ on disk. Then the backward adjoint integration from t_T to t_0 consists of the following operations for each step from t_l to t_{l-1}, $l = T, \ldots, 1$: a forward integration from t_l to t_{l+1} with storage of all operator split intermediate states, in main memory, followed by a backward integration from t_l to t_{l-1}. Each nonlinear adjoint operator requires the calculation of its forward version with internal storage of intermediate states. Finally, a call of the L–BFGS minimization routine ([9]) within the parallel implementation completes the iteration step. By this procedure, the nonlinear advection and vertical diffusion operators and the chemistry solver are integrated three times in forward direction per a single backward integration. A schematic of this procedure in a parallel application is presented in Fig. 2.

4. Results of identical twin experiments

The capability of the adjoint method is taxed within the framework of identical twin experiments. With this method artificial 'observations' are produced by a preceeding reference model integration based on initial values to be analysed by the subsequent assimilation procedure. The skill of the 4D–var method can then be estimated by comparing the analysis with initial values withheld from the reference run. The assimilation time interval spans six hours, starting at 06:00 local time in the centre of the integration domain at midsummer conditions. The length of the assimilation interval is limited by available computing time. The wind field is defined to form two vortices, a cyclonic one in the eastern, and an anticyclonic one in the western integration domain, with logarithmically increasing wind speed aloft.

In the experiment two circular emission areas are introduced, in the north–western and the south–eastern part of the integration domain. Only ozone observations of the lowest level are provided to the algorithm. Fig. 3 exhibits the the reference state, providing 'observations' (upper left panel), the assumed first guess state (upper right panel) for surface ozone, and the analysis result (lower panel) when the minimization process is stopped after 30 iterations. The emission areas of anthropogenic pollutants and the horizontal distribution of their plumes is clearly visible. A satisfying analysis is obtained with preconditioning by a factor 10 of ozone with respect to all other constituents, additionally to logarithmic scaling to preserve positive concentration levels.

In Fig. 4 the minimization progress is given in terms of the cost functions of different prescaling runs. The highest convergence rate is given with prescaling factor 10. For the other cases minimization proceeds at nearly the same rate, irrespective of a skillful analysis, as in the case of factor 1, or a false analysis, as in the cases of factors 100 and 1000. A closer look into the analyses of the NO_x concentrations reveals, that in the latter case high NO levels reduce the surface ozone very shortly after model start and hence reduce the cost function, while the ozone concentrations at the analysis time 06:00 UTC are erroneous. In fact it turns out that only the concentrations of $[O_x] = [O_3] + [NO_2]$ are skillfully analysed from the very start of the assimilation interval. A straightforward remedy of this problem would be a small shift to an earlier initial state, while maintaining the analysis time.

In Fig. 5 time series of the assimilation run with prescaling factor 10 are presented for a location in the plume of the south–eastern emission area at surface grid point (20,13). Each panel includes the 'true' state in terms of observations (dashed lines), first guess run (dotted lines), and the analysis results after 30 iterations (bold lines). The upper left panel displays the time series of the surface ozone, which is, recall, the only observed quantity. As expected it is clearly visible that the reference curve and the analysis curve are in very close agreement. A nearly similar picture is presented in the upper right panel, where the ozone analysis for the third layer (230m) is presented. All information obtained from the surface observations is propagated upwards by vertical transport and diffusion, which suffices to analyze ozone properly. The lower panels exhibit the analyses for NO (left) and NO_2 (right) at the surface on the basis of ozone observations mentioned above. Major discrepancies are visible for the first hour of the assimilation window, where the concentration levels are significantly off the chemical equilibrium. Taking notice of the different scale in the upper left panel, it may be observed that the analysis of ozone is about 1 ppbV higher than the 'measurement' at 06:00 UTC. Misfits between the three surface analyses may therefore be partly interpreted as an initialization problem with correctly analyzed O_x levels, as mentioned above.

5. Conclusions

In general the adjoint method appears to be a valuable tool for chemistry data assimilation also in the realm of atmospheric chemistry transport modeling. However, several precautionary remarks should be kept in mind, when interpreting the assimilation results. Minimization should be accelerated by preconditioning to attain a better optimization. The analysis at the starting point of the assimilation interval is not necessarily a good approximation to the true state, even if the cost function is small and the fit to observations later in the assimilation window is good. In this case there is an undue chemical imbalance at the starting point, which relaxes towards the true state very quickly. This is the chemical analog to the 'initialization problem' in meteorological data analysis, where fast gravity waves obstruct direct usage of the assimilation product. In the case of only ozone observations a further pitfall is given by ambiguous analysis results of precursor species like NO_x and VOCs, where various combinations may account for observed ozone variations ([10]). Hence, a good first guess or a priori knowledge is required. This is the reason why a sound suite of experiments are required to form a statistical basis for reliable background estimates. Nevertheless, the ability of the 4D-var to extract information from observations is unprecedented as compared to other methods within the limits of today's computing resources.

160

Acknowledgments

The authors would like to thank O. Talagrand, LMD–ENS, for many helpful discussions. We gratefully acknowledge the computational support by the Central Institute for Applied Mathematics (ZAM) of the Research Centre Jülich and the Regional Computing Centre of the University of Cologne (RRZK). Financial support was provided by the European Union's Climate and Environmental program project RIFTOZ under grant ENV4–CT95–0025, the Troposphärenforschungsschwerpunkt of the German Ministry for Education, Science, Research, and Technology (BMBF) under grant 07TFS 10/LT1–C.1, and the Ministry for Science and Research (MWF) of Northrhine–Westfalia.

References

1. Bott, A. (1989) A positive definite advection scheme obtained by nonlinear renormalization of advective fluxes. *Mon. Wea. Rev.*, **117**, 1006–1015.
2. Chang, J. S., Brost, R. A., Isaksen, I. S. A., Madronich, S, Middleton, P., Stockwell, W. R. and Walcek, C. J. (1987), A three-dimensional Eulerian acid deposition model: physical concepts and formulation, *J. Geophys. Res.*, **92**, 14618–14700.
3. Elbern, H. (1997), Parallelization and load balancing of a comprehensive atmospheric chemistry transport model. *Atmos. Env.*, **31**, 3561–3574.
4. Elbern H., Schmidt, H. and Ebel, A. (1997), Variational data assimilation for tropospheric chemistry modeling, *J. Geophys. R.*, **102**, 15967–15985.
5. Hass, H. (1991), Description of the EURAD Chemistry–Transport–Model Version 2 (CTM2), in A. Ebel, F.M. Neubauer, P. Speth (eds), *Mitteilungen aus dem Institut für Geophysik und Meteorologie der Universität zu Köln*, **Nr. 83**.
6. Hesstvedt, E., Hov, Ø., Isaksen, I. S. A. (1978), Quasi–steady state approximation in air pollution modeling: Comparison of two numerical schemes for oxidant prediction. *Int. J. Chem. Kinet.*, **10**, 971–994.
7. Le Dimet, F.–X., and Talagrand, O. (1986), Variational algorithms for analysis and assimilation of meteorological observations: theoretical aspects, *Tellus*, **38A**, 97–110.
8. Marchuk, G. (1974), *Numerical Solution of Problems of the Dynamics of the Atmosphere and Ocean*, Gidrometeoizadat, Leningrad, 1974, (in Russian).
9. Nocedal, J. (1980), Updating quasi–Newton matrices with limited storager,. *Mathem. of. Comp.*, **35**, 773–782.
10. Sillman, S., Logan, J. and Wofsy, S. (1990), The sensitivity of ozone to nitrogen oxides and hydrocarbons in regional ozone episodes. *J. Geophys. Res.*, **95**, 1837–1851.
11. Talagrand, O. and Courtier, P. (1987), Variational assimilation of meteorological observations with the adjoint vorticity equation. I: Theory, *Q. J. R. Meteorol. Soc.*, **113**, 1311–1328.

APPROACHES FOR IMPROVING THE NUMERICAL SOLUTION OF THE ADVECTION EQUATION

M. V. GALPERIN
Department of Biophysics, Radiation Physics and Ecology,
Moscow Physical - Engineering Institute,
Studencheskaya street, 38 - 31, Moscow, 121165, Russia

1. Introduction

One of the key problems in airborne pollution modelling is accuracy and fast solving of the advection equation. The best chemical schemes and boundary layer parametrization will be useless due to errors in calculation of advection. A lot of methods and schemes have been proposed for attacking the task [1,2,3], but only a few of them are suitable in practice, because the compromise between accuracy and computation speed is an open problem. The important sources of errors in advection computation are considered below and two numerical advection schemes developed on this base are presented.

2. Formulation of the problem

Since the time splitting of spatial co-ordinates is used, the one-dimensional form of advection equation may be considered:

$$\partial c / \partial t = - \partial (uc) / \partial x, \tag{1}$$

where c is a concentration, u is a wind velocity, t is time, x is a spatial co-ordinate. The both members of equation (1) will be integrated over t and x within the time step τ and one - dimensional grid cell Δx to get the finite-difference form:

$$\Delta Q = \int_0^\tau u_L(t)\, c_L(t)\, dt - \int_0^\tau u_R(t)\, c_R(t)\, dt, \tag{2}$$

161

Z. Zlatev et al. (eds.), Large-Scale Computations in Air Pollution Modelling, 161–172.
© *1999 Kluwer Academic Publishers. Printed in the Netherlands.*

where ΔQ is a difference of mass in the cell Q and the subscripts R è L correspond to the right and left boundaries of the cell. There is a difference between fluxes crossing the boundaries of the cell in right part of (2). Setting $\tau=1$, $\Delta x=1$, u=const. within τ and Δx and dimensionless velocity $U = u\tau /\Delta x$, the flux across every boundary can be expressed by equation:

$$F=\int_{-U}^{0} c(x)\, dx, \tag{3}$$

where $x=0$ corresponds to the boundary and $c(x)$ is a spatial continuous concentration profile. However, $c(x)$ is unknown, and only masses in cells Q_i $(i=1,...,N$ is a cell number) are known at each instant of time t. The numerical schemes are distinguished from each other by the way of restoration of $c(x,t)$ from the known Q_i and by the additionally used information, even though the fluxes do not appear explicitly in the scheme. In Lagrangian and semi-Lagrangian models [4 - 8] $c(x,t)$ is presented as a lattice constructed from δ-functions in centres of corresponding masses (the improved modifications are represented by methods of heavy points, among them random walk models). In spectral models $c(x,t)$ is approximated by Fourier series [9-12]. In finite-difference (and «slope») schemes [1, 13, 14] the approximation is constructed for central points of cells. Bott's scheme [15-17] is based on polynomial interpolation (a Lagrange polynomial of 4-th degree). It seems that the accuracy can be improved by increasing the order of the approximations. However, this does not take place due to the following reasons:

1. Values of $c(x,t)$ in far points can be used for approximation of $c(x,t)$ in a certain point under existence of high derivative of $c(x,t)$ with respect to x and t. In fact the last condition does not take place and, moreover, atmospheric fronts, precipitation and pollution sources produce the jump discontinuities of $c(x,t)$. As a result the approximation orders more than 3 or 4 accomplish nothing and even can impair the situation.

2. When the approximation of $c(x,t)$ is used, the mass in the cell differs in general

from $Q=\int_{x_L}^{x_R} c(x)dx$, where x_L and x_R are the co-ordinates of left

and right cell boundaries. This circumstance is not taken into account in many schemes at all or forces to introduce rather crude normalisation, limitation, etc..

Hence numerical advection schemes include the additional ad hoc assumptions, which should be validated. There are no efficient analytical methods for estimation of the total scheme errors (and possibility of their creating is questionable). At most, the order of linear error can be determined

[18] under the assumption that high derivatives exist and the non-linear re-normalisation does not add the significant error (very questionable assumption). Therefore the set of tests should be applied, each given only part of scheme features. The testing of scheme should be always aimed at its **disproving** (falsifying or refutation in terms of Popper [19]), **but not at its validation.** The successful test result permits only to hope that the scheme can operate satisfactorily in conditions close to test and nothing more.

From the above reasoning the numerical schemes for advection calculation meeting the conflicting requirements of accuracy (monotony, positive definiteness, a little numerical diffusion and transportivity, i. e. conservation of the first moments, *etc*) and computer resource economy were developed.

3. Self - normalising polynomial flux schemes (SNP)

The Bott's idea to use polynomial approximation of $c(x)$ is applied in this scheme. In Bott scheme the concentrations in point centres of cells are taken to be equal to $Q_i/\Delta x$. An interpolation polynomial passing through these points is built up, but in so doing the cell masses may not be equal to Q_i. An example is shown in *Figure 1, a*. As a result the calculated flux should be normalised by one or another expedient (see [15,20]).

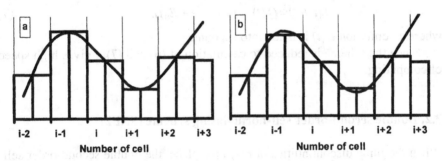

Figure 1. Two approaches to interpolation of concentration profile: a - with respect to mean cell concentrations, b - with respect to cell masses.

The approach to calculation of polynomial coefficients differs principally in presented self - normalising flux scheme. The assumption about concentrations $Q_i/\Delta x$ in cell centres is *ad hoc* hypothesis and it should not be fulfilled generally. At the same time the masses Q_i are known exactly and conservation of them is a necessary and sufficient condition of $c(x)$ restoration.

3.1. Third-order self-normalising polynomial scheme (SNP3)

Let

$$c(x) \approx f(x) = a_0 + a_1 x + a_2 x^2 + a_3 x^3 \tag{4}$$

and the flux between i-th and $(i+1)$-th cells should be determined. The origin of co-ordinates $x=0$ is chosen at boundary of the cells. The conditions of mass conservation in cells located symmetrically with the boundary are:

$$Q_{i-1} = \int_{-2}^{-1} f(x)dx; \quad Q_i = \int_{-1}^{0} f(x)dx; \quad Q_{i+1} = \int_{0}^{1} f(x)dx; \quad Q_{i+2} = \int_{1}^{2} f(x)dx. \tag{5}$$

Substitution of $f(x)$ from (4) in equations (5) gives the canonical algebraic equation set of the first degree for coefficients of polynomial (4). Then the polynomial and consequently the calculated flux will be normalised exactly to masses in cells (*Figure 1, b*). Of course the number of cells is taken at 1 more than polynomial degree and last one should be uneven. The substitution of $f(x)$ for $c(x)$ into (3) gives the flux F_i. For polynomial of 3-d degree it is found:

$$F_i = [7(Q_i + Q_{i+1}) - (Q_{i-1} + Q_{i+2})]U/12 + [5(Q_i - Q_{i+1})/8 - (Q_{i-1} - Q_{i+2})/24]U^2$$
$$- (Q_i + Q_{i+1} - Q_{i-1} - Q_{i+2})U^3/12 - [(Q_i - Q_{i+1})/8 - (Q_{i-1} - Q_{i+2})/24]U^4 \tag{6}$$

at limitations $Q_i \geq F_i \geq -Q_{i+1}$ and $F_i/U \geq 0$. After time step the value of Q_i is:

$$Q_i(t + \tau) = Q_i(t) - F_i + F_{i-1} + E_i(t), \tag{7}$$

where an emission $E_i(t)$ is taken into account.

Since the scheme is reduced to calculation of (6) and (7) it gives high speed of computing.

3.2. Double second-order self-normalasing scheme (SNP2)

When the jump discontinuities of $c(x)$ take place, the double second-order self-normalising scheme gives the somewhat better results. If $c(x) \approx f(x) = a_0 + a_1 x + a_2 x^2$ and the «upstream» approximation has been chosen, the conditions of mass conservation in $(i-1)$-th, i-th and $(i+1)$-th cells should be used for calculation of coefficients at $U \geq 0$ and in i-th, $(i+1)$-th and $(i+2)$-th cells - at $U \leq 0$. We obtain:

$a_1 = Q_{i+1} - Q_i$ in both cases;

$a_0 = 5\,Q_i/6 + Q_{i+1}/3 + Q_{i-1}/6$ and $a_2 = (Q_{i+1} + Q_{i-1})/2 - Q_i$, if $U \geq 0$;

$a_0 = 5\,Q_{i+1}/6 + Q_i/3 + Q_{i+2}/6$ and $a_2 = (Q_{i+2} + Q_i)/2 - Q_{i+1}$, if $U \leq 0$.

The flux through the boundary of i-th and $(i+1)$-th cells is

$$F_i = a_0 U - a_1 U^2 / 2 + a_2 U^3. \tag{8}$$

at limitations $Q_i \geq F_i \geq -Q_{i+1}$.

While the air pollution problem is solved, the initial conditions are taken to be zero or equal to emission value for both mentioned approximations. The calculation of cell row starts from $i = 0$ with $Q_{-1} = 0$ and $Q_0 = 2Q_1 - Q_2$ and stops at $i = N$ with $Q_{N+1} = 2Q_N - Q_{N-1}$ and $Q_{N+2} = 0$ (obviously $Q_0 \geq 0$ and $Q_{N+1} \geq 0$). The boundary effects are minimal under these conditions and the fluxes out of grid are determined correctly.

4. Scheme with step intracell concentration distribution (SCD)

The scheme discussed above has a significant numerical diffusion (the same as original Bott scheme), because the data of several adjacent cells are used for calculation in a given one. The mass movement of each cell should be computed independently from other ones for full elimination of pseudodiffusion as it is suggested by Egan and Mahoney [5]. The first and second moments of $c(x)$ in a cell are computed and conserved and the rectangular approximation of $c(x)$ is used in their pseudo-Lagrangian model. The accumulation and conservation of second moment result in cell mass may locate partially **outside the cell** before advection step and as a consequence a strong distortion of monotony can arise. In addition the second moment calculation requires a large computer time and store.

The presented scheme is based on calculation of the first moment only. The assumption about rectangular distribution of mass Q_i accords with conditions, when the first moment (e.g. mass centre) of distribution is known and the distribution is limited. Consequently, the mass Q_i should be initially distributed evenly **inside the cell** onto a maximum square at a known mass centre (and second moment conservation is not required). The movement and redistribution of Q_i within τ do not depend on masses in other cells and the new values of mass and first moment in cells must be obtained by superposition.

Let the mass $Q_i(t)$ and the first moment of mass $M_{xi}(t)$ in i-th cell be known at the moment t and

$$M_{xi} = x_{Ci} Q_i, \tag{9}$$

where x_{Ci} is a co-ordinate of mass centre. It is assumed that Q_i is distributed in the form of rectangular puff within the interval $(x_{Ci} - R,\ x_{Ci} + R)$, where $R = \min(x_{Ci} - i,\ i + 1 - x_{Ci})$ and cell numbers i and $i+1$ are equal to the co-ordinates of the left and right cell boundaries correspondingly (*Figure 2, a* and *b*). When the position of the cell puff after time step is calculated, the fractions of Q_i occurred in different cells can be determined as well as locations of their centres. Since all of $Q_i(t+\tau)$ and $M_{xi}(t+\tau)$ are results of superposition of all masses and moments, occurred at the moment $t+\tau$ in the i-th cell, $Q_i(t+\tau)$ and $M_{xi}(t+\tau)$ have been zeroized before routine time-step.

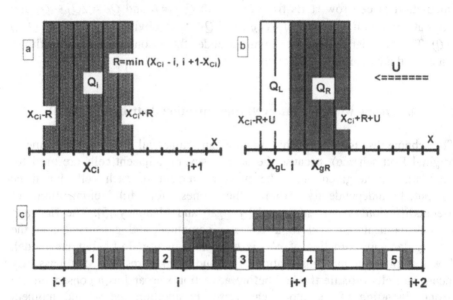

Figure 2. The SCD advection scheme: a - example of the cell mass distribution before time-step, b - same after time step, c - possible positions of cell puff: at the top and in the middle - before time-step, at the bottom - after it. Certainly, the puff may also cover the centre of cell before time step.

Examining the cells item-by-item (from $i=1$ to $i=N$) the co-ordinates of right and left puff boundaries of each cell after advection step are calculated: $B_R = x_{Ci} + R + U$; $B_L = x_{Ci} - R + U$. B_R and B_L are inside the cells with the numbers: $j_R = [B_R]$ and $j_L = [B_L]$, where square brackets is integer operator.

Let $K_i(j)$ be a fraction of $Q_i(t)$ occurred inside j-th cell and $x_{gi}(j)$ be a co-ordinate of its centre of mass. All possible cases can be described in the following way:

1. If $j_R = j_L$ (*Figure 2, c* positions 1,3 or 5) or R is small ($R \le \varepsilon$ and $0 < \varepsilon << 1$), then

$$K_i(j_R) = 1, \; x_{gi}(j_R) = x_{Ci} + U. \tag{10}$$

2. If $j_R \neq j_L$ (*Figure 2, c* positions 2 or 4), then

$K_i(j_R) = (B_R - j_R)/2R;\ x_{gi}(j_R) = (B_R + j_R)/2;\ K_i(j_L) = 1 - K_i(j_R);\ x_{gi}(j_L) = (B_L + j_R)/2.$

Values $K_i(j)Q_i(t)$ and $x_{gi}(j)K_i(j)Q_i(t)$ should be added to $Q_j(t + \tau)$ and $M_{xj}(t + \tau)$ respectively and calculation for i-th cell is finished. The emission $E_i(t)$ is prescribed as an additional flux discharged at the cell centre. As a result of time step routine the values:

$$Q_i(t + \tau) = \sum_{j=i-1}^{j=i+1} K_j(i)Q_j(t) + E_i(t);\quad M_{xi}(t + \tau) = \sum_{j=i-1}^{j=i+1} x_{gj}(i)K_j(i)Q_j(t) + (i+1/2)E_i(t). \quad (11)$$

are obtained for all cells. When calculation along axis x is performed in two-dimensional problem the fluxes of the first moment for y-axis should be taken into account:

$$M_{yi}(t + \tau) = \sum_{j=i-1}^{j=i+1} K_j(i)M_{yi}(t) + (i_y+1/2)E_i(t), \quad (12)$$

where i is cell co-ordinate on axis y. Calculation along axis y is performed in a similar way except the entering of emission.

The initial and boundary conditions are commonly assumed to be zero in air pollution problem. If it is necessary they can be pre-set in the same way as emission in corresponding grid cells and time steps. A horizontal diffusion may be included directly in the scheme by expanding the puff.

Additional examinations show that half-step time splitting does not change the calculation results. If the Courant number is less then 1 the accuracy does not depend on advection velocity.

5. Criteria for evaluation of the schemes and test results

The following criteria for evaluation of the schemes were used: *positivity* (negative values of mass or concentration should not arise in solution); *stability* at Courant number $-1 \leq U \leq 1$; *mass conservation* $(Q/E)100\%$, where Q is total mass within the calculation grid after last time step and E is the emitted mass; *transportivity* (bias of mass centre relative to the analytical solution); *relative error*; *non-monotony* (distortion of flat puff tops and continuous streams); numerical *pseudodiffusion* measure $(\sigma - \sigma_0)100\%/\sigma$, where σ_0 and σ are initial and final values of horizontal standard deviation; *non-additivity* (distortion of superposition, %); relative *computing time*; *using store* (the number of arrays). The testing included the rotating and linear movement of δ-function (point or one-cell puff), frontal, square flat, pyramidal and conic puffs and calculation of streams from continuous point sources in vortex and in uniform wind field with different disturbances (additional sources, precipitation).

The comparison was made for the schemes **SNP3, SNP2** and **SCD,** original and modifying Egan-Mahoney scheme **EMor** and **EMmd,** original Bott scheme **Bor** (Lagrange polynomial of 4-th degree, normalisation), modifying Bott scheme **Bmd** (Bessel polynomial of 3-d degree, flux normalisation), finite-difference MacCormack-Holmgren scheme **MCH** (approximation of 4-th order with respect to x and of 2-nd one - to t), simplest pseudo-Lagrangian scheme **LGN** (see Table 1).

Some examples are shown in *Figures 3* and *4. Figure 3* demonstrates the scheme response to impulse attack. **SCD** (and both **EM**) gives practically the analytical solution. **SNP3** gives the self-oscillating transient processes and significant errors (this is example of non-additivity). The same phenomena are inherent in Bott schemes and **MCH.** A reduction of the number of cells used for approximation in **SNP2** gives a significant decrease of oscillation. The 1-D tests, including movement of δ-function and square puff, show numerical pseudodiffusion and non-monotony. The numerical pseudodiffusion is measured by δ-function one-dimensional and rotating test. Since the mass centre of the one-cell puff does not commonly coincide with cell centre, the point mass should be distributed among four close cells. Hence the used pseudodiffusion measure should be within 50% (see *Figure 4, a*). The lesser values are accompanied by distortion of monotony or transportivity. The monotony is estimated by rotating and linear movement of square puffs (*Figure 4, b - d*) and calculation of «front» and continuous streams (*Figure 5*) in vortex. The mass conservation, stability and positivity are evaluated in every test.

Table 1. The integrated test comparison results.

Criterion	Emor	Emmd	Bor	Bmd	MCH	LGN	SNP3	NP2	SCD
Mass conservation, %	100	100	100	100	98-102	100	100	100	100
Transportivity, %	≤1	≤1	≤2	≤2	≤4	≤1	≤2	≤0.5	≤1
Relative error, %	≤20	≤20	≤10	≤10	≤40	≤100	≤12	≤12	≤8
Non-monotony, %	≤50	≤50	≤35	≤35	≤100	≤100	≤35	≤30	≤15
Pseudodiffusion, %	0-5	0-5	≈90	≈90	≈95	0	≈85	≈90	≤20
Non-additivity, %	≤1	≤1	≈50	≈50	≈50	≈10	≈50	≈50	≤1
Stability	+	+	+	+	+	+	+	+	+
Positivity	+	+	+	+	−	+	+	+	+
Computing time	5.3	5.2	3.8	3.4	1.7	1	1.4	1.5	1.8
Using store	10	10	2	2	2	6	2	2	6

Tests with the same schemes (excepting the new schemes from this paper) carried out in [21] and [20] have given similar results.

6. Conclusion

The mass conservation is a key principle in the development of advection schemes. Using information from a large number of cells in the computation of the concentration in a given cell leads to an increase of the pseudodiffusion, therefore the approximation order should in general be as low as possible.

The presented schemes, in which this principle is met, are stable, positive definite and mass conservative. Their monotony and transportivity are sufficient for most applications. The small numerical diffusion of the **SCD**-scheme makes it possible to calculate the transport of point mass. The **SCD** method makes it easy to check budget of mass and to calculate the export outside the grid limits. It is suitable for the 3-D problems and can be adapted for parallel calculation. The **SNP** - schemes are very fast and can be applied in relatively smooth problem. Both schemes can be simply harmonised with any models of vertical distribution, removal and transformation of substances.

Any comparive evaluation is approximate, because the speed of calculation depends on the implementation, software and hardware. Anyway, comparisons with results from other advection schemes show that the accuracy of the presented schemes is as good as that of the best known advection models. At the same time, the new schemes are often many times faster than the others.

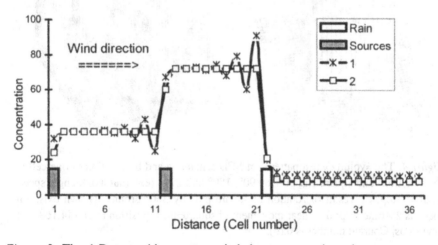

Figure 3. The 1-D test with constant wind demonstrates the scheme responses to impulse attack (one-dimensional streams from two continuous sources traverses a rain). The notation: 1 - SNP3 scheme, 2 - SCD scheme, the axis marks are cell boundaries.

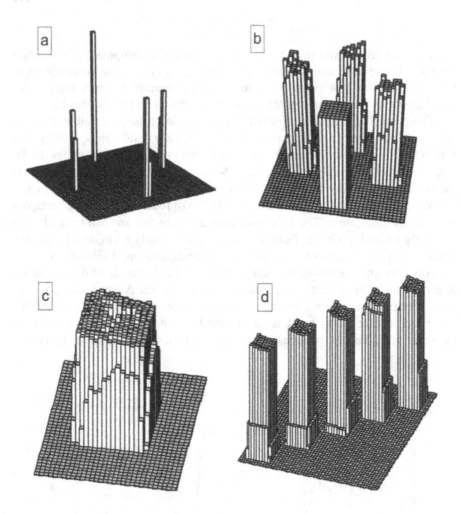

Figure 4. The «volume» test patterns of SCD scheme: a and b - rotation of one-cell pu
(δ - function) and square flat puff by 90°, 180° and 270° (external axis), angle speed
1°/ time step; c - result of rotation of square puff by 360° around its own axis, ang
speed is 2°/ time step; d - linear movement of square puff, position after 144,168,...,2
time steps, Courant number is 0,(6).

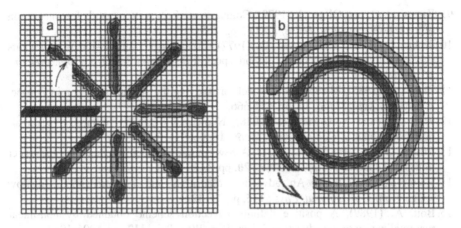

Figure 5. The «planar» test patterns of SCD scheme: a - front after rotation by 45°,...,315°; b - continuous streams from two point sources in vortex (rotation by 350°).

References

1. WMO-TCSU (1979), *Numerical methods used in atmospheric models*, Volumes I and II. - GARP Publication series No 17, Geneva.
2. Rood, R. B. (1987), Numerical advection algorithms and their role in atmospheric transport and chemistry models, *Rev. Geophys.*, **25**, pp. 71-100.
3. Peters, L. K., Berkowitz, C. M., Carmichael, G. R., Easter, R. C., Fairweather, G., Ghan, S. J., Hales, J. M., Leung, L. R., Pennell, W. R., Potra, F. A., Saylor, R. D., Tsang, T. T. (1995), The current state and future direction of Eulerian models in simulation the tropospheric chemistry and transport of trace species: a review, *Atmos. Envir.*, **29**, pp. 189-222.
4. Eliassen, A. (1978), The OECD study of long range transport of air pollutants: long range transport modelling, *Atmos. Environ.*, **12**, pp. 479-487.
5. Egan, A. B. and Mahoney, J. R. (1972), Numerical modeling of advection and diffusion of urban area source pollutants, *J. Appl. Meteor.*, **11**, pp. 312-322.
6. Pedersen, L. B. and Prahm, L. P. (1974), A method for numerical solution of the advection equation, *Tellus*, **XXYI**, pp. 594-602.
7. Pepper, D. W. and Long, P. E. (1978), A comparison of results using second - order moments with and without width correction to solve the advection equation, *J. Appl. Meteor.*, **17**, pp. 228-233.
8. Prather, M. J. (1986), Numerical advection by conservation of second-order moment, *J. Geophys. Res.*, **91**, pp. 6671-6681.

9. Kreiss, H. O. and Oliger, J. (1972), Comparison of accurate methods for integration of hyperbolic equation, *Tellus,* **XXIY,** pp. 199-215.

10. Prahm, L. P. and Christensen, O. (1977), Long range transmission of pollutants simulated by a two-dimensional pseudospectral dispersion model, *J. Appl. Meteor.,* **16,** pp. 896-910.

11. Zlatev, Z. (1985), Mathematical model for studying the sulphur pollution in Europe, *J. Comput. Appl. Math.,* **12,** pp. 651-666.

12. Zlatev, Z. and Berkowicz, R. (1988), Numerical treatment of large - scale air pollution model, *J. Comput. Appl. Math.,* **16,** pp. 93-109.

13. Russel, G. L., Lerner, J. A. (1981), A new finite-differencing scheme for the tracer transport equation, *J. Appl. Meteor.,* **20,** pp. 1483-1498.

14. Holmgren, P. (1994), An advection algorithm and an atmospheric airflow application, *J. of Comput. Phys.,* **115,** pp. 27-42.

15. Bott, A. (1989), A positive definite advection scheme obtained by non-linear renormalization of the advection fluxes, *Mon. Wea. Rev.,* **117,** pp. 1006-1012.

16. Bott, A. (1992), Monotone flux limitation in the area - preserving flux form advection algorithm, *Mon. Wea. Rev.,* **120,** pp. 2592-2602.

17. Bott, A. (1993), The monotone area - preserving flux - form advection algorithm: reducing the time - splitting error in two - dimensional flow fields, *Mon. Wea. Rev.,* **121,** pp. 2637-2641.

18. Odman, M. T. (1997), A quantitative analysis of numerical diffusion introduced by advection algorithms in air quality models, *Atmos. Environ.,* **31,** pp. 1933-1940.

19. Popper, K. (1959), *The Logic of Scientific Discovery,* Hutchinson, London.

20. Syrakov, D. and Andreev, V. (eds.) (1995), *Bulgarian contribution for EMEP. 1994 Annual Report,* NIMN, EMEP/MSC-E, Sofia - Moscow.

21. Berge, E. and Tarrason, E. (1992), *An evaluation of Eulerian advection methods for the modelling of long range transport of air pollution,* EMEP/MSC-W Note 2/92, DNMI, Norway.

APPLICATION OF PARALLEL ALGORITHMS IN AN AIR POLLUTION MODEL

K. GEORGIEV
Bulgarian Academy of Sciences,
Acad. G. Bonchev str., Bl.25A, 1113 Sofia, Bulgaria
E-mail: georgiev@parallel.bas.bg
Fax: +359 707 273

Z. ZLATEV
National Environmental Research Institute,
Frederiksborgvej 399, DK-4000 Roskilde, Denmark
E-mail: luzz@sun2.dmu.dk
Fax: +45 3630 1214

Abstract

One of the most important tasks has to be solved in the modern society is to find reliable and robust control strategies for keeping the pollution under certain safe levels and to use these strategies in a routine way. Large mathematical models, in which all physical and chemical processes are adequately described, in cooperation with the modern high-performance computers can successfully be used to solve this task. One such a model is the Danish Eulerian Model developed in the National Environmental Research Institute in Roskilde, Denmark.

The size of the computational task obtained after the appropriate splitting and discretization procedures of the system of partial differential equations describing the phenomenon is enormous and even when the computers with several GFlops top performance are in use, it is difficult to solve this problem efficiently and moreover, to prepare codes which may be used for operating purposes in estimating the pollution levels in different parts in Europe.

An new algorithm for implementation of the Danish Eulerian Model on parallel computers, both with distributed and shared memory, is discussed in this paper. The algorithm is based on partitioning of the computational domain in several subdomains. The number of the subdomains is equal to the number of the processors which are available. The splitting procedure used in the model is a splitting according to the different physical processes that are involved in it. As different numerical algorithms are used in the different submodels to types of subdomains are used: overlapping - into the advection-diffusion sub-

Z. Zlatev et al. (eds.), Large-Scale Computations in Air Pollution Modelling, 173–184.
© 1999 Kluwer Academic Publishers. Printed in the Netherlands.

models and nonoverlapping - into the chemistry-deposition submodels. The new algorithm is highly parallelizable. The Message Passing Interface (MPI) standard is used on both shared and distributed memory parallel computer platforms.

Numerical results obtained for an important module of the model (a transport-reaction scheme) on the parallel computer with distributed memory IBM SP (up to 32 processors) as well as concluding remarks explaining how the performance could be further improved are presented.

Key words: air pollution modelling, finite element method, domain decomposition, high-speed computers
AMS subject classifications: 65M20, 65N40, 65N55, 86A35

1. Mathematical description of the Danish Eulerian Model

The physical phenomenon, which is well known under the name "*Long-Range Transport of Air Pollution*" (the abbreviation LRTAP is often used) consists of three major stages:

1. **Emission.** During this stage different pollutants are emitted in the atmosphere from different emission sources. Many of the emission sources are anthropogenic, but some of the air pollutants are emitted also from natural emission sources.

2. **Transport.** The actual transport of the air pollutants is due to the wind. This is normally called "advection of the air pollutants".

3. **Transformations during the transport.** Three major physical processes take place during the transport of pollutants in the atmosphere:

   (a) **Diffusion.** The air pollutants are widely dispersed in the atmosphere (both in horizontal directions and in a vertical direction).

   (b) **Deposition.** Some of the pollutants are deposited to the surface of the Earth (soil, water and vegetation). Two different kinds of a deposition is considered: *dry* deposition (continues throughout the long-range transport) and *wet* deposition (takes place only when it rains).

   (c) **Chemical reactions.** Due to different chemical reactions which take place during the transport of the pollutants many *secondary* pollutants are created (the pollutants directly emitted

from the emission sources in the atmosphere are often called
primary pollutants).

There are two different ways to describe mathematically the air
pollution phenomena. The first is the Eulerian approach in which the
behavior of the species is described relative to a fixed coordinate system.
This description is a common way of treating heat and mass transfer.
The second approach is the Lagrangian in which the changes of the
concentrations are described relative to the moving fluid. The Eulerian
approach the Eulerian approach will be used. It is very useful because
the Eulerian statistics are readily measurable and the mathematical
expressions are directly applicable to situations in which the chemical
reactions take place.

Let us denote with q the number of the pollutants under consideration. The numerical experiments discussed in this paper have been
carried out by using a chemical scheme with 56 compounds, i.e. $q = 56$.
Let Ω be a bounded domain in the space with a boundary $\Gamma \equiv \partial\Omega$.
The Eulerian model for an adequately description of the physical and
chemical processes involved in the long-range transport of air pollutants
is able to be represented by the following system of partial differential
equations (PDE's) [11, 13]:

$$\frac{\partial c_s}{\partial t} = -\frac{\partial(uc_s)}{\partial x} - \frac{\partial(vc_s)}{\partial y} + -\frac{\partial(wc_s)}{\partial z} +$$

$$+\frac{\partial}{\partial x}\left(K_x \frac{\partial c_s}{\partial x}\right) + \frac{\partial}{\partial y}\left(K_y \frac{\partial c_s}{\partial y}\right) + \frac{\partial}{\partial z}\left(K_z \frac{\partial c_s}{\partial z}\right) + \quad (1)$$

$$+E_s + Q_s(c_1, c_2, \ldots, c_q) - (k_{1s} + k_{2s})c_s, s = 1, 2, \ldots, q,$$

where c_s are the concentrations; u, v and w are the wind velocities; K_x,
K_y and K_z are the diffusion coefficients; the sources are described by
the function E_s; k_{1s} and k_{2s} are the deposition coefficients; the chemical
reactions are described by the function Q_s.

The **advection** is described by the first two terms in (1), the **diffusion** - by the terms in the second line in (1). Very often advection part
of the equations can be simplified by assuming that a conservation law
is satisfied for the wind velocities in the lower parts of the atmosphere.
The **deposition** processes are presented by $- (k_{1s} + k_{2s})c_s$ (dry and
wet deposition, respectively). Sometimes, the dry deposition is put into
the boundary conditions. The terms E_s describe the **emission** sources
in the space domain under consideration. The **chemical processes**,
presented by the terms Q_s, play a special role in the model. The
equations in the system (1) are coupled only through the chemical

reactions. From the other hand the chemistry introduces nonlinearity in the model. As a rule, the chemical reactions are represented by the following formulae:

$$Q_s (c_1, c_2, \ldots, c_q) = -\sum_{i=1}^{q} \alpha_i c_i + \sum_{i=1}^{q}\sum_{j=1}^{q} \beta_{ij} c_i c_j \qquad (2)$$

Initial and boundary conditions are added to the system of PDE's (1).

2. Numerical treatment and computational complexity

It is difficult to treat numerically the PDE's system (1) directly due to the difficulties in finding a common, effective and fast, numerical method. The main reason is that the different terms of the right-hand side of the system come from very different physical processes, and they have different properties from a mathematical point of view. Therefore, it is very difficult to satisfy all requirements of the different terms simultaneously. The way to solve this problem is to apply some splitting procedure. Following ideas in Marchuk [6] and McRae et al. [7] the model is splited in five submodels according to the different physical processes that are involved in: *horizontal advection, horizontal diffusion, deposition, chemistry* (together with *source terms*) and *vertical exchange*.

$$\frac{\partial c_s^{(1)}}{\partial t} = -\frac{\partial(uc_s^{(1)})}{\partial x} - \frac{\partial(vc_s^{(1)})}{\partial y}, \qquad s = 1, 2, \ldots, q, \quad (3)$$

$$\frac{\partial c_s^{(2)}}{\partial t} = \frac{\partial}{\partial x}\left(K_x \frac{\partial c_s^{(2)}}{\partial x}\right) + \frac{\partial}{\partial y}\left(K_y \frac{\partial c_s^{(2)}}{\partial y}\right), \qquad s = 1, 2, \ldots, q, \quad (4)$$

$$\frac{\partial c_s^{(3)}}{\partial t} = E_s + Q_s \left(c_1^{(3)}, c_2^{(3)}, \ldots, c_q^{(3)}\right), \qquad s = 1, 2, \ldots, q, \quad (5)$$

$$\frac{\partial c_s^{(4)}}{\partial t} = -(k_{1s} + k_{2s}) c_s^{(4)}, \qquad s = 1, 2, \ldots, q, \quad (6)$$

$$\frac{\partial c_s^{(5)}}{\partial t} = -\frac{\partial(wc_s^{(5)})}{\partial z} + \frac{\partial}{\partial z}\left(K_z \frac{\partial c_s^{(5)}}{\partial z}\right), \qquad s = 1, 2, \ldots, q. \quad (7)$$

These submodels are discretized using different discretization algorithms and treated successively at each time-step (see for example

[12, 11]). In the algorithm considered in this paper, the spatial derivatives in (3) and (4) are discretized by using onedimensional linear finite elements. This descritization leads to systems of ordinary differential equations (ODE's) which are solved by using predictor-corrector schemes with several correctors (see [10]). The chemistry is treated numerically by an improved Quasi- Steady-State-Approximation (QSSA) algorithm (see: [1]) but the algorithm described in [4] is also possible to be used.

In the last few years the most often space discretization on a 96 × 96 × 10 grid (for the 3D version of the model, 10 layers in hight) is used. The grid is regular in xy planes and it is not regular in z direction. It is well seen that using this discretization one is to solve for each submodel systems of ODE's containing 5160960 equations at each time-step. More often the time-period for the model is one month + five days (to start-up the model, see [11]) and then 3456 time-steps are needed, assuming that the time-step for the advection submodel is $900sec$. Moreover, the systems of ODE's arising into the chemistry sub-model are stiff and therefore they have to be treated with smaller time-step ($150sec$. is actually used). It is clear that the computational task has to be solved is enormous and it causes great difficulties even when big and fast modern supercomputers are used. Therefore it is essential to select both sufficiently accurate and fast numerical algorithms and optimize the code realizing these algorithms for runs on parallel computers (with shared and distributed memory) and vector computers. Hereafter we will present an algorithm based on partitioning of the space domain for the 2D version of the Danish Eulerian Model for use on parallel computers.

3. Creating parallelism by using partitioning of the computational domain

Let us consider a parallel computer with p processors. Then we divide the computational domain Ω, which is a square now, in p subdomains in Y-direction. Similar algorithm one can get if the dividing is in X-direction. The combination of these two approaches will lead to a new algorithm. As was mentioned above, different numerical algorithms are used in the different submodels. Therefore, two types of subdomains are appeared. *Nonoverlapping* subdomains - Ω_i^{ch}, for the chemistry and deposition submodels and *overlapping* - Ω_i^{ad}, subdomains for advection and diffusion submodels (see Fig.1 for the case $p = 4$).

During the advection and diffusion processes the values of the concentrations inside Ω_i^{ad} are updated but we need the values of the con-

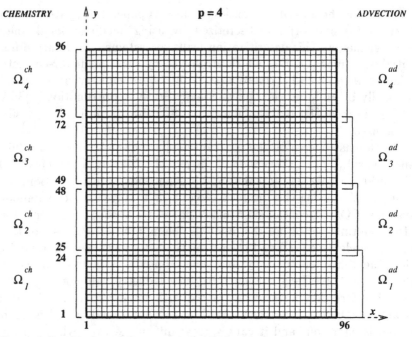

Figure 1. Partitioning of the computational domain in the case $p = 4$

centrations on the neighbor rows (24, 25, 48, 49, 72 and 53 for the four subdomains case on Fig.1) which are used as inner boundary conditions. In the chemistry and deposition submodels we do not need any boundary conditions and therefore there is no overlapping of the subdomains in these two submodels. Hence some communications are needed after the chemical and deposition part of the model. The number of communications is very small and communication time does not depend on the number of the subdomains when $p > 2$. The number of real numbers have to be communicated among the neighbor processors is two times the number of greed points belong to one grid lines. If the processor i is not the first or the last one then during the communication step it has to send to its neighbors in "north" and "south" directions the values of the concentrations in the grid-points belong to the second and last but one grid-line of the subdomain it is responsible. At the same time, it has to receive from the same neighbors the corresponding data and put these data on the grid-points belong to the first and last grid-line of its own subdomain. These updated values of the concentrations will be used as boundary conditions for the next advection time-step. It is clear that the first processor has not a neighbor from bellow and the last processor has not a neighbor from above. Therefore they will send and receive information only in "north" or "south" direction, correspondingly. Some additional communications are needed in the end of the algorithm when the values of the concentrations of all pollutants computed in the different processors have to be sent to the root processor which will prepare needed output data in whole computational domain.

For the communication part of the algorithm the Message Passing Interface (MPI) standard (see i.e. [3, 9]) is used on both shared and distributed memory parallel computer platforms. The use of MPI is a key component in the development of concurrent computing environment in which the applications and tools can be transparently ported between different computers.

At the current Fortran version of the code implementation of our algorithm the following primitives from MPI have been used:

— from the *Point-to-Point communications* in a *standard mode*:

  • **blocking send** (MPI_SEND) - for sending data;
  • **unblocking receive** (MPI_IRECV) - for receiving data;

— from the *Collective communications*:

  • MPI_BARRIER for *barrier synchronization*;
  • MPI_BCAST for *broadcasting* data;
  • MPI_GATHER and MPI_GATHERV (a vector variant of MPI_GATHER) for *gathering* data;
  • MPI_ALLREDUCE for *reduction* operations;
  • MPI_WTIME for *timing* (elapsed time) the codes or sections of the codes.

The most time consuming parts of the Danish Eulerian Model are *advection* and *chemistry* submodels. A module consisting of these two submodels was constructed and tested on different parallel computers (see [5, 11]). Such an advection-chemistry module is a good tool for testing the new algorithms according to the performance and accuracy in each of the processes and both together as well as some effects of the splitting procedure used in the last case. The same module as in [5, 11] will be used in this paper. The governing PDE's system is:

$$\frac{\partial c_s}{\partial t} = -(1-y)\frac{\partial c_s}{\partial x} - (x-1)\frac{\partial c_s}{\partial y} + Q_s(c_1, c_2, \ldots, c_q), \qquad (8)$$

where $s = 1, 2, \ldots, q$, $\quad 0 \le x \le 2$, $\quad 0 \le y \le 2$.

If the last term in the right-hand-side of (7) is skipped then (7) is transformed in the well known *rotation test* proposed in [2, 8]. The same splitting procedure as in the previous section is applied which leads to the same type of subproblems as (3) and (5). The algorithm for solving (7) after the discretization procedure is:

Initial stage
  - **read input data;**
  - **preparation of some matrices and other data which do not depend on the time;**

End of the initial stage

↓

Time cycle
  - **do the advection submodel;**
  - **do the chemistry submodel;**
  - **do the communications needed to update the boundary rows;**

End of the time cycle

↓

Final stage
  - **gathering the values of the concentrations into the root processor;**
  - **preparing data neede for the output;**
  - **output the results.**

End of the final stage

---

4. Numerical experiments

The numerical experiments for the transport-reaction scheme were performed on the parallel computer with distributed memory IBM SP Power 2 (up to 32 120 MHz RS/6000 processors). The top performance of each processor is 0.48 GFlops (billions of floating point computations per second). Let us call the direct use of the above described algorithm ALG-A. As it is well seen in the tables bellow this algorithm leads to a considerable reduction of the computational times. However, the efficiency which is measured by the ratio of the GFlops actually achieved and the maximal number of GFlops that can be achieved on the configuration used, is still very low. The main reason for these results is that the realization of this algorithm does not lead to an efficient use of the cache memory of the processors. From other hand it is seen from the numerical experiments that the computational time for the advection part is much less than the computational time for the chemical part. Therefore it is important to improve first and foremost the algorithm in its chemical part. The next three ideas were used and realized in this direction:

- Transposition of the arrays used in the chemical part (ALG-B);

- Using smaller arrays in the chemical part without any transposition of arrays (ALG-C);

- Using smaller arrays in the chemical part without any transposition of arrays and combining many loops into one loop (ALG-D).

The results from the numerical experiments when the *advection-chemistry* module was run on different number of processors and the above described four algorithms were applied can be found in the following tables. The *computing times - T_p* measured in seconds obtained for the four algorithms are given in Table I. The *speedups* defined as the computing time on p processors divided by the computing time on two processors - $S_p = T_p/T_2$ and *parallel efficiency - $E_p = 2S_p/p$* are presented in Table II. Finally, the *computational speeds* measured in GFlops and the *efficiency* defined as $100*GFlops/p*top-performance$ can be seen in Table III.

Table I. Computing times - T_p

Algorithm	2 proc.	4 proc.	8 proc.	16 proc.	32 proc.
ALG-A	3977	1703	869	474	272
ALG-B	3344	1598	838	439	257
ALG-C	2915	1457	756	412	249
ALG-D	2683	1321	689	389	218

The superlinear speedup which can be seen in some of the runs (see Table II) can be explained with *the increased memory locality, the cash locality* and *the communication overhead*.

The code realizing the same algorithm was run on the vector computer CRAY C92A with the top performance of the processor of 0.9

Table II. Speedups - S_p and parallel efficiency - E_p (in the brackets)

Algorithm	4 proc.	8 proc.	16 proc.	32 proc.
ALG-A	2.33 (1.17)	4.57 (1.14)	8.39 (1.05)	14.62 (0.91)
ALG-B	2.09 (1.05)	4.00 (1.00)	7.61 (0.95)	13.01 (0.81)
ALG-C	2.00 (1.00)	3.85 (0.96)	7.09 (0.88)	11.70 (0.73)
ALG-D	2.03 (1.02)	3.89 (0.97)	6.89 (0.85)	12.30 (0.77)

Table III. Computational speeds in GFlops and efficiency in percent (in brackets)

Algorithm	2 proc.	4 proc.	8 proc.	16 proc.	32 proc.
ALG-A	0.122 (12.2)	0.284 (14.2)	0.566 (13.9)	1.02 (12.8)	1.77 (11.1)
ALG-B	0.145 (14.7)	0.303 (15.1)	0.577 (14.4)	1.10 (13.8)	1.88 (11.8
ALG-C	0.166 (16.6)	0.322 (16.6)	0.640 (16.0)	1.17 (14.6)	1.94 (12.1)
ALG-D	0.180 (18.0)	0.366 (18.3)	0.702 (17.6)	1.24 (15.5)	2.22 (13.9)

GFlops. The results from this run are given in Table IV. It is seen that the best computing time achieved on the IBM SP computer platform is about five times less than the the computing time obtained on the IBM SP computer (compare Table I and Table IV). However, the efficiency of the vector variant of the algorithm, achieved on the CRAY C92A computer is much better than the efficiency of the parallel variant on the IBM SP computer.

Table IV. Results obtained on CRAY C92A computer

Parameters measured	Results
Computing time (CPU in seconds)	1050
Computational speed (in GFlops)	0.4601
Computational efficiency (in percent)	51

5. Conclusions and outlook

New algorithm for an implementation of the Danish Eulerian Model on parallel computers is discussed in this paper. The algorithm and the numerical tests performed are for only the main and the most time consuming parts of the model: advection and chemistry submodels. The same ideas are applicable to the big two- and three-dimensional Danish Eulerian Models and some work with parallelization of these models is in progress. The partitioning of the computational domain in several subdomains (as a rule the number of subdomains is equal to the number of the processors used) allows to solve efficiently large-scale problems in air pollution modelling. There are two main advantages of the algorithms based on the domain partitioning. Firstly, it is possible store different parts of the input and output data as well as different parts of the coefficient matrices after the discretization procedures (space and time discretizations) into the memories of the different processors, where distributed memory parallel computers are used and, hence, to solve efficiently huge real-life problems, when the problem does not fit in the memory of one processor. Secondly, the computational work may be done in parallel and, hence, to reduce considerably the computing time (proportional to the number of the processors used) which is very important in order to use the computer implementation of the Danish Eulerian Model for operational purposes even in its three dimensional version. The use of the standard message passing interface (MPI) is a key component in the development of this concurrent computing environment in which applications and tools can be transparently ported between different computers. The developed code is a portable code for parallel computers with distributed and shared memories, symmetric multiprocessor computers and clusters of workstations with MPI installation. The results for the numerical experiments show that our approach is an effective tool to determine a high-performance parallel algorithm for implementation of the Danish Eulerian Model on above mentioned parallel computers.

Acknowledgments

This research was partly supported by NATO (Grants ENVIR.CGR 930449 and OUTS.CRG.960312) and the Ministry of Science and Education of Bulgaria (Grant I-505/95)

184

References

1. Alexandrov, V., Sameh, A., Siddique, Y. and Zlatev, Z. (1997), Numerical integration of chemical ODE problems arising in air pollution models, *Environmental Modeling & Assessment*, **2**, pp. 365-377.

2. Crowley, W. P. (1968), Numerical advection experiments, *Monthly Weather Review*, **96**, pp. 1-11.

3. Gropp, W., Lusk, E. and Skjellum, A. (1994), *Using MPI: Portable programming with the message passing interface*, MIT Press, Cambridge, Massachusetts.

4. Hertel, O., Berkowicz, R., Christensen, J., Ø. Hov,Ø. (1993), Test of two numerical schemes for use in atmospheric transport-chemistry models, *Atmos. Envir.*, **27A**, pp. 2591-2611

5. Hov, Ø., Zlatev, Z., Berkowicz, R., Eliassen, A. and Prahm, L. P. (1988), Comparison of numerical techniques for use in air pollution models with nonlinear chemical reactions , *Atmospheric Environment*, **23**, pp. 967-983.

6. Marchuk, G. I. (1985), *Mathematical modeling for the problem of the environment*, North-Holland, Amsterdam.

7. McRae, G. J., W. R., Goodin, W. R. and Seinfeld, J. H. (1982), Numerical solution of the atmospheric diffusion equations for chemically reacting flows, *Journal of Computational Physics*, **45**, pp. 1-42.

8. Molenkampf, C. R. (1968), Accuracy of finite-difference methods applied to the advection equation, *Journal of Applied Meteorology*, **7**, pp. 160-167.

9. Snir, M., Otto, St., Huss-Lederman, St., Walker, D. and Dongara, J. (1996), *MPI: The Comploete Reference*, MIT Press, Cambridge, MA.

10. Zlatev, Z. (1984), Application of predictor-corrector schemes with several correctors in solving air pollution problems, *BIT*, **24**, pp. 700-715.

11. Zlatev, Z. (1995), *"Computer treatment of large air pollution models*, Kluwer Academic Publishers, Dordrecht-Boston-London.

12. Zlatev, Z., Dimov, I. and Georgiev K. (1994), Studying long-range transport of air pollutants, *Computational Science and Engineering*, **1**, pp. 45-52.

13. Zlatev, Z., Dimov, I. and Georgiev K. (1996), Three-dimensional version of the Danish Eulerian Model, *Zeitschrift für Angewandte Mathematik und Mechanik*, **76**, pp. 473-476.

AEROSOL MODELLING WITHIN THE EURAD MODEL SYSTEM: DEVELOPMENTS AND APPLICATIONS

H. HASS, I. J. ACKERMANN and B. SCHELL
Ford Forschungszentrum Aachen
Dennewartstr. 25, 52068 Aachen, Germany

F.S. BINKOWSKI
NOAA, on assignment to EPA, NERL
Research Triangle Park, NC, USA

1. Introduction

In order to fullfill their scientific tasks state-of-the-art air quality models should be capable of predicting particulate matter in addition to the gas-phase concentrations. A suitable aerosol model for the application in complex regional transport models has to provide sufficient information on the chemical composition as well as on the size distribution of the particles. Furthermore it has to be coupled to a photochemical model in order to be able to represent the interactions between the gas-phase and the particle-phase.

The Modal Aerosol Dynamics Model for Europe (MADE) has been developed as such an aerosol model based on the Regional Particulate Model (RPM) as described in [1] and successfully applied within the EURAD model system to the simulation of tropospheric aerosols over Europe [2]. However this model version was limited to submicron particles consisting of inorganic ions and water.

This paper describes the further development of the model, among others the extension to include the coarse particle size range and a more complete fine particle chemistry. These modifications have been performed by the incorporation of the aerosol portion of the EPA's Models3 system as described in [3]. Also a first three-dimensional application of the extended system is shortly illustrated.

2. Model Description

A detailed description of the MADE model can be found in references [2] and [1], therefore we will focus here on the major modifications

185

Z. Zlatev et al. (eds.), Large-Scale Computations in Air Pollution Modelling, 185–194.
© *1999 Kluwer Academic Publishers. Printed in the Netherlands.*

made upon that. The aerosol dynamics model MADE is coupled to the chemistry transport model CTM2 of the EURAD model system which is described in [4] and is called at every transport time step of the gas-phase model.

MADE solves the following conservation equations that are formulated in terms of integral moments M_k, defined as

$$M_k = \int_{-\infty}^{\infty} d_p^k n(ln\ d_p) d(ln\ d_p), \tag{1}$$

of three particle modes:

$$\frac{\partial}{\partial t} M_{k_i}^* = -\nabla \cdot (\vec{v} M_{k_i}^*) - \frac{\partial}{\partial \sigma}(\dot{\sigma} M_{k_i}^*) + \left(\frac{\partial M_{k_i}^*}{\partial t}\right)_{dif} \tag{2}$$

$$+ coag_{kii} + coag_{kij} + cond_{ki} + nuc_{ki} + e_{ki}$$

$$\frac{\partial}{\partial t} M_{k_j}^* = -\nabla \cdot (\vec{v} M_{k_j}^*) - \frac{\partial}{\partial \sigma}(\dot{\sigma} M_{k_j}^*) + \left(\frac{\partial M_{k_j}^*}{\partial t}\right)_{dif}$$

$$+ coag_{kjj} + coag_{kji} + cond_{kj} + e_{kj}$$

$$\frac{\partial}{\partial t} M_{k_c}^* = -\nabla \cdot (\vec{v} M_{k_c}^*) - \frac{\partial}{\partial \sigma}(\dot{\sigma} M_{k_c}^*) + \left(\frac{\partial M_{k_j}^*}{\partial t}\right)_{dif}$$

$$- \frac{\partial}{\partial \sigma} ((V_{k_c}^*)_{Sed} M_{k_c}^*) + e_{kc},$$

where the index i represents the Aitken mode, j the accumulation mode, c the coarse mode and $\vec{v}$ the horizontal wind vector. The equations are formulated in a σ-coordinate system defined as

$$\sigma = \frac{p - p_{top}}{p_{surf} - p_{top}}, \tag{3}$$

with p the pressure at layer height and p_{surf} and p_{top} the pressure at the lower and upper model boundary respectively. Within each mode a lognormal size distribution

$$n(ln\ d_p) = \frac{N}{\sqrt{2\pi} ln\ \sigma_g} exp\left[-\frac{1}{2}\frac{(ln\ d_p - ln\ d_{pg})^2}{ln^2 \sigma_g}\right] \tag{4}$$

is assumed, where N is the number concentration, d_p the particle diameter and d_{pg} the median diameter and σ_g the standard deviation of the distribution. The time rate of change of a given moment M_k is therefore given by the transport terms of horizontal and vertical advection and vertical diffusion (additionally horizontal diffusion can be activated if required), the aerosol dynamics terms for coagulation ($coag_k$) within and between the modes, condensation ($cond_k$) and nucleation (nuc_k) for the Aitken and accumulation mode and sedimentation for the coarse mode as well as primary particle emissions (e_k). Size dependent dry deposition is treated as lower boundary condition for vertical diffusion.

Equations 2 are solved for $k = 0$ (i.e. the total number concentration per mode) and for $k = 3$ (which is proportional to the aerosol volume concentration in the mode). In contrast to the formulation given in [2] the standard deviations σ_g of the modes are assumed to be constant. Consequently the prediction of two moments is sufficient to define the size distribution functions.

The equations for the third moment are further subdivided into the submoments of the substances contributing to the chemical composition of the aerosol particles M_3^n:

$$M_{3l} = \sum_{n=1}^{n=spec} M_{3l}^n = \sum_{n=1}^{n=spec} \frac{m_l^n}{\frac{\pi}{6}\rho_l^n} \qquad l = i, j, c \qquad (5)$$

where m_l^n is the mass concentration of species n and ρ_l^n the bulk density of the species in mode l (currently $spec = 3$ for the coarse mode and $spec = 8$ for each of the two fine particle modes, water is not transported but locally equilibrated, moment index k is neglected here). This results in a total of 22 prognostic variables in the aerosol model.

A description of the process parametrizations and the derivation of the according terms of equations (2) can be found in references [1] and [2]. In addition to the nucleation scheme described in [1] and [2] an optional nucleation treatment has been implemented into the model [5]. Sedimentation is treated as a vertical advection process using the implemented advection schemes of the model. This requires interpolation of the sedimentation velocities to the layer boundaries and a subsequent unit conversion according to

$$\left(V_{k_c}^*\right)_{Sed} = \left(V_{k_c}\right)_{Sed} * \rho_{air} * g / \left(p_{surf} - p_{top}\right) \qquad (6)$$

where ρ denotes the air density and g the mean gravitational acceleration. The sedimentation velocities $\left(V_k\right)_{Sed}$ are calculated by solving

$$V_{Sed} = \left[\frac{g}{18\nu} \left(\frac{\rho_p}{\rho_{air}}\right) d_p^2\right] C_C \qquad (7)$$

where ν is the kinematic viscosity, ρ_p the particle density, d_p the particle diameter and C_C the Cunningham correction factor, in the appropriate form for the moments of a lognormal distribution.

Figure 1 shows a schematic representation of the chemical composition of the fine particle modes together with gas-phase precursors. Secondary inorganic ions are represented by sulfate, nitrate and ammonium. Emissions of sulfate that are given from the emission inventory are no longer treated as particle emissions -as in the original model formulation- but assigned to the gas-phase species H_2SO_4-vapour which has been added to the RADM2 formulation. This vapor is then -depending on the actual conditions- either condensing on existing particles or nucleating to form new particles. Precursor concentrations of ammonia, nitric acid vapor and sulfuric acid vapor are treated by the gas-phase model.

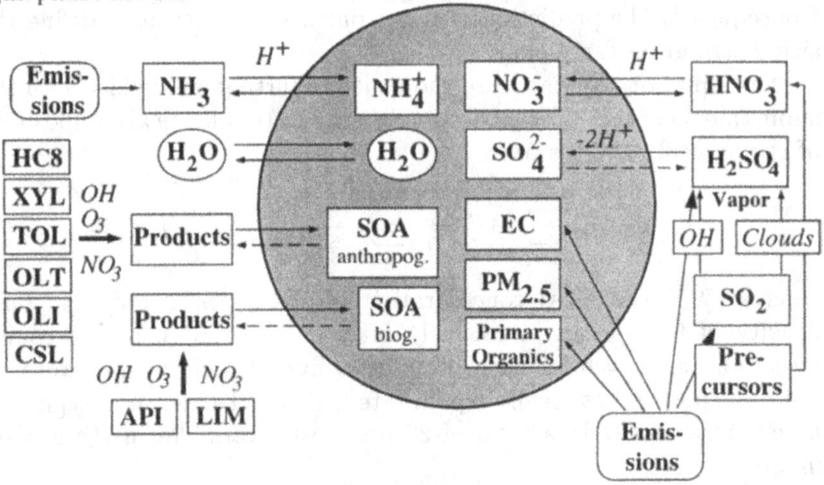

Figure 1. Schematical representation of the fine particle (Aitken and accumulation mode) chemical composition with gas-phase precursors in MADE.

Three groups of primary aerosol particles have been added to the fine particle composition, elemental carbon (EC), primary organics and PM2.5, which is a lumped class of other primary particles.

In the troposphere reactive organic gases (ROG) are mainly oxidised by the OH radical, the nitrate radical NO_3 and O_3. Some of the oxidation products have low volatility products and might condense on available particles forming secondary organic aerosol (SOA). Following the approach of [6] these reaction products accumulate in the gas phase. If the saturation concentration is exceeded they condense on the available aerosol surface establishing a thermodynamic equilibrium between the gas phase and the aerosol phase. Therefore a condensable reaction product G_i which is the product from the oxidation of a ROG by e.g. the OH radical is partitioned between the gas and the aerosol phase:

$$[G_i]_{tot} = [G_i]_{gas} + [G_i]_{aer}. \qquad (8)$$

The measured quantity available is the aerosol mass yield

$$Y_i = [G_i]_{aer}/\Delta ROG \qquad (9)$$

where ΔROG represents the reacted precursor concentration. The aerosol yields Y_i in $[\mu g/m^3/ppb]$ are adapted from [6] for the use with RADM2 classes of lumped organic species given in figure 1. A more detailed description of the scheme and an outline of future developments is given in [7].

The gas phase chemical mechanism RADM2 was extended to include a more detailed treatment of biogenic compounds. Hence the biogenic model species API and LIM and their reactions were adapted from the new RACM mechanism [8]. API represents α-pinene and other cyclic terpenes with one double bond, LIM stands for d-limonene and other cyclic diene terpenes. In the aerosol phase SOA is distinguished betweem those of biogenic and those of anthropogenic origin.

The coarse mode particles currently are segragated into three contributions, which are soil-derived particles, particles of marine origin and anthropogenic coarse mode particles. However, as for the fine mode primary particles, the number of contributors and type of segragation can be adjusted to the emission inventory available.

3. Model Application

Figure 2 shows the model domains and the nesting scheme used for the simulations with the coupled system of the CTM and MADE. Three subdomains are nested with a one-way nesting scheme into the mother domain that covers Central Europe, whereas the highest resolution grid covers the greater Cologne area in Germany with a resolution of 1 km. Nesting is realized by using output from the higher level domain for the initial values at the beginning of the simulation and the boundary values throughout the period for the nested domain. This is done for both, the gas-phase as well as the aerosol fields.

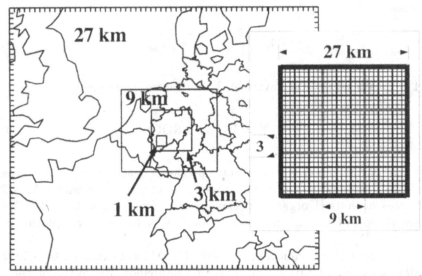

Figure 2. Model domains and different nest levels used for the aerosol model in a one-way nesting mode.

A ten day episode in July 1994 has been simulated in order to test the model system. Figure 3 shows the calculated number concentration of Aitken mode particles for the 9 km grid and the according subdomain of the 27 km simulation as an example of model results. Although the general features -with a dominating inflow from the the eastern boundary- as well as the maximum and minimum concentrations in the domain are similar between the two simulations, the results from the finer resolution show greater horizontal gradients, e.g. in Belgium and along the coast of the Netherlands. Another significant difference is the location of the central minimum which is shifted to the South-West in case of the simulation with finer resolution. In case of nucleation events number concentrations in the Aitken mode reach up to values of several 10^5 $1/m^3$. However these high number concentrations are rather rapidly reduced by efficient coagulation processes.

In figure 4 concentrations of secondary organic aerosols resulting from biogenic and anthropogenic sources are given at the same point of time for the coarse level domain. In this case both source categories contribute about equally to the total SOA load, whereas the spatial distribution shows some differences. Simulations over the ten day period show that the spatial distribution of SOA is highly variable and that the maximum concentration can be dominated by anthropogenic as well as biogenic precursors, depending on the actual conditions (meteorology, time and temperature dependence of the emissions etc.) over the domain. Concentration ranges up to about 10 $\mu g/m^3$, thus providing a significant contribution to the total PM2.5 mass.

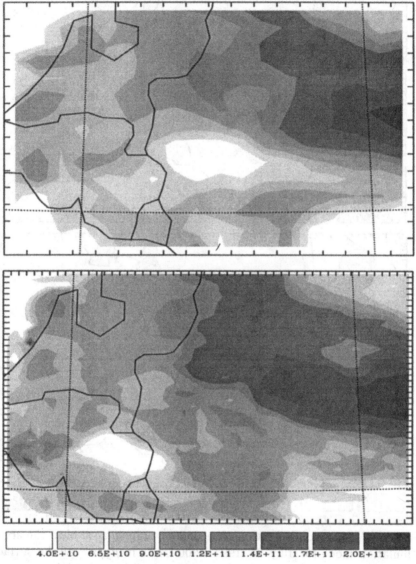

Figure 3. Number concentration in the Aitken mode for July 24. 1994 12.00 GMT $[1/m^3]$ at the surface layer for a simulation with 9 km resolution (top) and for a subdomain of the simulation with 27 km resolution (bottom).

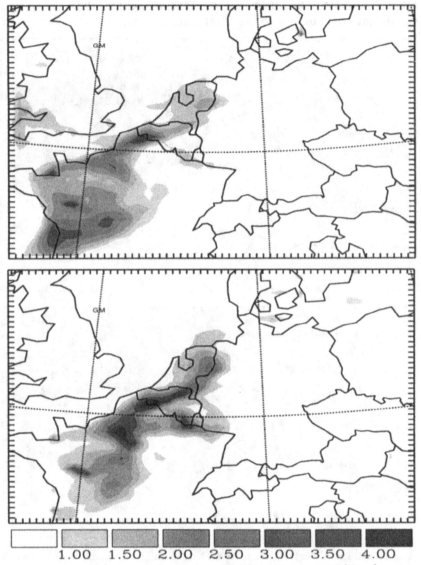

Figure 4. Mass concentration of secondary organic aerosol in the accumulation mode for July 24. 1994 12.00 GMT in $\mu g/m^3$ at the surface layer from anthropogenic (top) and biogenic (bottom) origin.

4. Summary and Outlook

An extended aerosol model system coupled to a sophisticated gas-phase chemistry model has been developed and applied to test simulations over Europe with variable horizontal resolutions. The nesting capability of the model system provides the opportunity to study the importance of aerosol effects on different horizontal scales as well as the impact of horizontal resolution on the results of the process parametrisations.

It has been shown that secondary organics provide a potentially significant aerosol source over Europe which has to be considered in atmospheric models although knowledge on the formation processes and ambient measurements are still limited.

Since particulate matter in the troposphere is a primary as well as secondary pollutant, the successfull application of the model system is crucially dependent on the availability of sophisticated PM emission inventories.

Future developments of the model will -among others- include parametrisations of aerosol-cloud interactions [9] and an updated formation mechanism for SOA [10]. Field measurements of the relevant aerosol parameters are required to evaluate the aerosol model.

References

1. Binkowski, F. and Shankar U. (1995), The regional particulate model 1. Model description and preliminary results, *J. Geophys. Res.*,**100**, pp. 26191-26209.
2. Ackermann, I. J., Hass, H., Memmesheimer, M., Ebel, A., Binkowski, F. S., and Shankar, U. (1998) Modal aerosol dynamics model for Europe: Development and first applications, *Atmos. Environm.***32**, pp. 2981-2999.
3. Binkowski, F.S. (1998), The aerosol portion of Models-3 CMAQ. Models3-CMAQ Science Document, Unpublished manuscript, *U.S. Environmental Protection Agency, RTP, North Carolina.*
4. Hass, H., Jakobs, H. J., and Memmesheimer M. (1995), Analysis of a regional model (EURAD) near surface gas concentration predictions using observations from networks, *Meteorol. Atmos. Phys.*, **57**, pp. 173-200.
5. Youngblood, D. A., and Kreidenweiss, S. M. (1994), *Further development and testing of a bimodal aerosol dynamics model*, Report. no. 550, Colorado State University, Dept. Atmos. Science.
6. Pandis, S. N., Harley, R. A., Cass, G. R., and Seinfeld, J. H. (1992), Secondary organic aerosol formation and transport, *Atmos. Environm.*,**26A**, pp. 2269-2282.
7. Schell, B., Ackermann, I. J., Hass, H., and Ebel, A. (1998), Secondary organic aerosol modelling with MADE: Biogenic and anthropogenic contributions, in Borrell, P.M. and Borrell, P. et al. (eds.), *Proc. of EUROTRAC Symposium 1998*, Computational Mechanics Publications, Southampton.

8. Stockwell, W. R., Kirchner, F., and Kuhn, M. (1997), A new mechanism for regional atmospheric chemistry modeling *J. Geophys. Res.*,**102**, pp. 25847-25879.

9. Meyer, R. K., Ackermann, I. J. , Hass, H. and Ebel, A. (1998), Modelling of Aerosol Dynamics with MADE: CLOUD Contributions, in Borrell, P.M. and Borrell, P. (eds.) *Proc. of EUROTRAC Symposium 1998*, Computational Mechanics Publications, Southampton.

10. Schell, B., Ackermann, I. J., Hass, H., and Ebel, A. (1998), Modeling secondary organic aerosol with MADE: a 3-D application to study the formation, transport, and impact on aerosol properties, In: Abstracts of the 17th Annual AAAR Conference, June 22-26, Cincinnati, Ohio, p. 220.

CALCULATION OF OZONE AND OTHER POLLUTANTS FOR THE SUMMER 1996:

The Influence of Lateral Boundary Concentrations on Ozone Levels.

J.E. JONSON and L. TARRASON
The Norwegian Meteorological Institute
P.O. Box43 Blindern, N-0313 Oslo, Norway

J. SUNDET
Department of Geophysics, University of Oslo
P.O. Box 1022 Blindern, N-0315 Oslo Norway

Abstract

Results for the summer months April - September 1996 from the MA-CHO model (Multi-layer Atmospheric CHemistry model, Oslo), a regional scale photo-chemistry model, are presented and compared to measurements. Initial and lateral boundary concentrations are provided by a global CTM (Chemical Tracer Model). As a base run monthly averaged concentrations for June 1996 are used as initial and lateral boundary concentrations.

The sensitivity of the interior model domain to the lateral boundaries is analyzed by comparing the base run with a model run with lateral boundary concentrations updated at six hour intervals. In the boundary layer over Europe differences between the two runs are small both for accumulated exceedances and for concentrations on individual days, indicating that ozone levels here are determined predominantly by local European scale sources. In the free troposphere however, the effects from the lateral boundary concentrations are shown to be significant.

1. Introduction

In 1997 results from the MACHO model (Multi-layer Atmospheric CHemistry Model, Oslo) were presented for the first time (Jonson et al. [12]). Comparisons with measurements were made, predominantly for ozone. It was shown that the model was able to reproduce major ozone events over Europe for May 1992. In Jonson et al. [13] calculations for the summer 1996 were presented, and in Simpson and Jonson [20]

Z. Zlatev et al. (eds.), Large-Scale Computations in Air Pollution Modelling, 195–206.
© *1999 Kluwer Academic Publishers. Printed in the Netherlands.*

comparisons were made between the MACHO model and the EMEP Lagrangian photo-chemistry model.

All chemical components calculated in the EMEP Eulerian acid deposition model (Bartnicki et al. [1]) are now also included in the MACHO model. Thus the chemistry has been extended so that sulphur and ammonia components are also included. The aqueous phase chemistry and wet deposition parameterization from the EMEP Eulerian acid deposition model is also included in the MACHO model.

The year 1996 was chosen for this model study as data from the Oslo CTM2, a global chemical tracer model with T21 resolution (Sundet [22]), were available for this year. In the base run monthly averaged concentrations for June 1996 from the Oslo CTM2 are used in the MACHO model as initial and lateral boundary concentrations.

The main focus in this paper is on the boundary layer over Europe, and to what extent it is affected by the innflux across the lateral boundaries. Transport to Europe through the lateral boundaries is predominantly from the north American continent. Ozone and other pollutants, lifted into the free troposphere, may be advected across the Atlantic with a characteristic transport time of less than a week. Furthermore, it is likely that additional ozone production will take place as the plume is advected. Elevated ozone in the free troposphere may be mixed into the European boundary layer, contributing to the exceedances of the critical levels for ozone in Europe.

2. Model formulation

A detailed description of the MACHO model is given in [13]. In Berge and Jakobsen [4] a sulphur only version of the the model is described. Only a brief description of the model will be given here.

2.1. ADVECTION

The model solves the mass continuity equations for trace gases in the same horizontal and vertical grid as the meteorological data (Berge [3], Jakobsen et al.[11], Berge and Jakobsen [4]). Hence σ (normalized vertical pressure coordinate), together with a polar stereo-graphic projection true at $60°N$ is employed.

The continuity equation can be formulated as follows (Berge and Jakobsen, [4]):

$$\frac{\partial}{\partial t}(qp^*) + m^2 \nabla_h \cdot (qp^* \vec{v}_h/m) + \frac{\partial}{\partial \sigma}(qp^* \dot{\sigma})$$
$$= \left[\frac{g}{p^*}\right]^2 \frac{\partial}{\partial t}(\rho^2 K_z \frac{\partial}{\partial \sigma}(qp^*)) + \frac{p^*}{\rho}S$$

The first term on the left hand side is the transient term, the second and third terms represent the horizontal and the vertical flux divergence formulation of the advective transport respectively. The first term on the right hand side in the equation represents the vertical eddy diffusion where g, ρ and K_Z are the gravitational acceleration, air density and vertical diffusion coefficient respectively. S describe the chemical and physical source or sink terms. q is the mixing ratio of the pollutant. $\vec{v}_h$ and ∇_h are the horizontal wind vector and the horizontal del operator respectively, and m is the map correction factor on a polar stereographic map projection. The vertical coordinate, σ, is defined as $\sigma = \frac{p-p_T}{p^*}$ where $p^* = p_s - p_T$. p, p_S and p_T are the pressure at the σ surface, the surface pressure and the pressure at the top of the model atmosphere (100 hPa) respectively. $\dot{\sigma}$ $(\frac{\partial p}{\partial z})$ is the vertical velocity.

Time splitting is used in order to solve the system of equations of the Eulerian model. In the horizontal a 4'th order Bott scheme is used for space discretization of the advection, and a 2'nd order scheme in the vertical (Bott, [6], [7]). An implicit central-difference scheme is applied for space discretization of the vertical diffusion (Berge and Jakobsen, [4]). The vertical sub-grid scale turbulent transport is modeled as a diffusive effect. In the air pollution model the diffusivity coefficient, K_Z, was derived from the basic meteorological parameters following the same procedure as utilized in the NWP- model described by Nordeng [18]. In the model the time-step is 10 minutes, but in the vertical advection the the time-step is reduced to 5 minutes.

2.2. EMISSIONS

Emission field estimates for SO_2, NH_3, NO_x, CO and VOC (volatile organic compounds) are based on data submitted officially to EMEP for 1996 from the participating countries. For most countries the emission data are available in the 50 km EMEP grid. A distinction is made between surface sources (below 100 m) and high stacks sources (above 100 m). Monthly averaged lightning emissions, and seasonally averaged aircraft emissions, are included for NO_x (Gardner et al.[10]). Emissions of NO_x from lightning is included as monthly averages on a T21 resolution (Köhler et al. [14]). Biogenic emissions of isoprene are calculated, based on the E-94 inventory (Simpson et al. [19]).

2.3. CHEMISTRY

The present chemical mechanism is based on the Oslo CTM1 (Berntsen and Isaksen [5]), and the Oslo CTM2 (Sundet [22]) models. The mechanism applied in the MACHO model has been extended to also include sulphur and ammonium chemistry. As part of this extension the aqueous phase chemistry already applied in the EMEP Eulerian acid deposition model (Jakobsen et al. [11]) is also implemented in the MACHO model. Acetone and DMS are not calculated in the MACHO model as is done in the Oslo CTM1 and CTM2. As concentrations from the Oslo CTM2 are used as lateral boundary values for the MACHO model, it is highly beneficial to have a virtually coherent chemistry parameterization in the two models.

2.4. PHOTO-DISSOCIATION RATES.

The photodissosiation rates (j-values) are calculated for clear sky conditions and for two predefined clouds using the phodis routine (Kylling, [15], Kylling et al. [16]). Ozone concentrations from a 2-D global model, extending from the surface to 50 km (Stordal et al. [21]) are scaled by observed total ozone columns Dutsch [9]. Cloud base for both the predefined clouds is at 1 km above the ground. The first predefined cloud is 3 km deep, with a water content of 0.7 g cm^{-3} and a mean droplet radius of 10 μm. The second predefined cloud is 1 km deep, with water content of 0.3 g cm^{-3} and a mean droplet radius of 10 μm. The j-values are calculated using the new recommendations for absorption cross sections and quantum yields from DeMore et al. [8]. For most components the changes are small compared to earlier recommendations. However the photolysis of O_3 to $O'D$ is now approximately 30% higher.

2.5. INITIAL AND LATERAL BOUNDARY CONCENTRATIONS.

For most of the chemical components the initial concentrations and the mass advected into the model domain across the lateral boundaries are calculated with the Oslo CTM2 (Sundet [22] with a T21 resolution (5.625 degrees). In the base run mean concentrations for June 1996 are used. For SO_2, and SO_4 lateral boundary concentrations are based on the results from a Hemispheric model (Leonor Tarrason, personal communication), and for ammonia they are zero.

2.5.1. *Potential vorticity adjustment of O_3.*
Across the tropopause there are large gradients in the concentrations of O_3. In an attempt to reproduce these gradients, initial and lateral boundary concentrations of O_3 at the four uppermost layers (level four corresponds to the approximate level of the tropopause) are scaled according to the potential vorticity (PV). It has been shown that there is a good correlation between O_3 and PV in the stratosphere and lower troposphere (Lary et al. [17], Beckmann et al. [2]). Thus, at these levels lateral boundary concentrations of O_3 (in ppb_v) are scaled to the PV

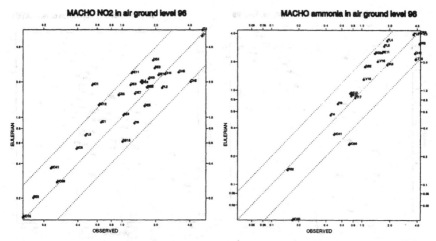

Figure 1. Scatter plots for calculated versus measured concentrations $\mu g\ m^{-3}$) as N for NO_2 (upper left), ammonium + ammonium (upper right). For NO_2: Calculated mean= 1.75, measured mean= 1.59. For ammonium + ammonia: Calculated mean= 1.75, measured mean= 1.68.

as:

$$O_3 = 4.5 \times 10^{-8} \times PV \qquad (1)$$

2.5.2. *Updating the lateral boundary concentrations at 6 hour intervals for June 1996.*

In order to test the model sensitivity to transient lateral boundary concentrations, 15 key chemical components are updated at 6 hour intervals (as for the meteorological input data for the MACHO model) with concentrations calculated with the Oslo CTM2. The 15 components are: O_3, HNO_3, PAN, CO, C_2H_4, C_2H_6, C_3H_6, C_4H_10, C_6H_14, m-Xylene, CH_2O, H_2O_2, NO and NO_2.

3. Model results for the summer 1996.

In Jonson et al. [13] results, including comparisons with measurements, for the summer 1996 are presented, and in Simpson and Jonson the performance of the MACHO model is compared to the EMEP Lagrangian photochemistry model. A few examples only of the model performance will be shown here.

In Figure 1 and 2 scatter plots of calculated versus measured concentrations of NO_2, ammonia + ammonium, total nitrate (HNO_3 and ammonium nitrite) and wet deposition of total nitrate are shown for the six summer months. There is a considerable difference between

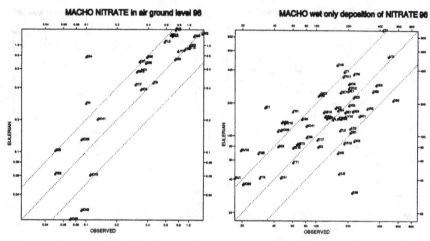

Figure 2. Scatter plots for calculated versus measured concentrations μg m^{-3}) as N for total nitrate (lower left), and calculated versus measured wet deposition of oxidized nitrogen (lower right) in mg m^{-2} as N. For total nitrate: Calculated mean= 0.67, measured mean= 0.48. For wet deposition of total nitrate: Calculated accumulated deposition= 172, measured accumulated deposition= 156.

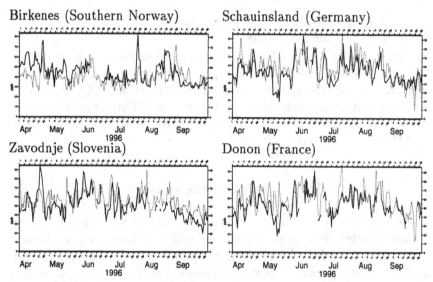

Figure 3. Measured (full line) versus calculated (dotted line) daily maximum concentrations of ozone inn ppb_v.

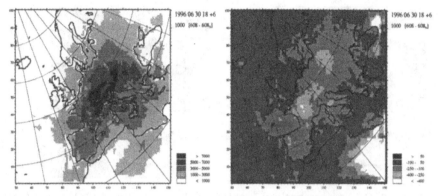

Figure 4. aot40 for June 1996 (left) and aot40$_{bou}$ − aot40$_{std}$ (right), where aot40$_{bou}$ is calculated with updated lateral boundary concentrations and aot40$_{std}$ is the standard model run.

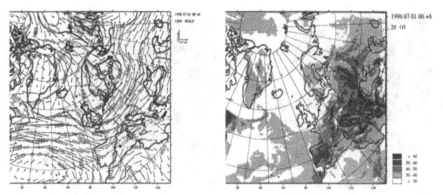

Figure 5. Mean sea level pressure and winds in the lower free troposphere (left) and ozone in ppb in the lowest model layer (right).

calculated and measured values. However these components all have a characteristic residence time in the atmosphere of the order of a day or less. Combined with large variabilities in the source strengths of NO_x and NH_3 within the grid boxes, differences between measurements at a point and model calculated grid averages are not surprising.

For total nitrate mean calculated concentrations are about 50% to high compared to measurements. The reason for this overestimation is unclear.

In Figure 3 calculated and measured daily maximum ozone concentrations are shown for four sites representing northern Europe (Birkenes), central Europe (Schauinsland), southern Europe (Zavodnje) and southwest Europe (Donon). In general the model is able to reproduce the day by day variations in ozone. However, in northern Europe ozone levels tends to be underpredicted.

202

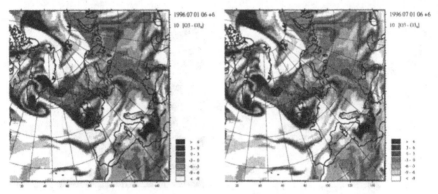

Figure 6. Ozone concentrations in ppb from the standard model subtracted from the model run with updated boundary concentrations. Lowest model layer (left) and lower free troposphere (right).

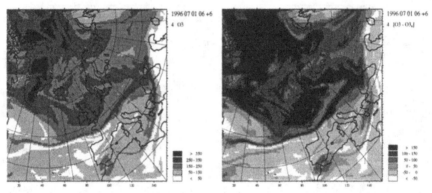

Figure 7. Ozone concentrations at model level 4 (approximately 230 hpa) (left). Ozone concentrations in ppb from the standard model subtracted from the model run with updated boundary concentrations also for model level 4 (right).

3.1. UPDATING THE LATERAL BOUNDARY CONCENTRATIONS.

Compared to the base run (with fixed lateral boundary concentrations) the effect of updating the lateral boundary concentrations every 6 hours is small in the boundary layer over Europe. In Figure 4 (left) aot40 for June is shown for the base run. aot40 is defined as the accumulated daytime ozone above the threshold value 40 ppb in ppb hours (50 ppb ozone for one hour will add 10 ppb hours to the accumulated value). With updated lateral boundary concentrations aot40 is lower over central parts of Europe, but the difference is small (Figure 4 right), in particular when considering the sensitivity of this parameter, as changes in average ozone concentrations are likely to be enhanced when

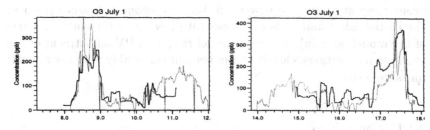

Figure 8. Measured (dotted line) versus calculated (full lines) concentrations inn ppb$_v$ of O$_3$ for a morning flight from Shannon airfield to the Canary islands (outside the model domain) and the return flight in the afternoon on the 1'th of July 1996. Time in UTC hours along the x-axis.

ozone concentrations are interpreted as aot40 (Simpson and Jonson [20]).

For individual days we found no large differences in boundary layer ozone over Europe between the two model runs, even though marked differences were often found in the lower free troposphere. In the upper troposphere the differences were often large. In Figures 5 to 7 this is visualized for July 1. Figure 5 (left panel) depicts the meteorological situation at noon. A westerly air current is set up between a high pressure system over the Azores and a low over Scandinavia. Ozone concentrations, also at noon, are shown in the right panel. Ozone concentrations above 50 ppb are mostly confined to southern Europe. Figure 6 shows differences in ozone between the model run with transient lateral boundary concentrations and the base model run in the lowest model layer (left) and in the lower free troposphere (right). The difference between the two model runs is negative in the lower left corner, indicating that the contribution from the north American continent is less than the average for June. In the lower free troposphere low ozone compared to the base run, originating from the north American continent, is seen in the westerly air current across the Atlantic and over central Europe. However, in the boundary layer differences are small, indicating that ozone concentrations here to a large extent are determined by local chemistry.

In Figure 7 (left panel) ozone at level 4 (approximately 230 hpa) is shown. High ozone concentrations (around 300 ppb), indicating stratospheric air-masses, are found over the British isles. This is also seen in Figure 8, comparing model results from the base run with aircraft measurement at flight altitude (around 10 km) in a flight from Shannon airfield, in northern England, to the Canary islands in the morning, and a return flight in the afternoon. Figure 7 (right panel) shows the difference between the two model runs at model level 4. With transient lateral boundaries (and without PV adjustment of ozone)

ozone concentrations are now much lower in ozone rich areas (as over the British isles) and higher in areas with low ozone (in southern parts of the model domain). The base model run, with PV adjustment of the ozone concentrations, clearly compares more favorably with the aircraft measurements.

3.1.1. *Conclusions.*

Although, as demonstrated above, there are strong indications that pollution from the north American continent will reach the lower free troposphere over Europe, calculated ozone levels in the boundary layer over Europe in June 1996 are dominated by local European scale production. Even though average June concentrations are used as lateral boundary concentrations throughout the six months period, the correspondence between calculated and measured ozone concentrations is about the same for all the months (A larger set of time-series for ozone is included in Jonson et al. [13]), suggesting that ozone levels in summer are determined predominantly by sources within the model domain.

At present cumulus convection is not fully accounted for in the model. In the boundary layer calculated effects of changes in lateral boundary concentrations are likely to be larger as the parameterization of this process is improved.

In the period of main concern with regard to exceedances of critical levels of ozone in the boundary layer (critical levels are rarely exceeded in winter) the influence from the across the lateral boundaries is likely to be largest in spring and autumn/late summer when the solar zenith angle is lower and thus the local photochemical production of ozone is less effective.

In the free troposphere the model calculations indicates that ozone concentrations scaled to the potential vorticity gives more realistic ozone levels than with concentrations from the global model alone, updated every 6 hours. This is not surprising as the correlation between potential vorticity and ozone is remarkably high (Spoerre Frode).

Acknowledgment

We would like to thank Dr. Hans Schlager from DLR, Oberpfaffenhofen, Germany for the permission to show the aircraft measurements from the Schadstoffe in der luft project.

References

1. Bartnicki, J., Olendrzynski, K. and Jonson, J. E. (1998), Description of the Eulerian acid deposition model. In *Transboundary acidifying air pollution in Europe*. EMEP/MSC-W, Status Report 1/98. The Norwegian Meteorological Institute, Oslo, Norway.

2. Beckmann, M., Ancellet, G. and Megie, G. (1994), Climatology of trpospheric ozone in southern Europe and its relation to potential vorticity. *J. Geophys. Res.* **99**, pp. 12841–12853.

3. Berge, E. (1993) Preliminary estimates of sulphur transport and deposition in Europe with a regional scale multi-layer Eulerian model. EMEP/MSC-W, Note 1/93. The Norwegian Meteorological Institute, Oslo, Norway.

4. Berge, E. and Jakobsen, H.A. (1998) A regional scale multi-layer model for the calculation of long-term transport and deposition of air pollution in Europe, *Tellus B* **50**, pp. 205–223.

5. Berntsen, T. and Isaksen, I. S. A. (1997), A global 3D chemical transport model for the troposphere: Model description and CO and O_3 results, *J. Geophys. Res.*, **102**, pp. 21239–21280

6. Bott, A. (1989), A positive definite advection scheme obtained by non-linear re-normalization of the advection fluxes, *Monthly Weather Rev.*, **117**, pp. 1006–1015

7. Bott, A. (1989), Reply *Monthly Weather Rev.*, **117**, pp. 2633–2636

8. DeMore, W. B., Sander, D. M, Golden, D. M., Hampson, R. F., Kurylo, M. J., Howard, C. J., Ravishankara, A. R., Kolb, C. E. and Molina, M. J. (1997), *Chemical kinetics and photochemical data for use in stratospheric modeling*, NASA/jpl Publ.

9. Dutsch, M. U. (1974), The ozone distribution in the atmospherer,. *Can.J. Chem.*, **52**, pp. 1491–1504.

10. Gardner, R. M., Adams, K., Cook, T. Deidewig, F., Ernedal, S., Falk, R. Fleuti, E., Herms, E, Johnson, C. E., Lecht, M., Lee, S. D., Leech, M., Lister, D., Masse, B., Metcalfe, M., Newton, P., Schmitt, A., Vandenbergh, C. and Van Drimmelen, R. (1997), The ANCAT/EC global inventory of NO_x emissions from aircraft, *Atmos. Environ.*, **31**, pp. 1751–1766.

11. Jakobsen, H. A., Berge, E., Iversen, T. and Skalin, R. (1995), Status of the development of the multi-layer Eulerian model: a) Model description, b) A new method of calculating mixing heights, c) Model results for sulphur transport and deposition in Europe for 1992 in the 50 km grid. EMEP/MSC-W, Note 3/95. The Norwegian Meteorological Institute, Oslo, Norway.

12. Jonson, J. E., Jakobsen H. A. and Berge E. (1997), Status of the development of the regional scale photo-chemical multi-layer model. EMEP/MSC-W, Note 2/97. The Norwegian Meteorological Institute, Oslo, Norway.

13. Jonson, J.E., Tarrason, L., Sundet, J., Berntsen, T. and Unger, S. (1998) The Eulerian 3-D oxidant model: Status and evaluation for summer 1996 results and case-studies, in *Transboundary photo-oxidant air pollution in Europe*. EMEP/MSC-W, Status Report 2/98. The Norwegian Meteorological Institute, Oslo, Norway.

14. Köhler, I., Sausen R. and Gallardo K. (1995), NO_x production from lightning, in U.Schumann, (ed.), *The impact of NO_x emissions from aircraft upon the atmosphere at flight altitudes 8-15 km (AERONOX)*, Final report to the Commission of the European Communities, Deutch Luft und Raumfart, Oberpfaffenhofen, Germany, pp. 343–345

206

15. Kylling, A. (1995), Phodis, a programme for calculation of photo-dissociation rates in the Earths atmosphere, available by anonymous ftp to pluto.itek.norut.no, cd pub/arve

16. Kylling, A., Stamnes, K. and Tsay S. C. (1995), A reliable and efficient two-stream algorithm for radiative transfer: Documentation of accuracy in realistic layered media. *J. Atmos. Chem.*, **21**, pp. 115–150.

17. Lary, D. J., Chipperfield, M. P., Pyle, L. A, Norton, W. A. and Riishøygard, L. P. (1995), Three-dimensional tracer initialization and general diagnostics using equivalent PV latitude-potential-temperature coordinates, *Quart. J. Roy. Meteor. Soc.*, **121**, pp. 187–210.

18. Nordeng, T. E. (1986), Parameterization of physical processes in a three-dimensional Numerical Weather prediction model. Tech. Rep. no 65, The Norwegian Meteorological Institute, Oslo, Norway.

19. Simpson, D., Guenther, A., Hewitt, C. N. and Steinberger, R. (1995), Biogenic emissions in Europe 1. Estimates and uncertainties, *J. Geophys. Res.*, **100**, pp. 22875–22890

20. Simpson, D. and Jonson, J. E. (1998), Comparison of Lagrangian and Eulerian models for summer of 1996. In *Transboundary photo-oxidant air pollution in Europe*, EMEP/MSC-W, Status Report 2/98, The Norwegian Meteorological Institute, Oslo, Norway.

21. Stordal, F., Isaksen I. S. A. and Horntveth, E. (1985), A diabatic circulation two-dimensional model with photo-chemistry. Simulations of ozone and long-lived tracers with surface sources, *J. Geophys. Res.*, **90**, pp. 5757–5776.

22. Sundet, J. (1997), *Model studies with a 3-d global CTM using ECMWF data*, Ph.D thesis, Dept. Geophys., Univ. Oslo.

ATMOSPHERIC MECHANISMS OF ADMIXTURES TRANSFER

K. A. KARIMOV and R. D. GAINUTDINOVA
Institute of Physics, National Academy of Sciences
265-A Chui Prosp., Bishkek 720071, Kyrgyz Republic

Introduction

One of the major problems of the atmospheric transfer of small gas admixtures in ionoshheric levels is assessment of its impact on variability of the main parametres of ionosphere, propagation and absorbtion of radiowaves in D-region of lower ionosphere.

It is known that D-region of ionosphere is a geophysical and meteorological system. It is forming as a result of an influence of geophysical agents such as Solar short-length radiation, flows of powerful protons and electrons, cosmic rays. However the dynamics of D-region depends on the meteorological factors. Investigations of the dynamics of D-region is very interesting for explanation of some variabilities depended on the meteorological factors.

A brief description of meteorological effects in dynamics of D-region and explanation of winter anomaly of radiowaves absorbtion connected with transfer of nitrogen oxide and atmospheric mechanisms of its transfer obtained on the base of many years observations of the atmospheric and ionospheric parameters have been presented in this paper. The proposed mechanisms could be used to interpret the sharp increase of radiowave's absorption in D-region of ionosphere within middle latitudes.

Data and Methods of Investigation

The analysis of atmospheric processes was carried out on the basis of many years data, obtained from meteorological stations network measurements, aerological and satellite sounding of the atmosphere over Middle Asia region, synoptical maps and radiolocation measurements of high atmospheric parametres (Institute of Physics NAS, Kyrgyz Republic).

Z. Zlatev et al. (eds.), Large-Scale Computations in Air Pollution Modelling, 207–210.
© 1999 *Kluwer Academic Publishers. Printed in the Netherlands.*

For analysis of dynamic processes the following parameters of atmosphere were used: ionospheric and meteorological parametres, pressure, direction and wind velocity, temperature. Daily measured atmospheric parameters was processed by various integration-, trend- and filtering techniques.

The Main Results

The atmospheric processes within Central Asia region have several regional peculiarities, connected with geographical position in the central part of the Eurasia. The atmospheric mechanisms of admixtures tranfer essentially depends on the physical-geographical and meteorological conditions in the region. The influence of geophysical conditions on the distribution of the small gas admixtures in ionosphere is one of the main objectives in the investigations of transfer mechanisms.

The atmospheric processes above middle latitudes in the Northern Hemisphere are defined by the heightened level of inhomogeneous of the atmospheric parameters in horizontal and vertical directions, characteristics of the latitudinal distribution of Hadley's vertical circulation cells and the structure of tropopause.

Long investigations have shown that the atmosphere over Kyrgyzstan is influenced by two circulation systems: polar and subtropical. In winter period anomalous variations of radiowaves absorption in D-region of ionosphere are caused by variations of electron concentrations in the lower ionosphere [1].

Nitrogen oxide NO has a significant role in ionization balance of lower ionosphere [2]. This was shown by analyzing the geophysical conditions in atmosphere and lower ionosphere and by using results of radiophysical observations in Kyrgyzstan in periods of winter anomaly of radiowaves absorbtion in D-region of ionosphere. It has been suggested that the effect of the winter anomaly in D-region of ionosphere can be caused by considerable increase of the NO concentrations in D- and E-regions of lower ionosphere. NO is an easy ionized component. It is formed in high latitudes of the Earth as a result of intrusion of high energy particles into the polar atmosphere. NO can be increased in lower latitudes because of meridional transfer from high latitudes and cause the effect of radiowave absorption in the middle latitudes.

The ionization and recombination cycle in the D-region of ionosphere is a balanced system which is sensitive to changes of temperature of gases and concentrations of, first of all, NO, O, H_2O and O_3. The impact of dynamic and thermodynamic processes on the parameters of D-region of ionosphere is explained by the temperature regime and the variation of the concentrations of the species mentioned above.

The problem of increasing the NO during winter anomaly radiowave absorbtion by meridional transfer was repeatedly discussed by scientists. The analysis of the data from [3], shows the existence of Northern wind directions on the altitudes up to 70-80 km in D-region within Central Asia which could provide the transfer of NO in lower ionosphere from high to low latitudes.

The analysis of global circulation maps for altitudes 20-95 km found by using geophysical and meteorological data including satellite measurements shows that horizontal transfer (advection) of NO, vertical transfer, changes of middle balanced temperature of gas and vertical propagation of planetary waves are the main factors of electron concentration changes [1]:

$$L(t) = -32 + 115AdvV - 2,4w + 0,25T.$$

Advection $AdvV$ is determined as a back proportional value to the time of air mass moving from auroral zone to the point of carrying out of radiowave absorption measurements. It is difficult to use this parameter for interpretation of the mechanism of increase NO concentration within local region because there is no sufficiently information about meridional changes of NO concentration. The vertical transfer of NO down from E-region or regular transfer and turbulent mixing could be an another significant mechanism of increase NO concentration in D-region of ionosphere.

The expression describing temporal changes of radiowave absorbtion $L(t)$ in dependence of mesospheric temperature T and vertical movement w could be obtain on the basis of correlation dependencies between «minimum frequency of radiowave absorbtion and meridional wind component» and «minimum frequency of radiowave absorbtion and vertical wind velocity»:

$$L(t) = C - cT - bw.$$

Coefficients C and c are determine from the equation of the empiric regression. For example, down vertical movement could to lead to supplementary transfer of NO to D-region from higher altitudes. When $w = 5$ cm/s transfer of NO from the altitude 110 km to 90 km will be carried out during 4 days. In this case maximum values of radiowaves absorbtion are observing later then the minimums values of vertical velocity.

In winter period the daily variations of $L(t)$ depend on the effective changes of temperature on the layer 60-90 km, on the advection of ionized component and on the vertical movement. In summer period daily variations of $L(t)$ depends on the effective temperature in the mesosphere and vertical transfer.

Experimental data show the connection between temporal variations of $L(t)$ and w in D-region above middle latitudes of Central Asia region. The beginning of development of regular part of winter anomaly of radiowaves absorbtion coincides with the reverse of vertical velocity direction connected with seasonal changes of w from rising to downward movements. Ending of the regular part of winter anomaly coincides with the beginning of short-time spring reconstruction of zonal circulation on the level 93 km (meteor zone) [2].

The phenomenon of regular part of winter anomaly radiowaves absorption in D-region of ionosphere is connected with complete reconstruction of whole system of circulation on the altitudes of high mesosphere - lower thermosphere. In this connection information about spring and autumn reconstruction of circulation regime in meteor zone could be use for prognosis of date of beginning regular part of winter anomaly. Presented data illustrated meteorological control of lower ionosphere by neutral atmosphere.

The rising level of radiowave absorption in period of winter anomaly is included irregular part of winter anomaly. In this period there take place intensive downward vertical movements with amplitudes higher in 5-8 times than middle season vertical velocities. Irregular part of winter anomaly has correlation with downward vertical velocities which are providing transfer of NO from the level 110-120 km to D-region of ionosphere during 3-5 days [1-2].

Conclusions

1. The regularities of winter anomaly radiowaves absorption in D-region of lower ionosphere above Central Asia and conditions of nitrogen oxcide NO transfer have been obtained:
   - beginning and ending of regular part of winter anomaly is connected with seasonal variations of vertical movements and zonal circulation in mesosphere-lower thermosphere;
   - the irregular part of winter anomaly is connected with intensive downward vertical velocities.
2. The proposed mechanism of transfer of small admixtures in mesosphere-lower thermosphere explains the sharp increase of the radiowaves absorption in periods of winter anomaly within Central Asia region.

References

1. Karimov, K. A. and Gainutdinova, R. D. (1986), *Disturbances in Lower Ionosphere in Connection with Dynamic Characteristics of Neutral Atmosphere*, Frunze, Ilim.
2. Gainutdinova, R. D., Karimov, K. A., Karimov A. K. and Zelenkova I. A. (1991), Long-Term Variations of Radiowaves Absorbtion in Connection with Dynamics of High Atmosphere, *Dynamics of Ionosphere*, Part I, Alma-Ata, pp.43-53.
3. Gainutdinova, R. D., Ivanovsky, A. I. and Karimov, K. A. (1988), On Dependence of Method of Estimation Vertical Movements by Radiometeor Measurements, *Izvestia of Acad. Sci. USSR, Atmospheric and Oceanic Physics*, **24**, pp. 1123-1133.

THE VARYING SCALE MODELLING
OF AIR POLLUTION TRANSPORT

V.K. KOUZNETSOV*, V.B. KISSELEV** AND V.B. MILJAEV*

*Institute for Atmosphere Protection,7, Karbysheva str., St.-Petersburg, Russia
e-mail: sriatm@mgo.main.rssi.ru
** Institute for Informatics and Automation RAS, 39,14-th linea, St.-Petersburg, Russia
e-mail: kisselev@epr.pu.ru

1. Introduction

Different models are used to study the transport and deposition of pollutants and their influence on natural ecosystems. These models have different spatial resolution according to the goals of investigations. For example, the influence of one European country on another is determined on a scale $(150 \times 150 \text{ km}^2)$ in [1]. In Russia, considering its natural and industrial heterogeneities, the mesoscale interregional model of pollution transport has been created [2]. This is necessary, because there are great variations of depositions in EMEP cells [3] when the emissions are big. There are models for computing the pollution in cities [4]. All models use emission data and different types of meteorological information: (i) 6-houres meteorological data [1], (ii) climatological data [2] and (iii) most unfavorable meteorological parameters [4].

The climatological model is analogous to EMEP model, but its main feature is the use of climatological characteristics of the considered region as input variables and, thus, statistical generation of the fields of meteorological elements, simulating in this way the real conditions of pollutant transport .

The comparison of sulphur deposition results, obtained with a help of a stochastic model and EMEP models in the same sells shows that the average difference is not higher than 13% [2].

Evidently, the total deposition of pollutants in some region consists of external deposition, that is import of pollutants from another areas, and internal deposition conditioned by the regional emission sources.

In order to calculate the total deposition in some region it is possible to use both global modelling for external deposition estimation and mesoscale modelling for internal deposition estimation.

Z. Zlatev et al. (eds.), Large-Scale Computations in Air Pollution Modelling, 211–214.
© 1999 Kluwer Academic Publishers. Printed in the Netherlands.

2. Results and discussion

The case study of Karelia, situated in the North-West part of Russia, demonstrates the usefulnes of the approach. Average (1991-1996) atmospheric fluxes of SO_2 in Karelia obtained from the EMEP model and stochastic model are shown in table 1. The share of external deposition is 67% of total deposition in this region and 62% of internal emission is exported outside the region.

Table 1. The average budget of sulphur in Karelia (1991-1996)

Sulphur budget parameters	1000 tonnes S
Deposition from external sources	49
Deposition from internal sources	24
Total deposition	73
Emissions from internal sources	63
Exportation outside the region	39

The index of soil acidification is the exceedance of total sulphur deposition over a value of critical load for different types of soil [5]. The performed calculations show that the exceedances (Ex) of total sulphur depositions (D_{tot}) over values of critical loads (CL), i.e. the acidification of soils, take place over the whole territory of the Southern Karelia and partly in Central Karelia due to deposition of sulphur both from external and internal emission sources (Fig.1). Calculation of Ex is performed on the basis of the balance equation:

$$D_{ex} + D_{in} - CL = Ex \qquad (1)$$

Total deposition is the sum of external (D_{ex}) and internal (D_{in}) depositions. D_{in} in (1) can be managed at a regional level, and the excess depositions can be avoided only by reducing emission of internal sources. It follows from (1) that if $D_{ex} > CL$, Ex cannot be avoided by reducing these emissions. In some parts of Karelia, where acidification is conditioned by external emission sources, the quality of soils can not be managed by regional authorities. On the other hand, on the territories where $Ex > 0$ and $D_{ex} < CL$, the exceedances can be avoided by reducing the internal emissions to a certain value D_{in}^0, which can be determined ($Q = D_{in}/D_{tot}$) from the equation:

$$D_{in}^0 = CL \times Q \qquad (2)$$

In the fig.1 there are the black cells where it is not possible to avoid exceedances, and the gray cells where it is possible to do it with a help of internal emission reduction

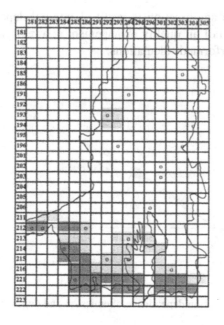

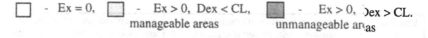

☐ - Ex = 0, ☐ - Ex > 0, Dex < CL, ▓ - Ex > 0, Dex > CL.
 manageable areas   unmanageable areas

Figure 1. The areas of Karelia with different opportunities for air pollution management.

Using regression dependence of emission and deposition values for different sources (fig. 2) it is possible to determine sulphur emission reduction which is necessary to achieve for the required deposition (D_{in}^0). Taking into account the tendency of sulphur emission reduction in other countries of Europe two scenarios have been created for reduction of internal emissions in Karelia (table 2). In the first case sulphur emissions in Karelia have to be reduced by 36-70% and in the second case - by 45-83% depending on the source.

3. Conclusion

1. In the North-West part of Russia the transboundary flux of acidifying substances is more than deposition conditioned by internal emission sources.

2. In this territory there are areas where the acidification of soils can not be avoided with a help of reduction of internal emission sources only.

214

3. The areas where acidification is result of internal deposition have to be included in the system of air quality management at a regional and state levels in accordance with performed calculations.

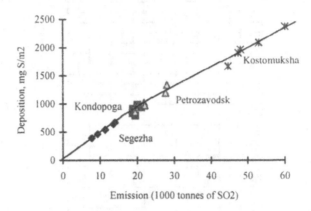

Figure 2. The regression dependence of emission and deposition of sulphur.

Table 2. Possible scenarios for SO2 emission reduction (1000 ton/yr) in Karelia

Emission source	Necessary reduction of internal SO$_2$ emissions under proportional reduction of external impact	Necessary reduction of internal SO$_2$ emissions in the absence of external impact reduction
Kostomuksha	from 51 to 14	from 51 to 11
Segezha	from 11 to 7	from 11 to 6
Kondopoga	from 20 to 8	from 20 to 6
Petrozavodsk	from 24 to 8	from 24 to 4

4. References

1. Berge, E. (ed.) (1997) *Transboundary Air Pollution in Europe*, MSC-W Status Report for 1997. Part 1, 2. DNMI, Oslo.
2. Kouznetsov, V. K., Kisselev, V. B., Miljaev, V. B. (1997), A comparison of mesoscale pollution transport models with varying spatial resolution. *Air Pollution V Modelling, Monitoring and Management*. Southamton Boston, pp. 129-139.
3. Galperin, M. V. and Cheshukina, T. V. (1996), *The estimation of subgrid inhomogeneity of atmospheric pollution on the basis of model calculations of sulphur compaund deposition in the North-Eastern Europe*, EMEP/MSC-E Report 8/96, MSC-E, Moscow.
4. Berljand, M.E. (ed.) (1987), *The technique of calculation of pollutant concentrations in ambient air due to industrial emissions*, OND-86, Leningrad.
5. Posch, M., de Smet, P. A. M., Hettelingh, J.-P., Downing, R. I. (1995), *Calculation and Mapping of Critical Thresholds in Europe*, Status Report for 1995, CCE/RIVM, Bilthoven, The Netherlands.

DETERIORATION ON HISTORIC BUILDINGS DUE TO AIR POLLUTION AND SOME DIFFICULTIES DURING THEIR RESTORATION WORKS

A. G. KUCUKKAYA
Trakya University, Faculty of Eng. & Architecture,
Department of Restoration, Edirne/Turkey

Abstract

The conservation of cultural heritage is one of the most important problems in our polluted environment. The new modern technology especially on computers are developing faster then ever. Interdisciplinary task of large scale computation in air pollution needs to be solved with combined efforts from specialists who are informed from several different fields.

To make clear the reflection of air pollution on historic buildings and difficulties on their restoration works and explanation of some faulty restorations are the main subject of this paper.

Air pollution, consisting of chemical and mechanical compositions, causes deterioration especially on stone materials which are mostly used on the facade of historic buildings. Effects caused by compounds of S, C, N will be reported.

1. Introduction

The conservation of cultural heritage, specifically historic buildings, monuments and sites, is one of the most important problems in our polluted environment. New modern computers and technology used to investigate and to solve deterioration problems in building facades resulting from air pollution offer us new approaches to conservation of historic architecture.

Air pollution, consisting of volatile chemical compositions, cause deterioration especially on stone materials which can clearly be seen on the facades of historic buildings, the use of chemical solutions intended for the preservation has created new problems for historic buildings: specifically, the

Z. Zlatev et al. (eds.), Large-Scale Computations in Air Pollution Modelling, 215–220.
© *1999 Kluwer Academic Publishers. Printed in the Netherlands.*

increase in chemical deterioration in building facades resulting from the use of such compounds. In addition, other cleaning processes generally create different and difficult problems which are not resolved yet.

The aim of this study is to focus on the weathering and deterioration of historic buildings caused by air pollution. This study will highlight current problems in the restoration of historic buildings in order to demonstrate the importance of preserving our historic building environment for the benefit of future generations as we enter the 21st century. It is the hope of this study to engage professionals working with computers and technology in this important undertaking.

2. Deterioration on the surface of building materials due to air pollution

The German "verwitterung" and the English "weathering" are terms used to describe the deterioration of buildings due to atmospheric effects, unlike the term "alteration" used in French [6]. Any other special term couldn't be found to describe the determination of deterioration due to the air pollution until the present.

Air pollution combined with atmospheric effects such as rain, fog, wind and snow cause deterioration on the surface of historic buildings. The SO_2 [sulphur dioxide], SO_3 [sulphur trioxide], CO_2 [carbon dioxide], and Cl_2 [chlorine] in polluted atmosphere; nitrates over the normal values, and humidity play an active role to deteriorate the surface of building materials.

2.1. Sulphates and SO_2 . Sulphur dioxide gas, formed when sulphur burns in air, is the most significant cause of acid gas degradation. When the gas is dissolved in water, from either moisture in the air or rain, dilute sulphurous acid is formed. Sulphurous acid combines with oxygen to form sulphuric acid. When the acid is deposited on calcium carbonate the surface of the stone is converted to calcium sulphate takes up water as it crystallises to the mineral gypsum [1]

$$S + O_2 \text{-------} > SO_2 \text{ [S is found mostly in industrial areas]}$$
$$SO_2 + H_2O \text{------} > H_2SO_3 \text{ [sulphurous acid]}$$
$$CaCO_3 + H_2SO_3 \text{------} > CaSO_3 + H_2CO_3$$
$$CaSO_4 + 2H_2O \text{ [gypsum formats Stone materials tears apart]}.$$

The grain size of stone materials, pollutant factor in atmosphere; SO_2, and relative humidity has to be known for the determination of the speed of these chemical reactions.

When limestone and sandstone are used together in polluted environment, then SO_2, H_2S and water vapour effect limestone. The dissolved $CaCO_3$ settles over sandstone as $CaSO_4$.

$$SO_4^{2-}$$
$$CaCO_3 \text{ --------------} > .CaSO_4$$

The volume of $CaSO_4$ is 1,7 times higher than volume of $CaCO_3$. This expansion of volume causes the breakage of sandstone.

$$2SO_2 + O_2 \text{------} > 2SO_3$$
$$SO_3 + H_2O \text{------} > H_2SO_4$$
$$CaCO_3 + H_2SO_4 \text{------} > CaSO_4 + H_2CO_3 \text{ [Crusts on surface]}$$

2.2. Carbonates. The most important reactions are:

$$CO_2 + H_2O \text{-------} > H_2CO_3 [acid]$$
and $\quad H_2CO_3 \text{-------} > CO_2 + H_2O$

$CaCO_3$ dissolves in a certain amount of water in polluted atmosphere

$$CaCO_3 \text{ ------} > Ca^{2+} + CO_3^{2-}$$
and $\quad Ca^{2+} + CO_3^{2-} \text{------} > CaCO_3$

Water dissolves the carbonates by bounding with carbon dioxide in the polluted atmosphere.

$$CaCO_3 + H_2O + CO_2 \text{------} > Ca(HCO_3)_2 \text{ (dissolves)}$$

2.3. Chlorates. Chlorates can easily be found in the see side and reacts with rain water .

$$Cl_2 + H_2O \text{------} > HCl + HOCl \qquad\qquad\qquad acid$$

As a result of the effect of acidic properties on the carbonates help to deteriorate especially limestone building materials.

$CaCO_3 + 2HCl \text{------} > CaCl_2 + H_2O + CO_2$ [erosion on the surface] $\qquad$ g a s

2.4. Nitrates. NO_3^- (the phase of NO_2 in the atmosphere saturated with water) is not met in ordinary environment when marbles rich in NO_2 are examined. Burned fossil fuels, especially coal produces gases and particles such as SO_2 and NO_2 that effect marble. These gases and some particles dissolve in water and form acid rain.

$$NO_2 + H_2O \text{------} > HNO_3 \text{ (acid)}$$

Briefly, the analysis of present environmental conditions indicates a decrease in the amount of atmospheric pollutants over the past few years. However, the major constituent was either SO_2, which can be absorbed by building materials, or converted to acid on moist surfaces in the presence of catalytic active material such as soot, iron or manganese, or $SO_4^{..}$ contained in the dry deposition. When the crust material is exposed to moisture, the SO_2 can go into solution, leading to the chemical deterioration of stone, or the $SO_4^{..}$ can come directly from rain.

3. Deterioration And Difficulties Encountered During Restoration Works

Although environmental control would appear to be a solution, the issue is complex. Air pollution can never be completely eliminated, nor can all historic material that is at risk be removed to a less hostile location. When alteration of environment is not a suitable option, protective treatments are often applied directly to the surface of the building material.

In the search for sympathetic surface treatment, many materials have been tried with varying degrees of success. Wax and oils, used extensively in the past, discolour the surface, attract particulate pollutants, are difficult to remove, and may render the surface impermeable. Solutions, such as lime water, believed to strengthen decaying stone by the addition of sound calcium carbonate, have not proven to be entirely effective. Shelter coat is intended to provide a sacrificial layer, decaying preferentially to the stone material itself. Many protective treatment are unsightly and some are often more damaging to the decay itself. Buildings and wall paintings, unlike much sculpture in the round are vulnerable to air pollution attack their surface. Water repellent coating have been favoured for their reversibility but may cause more damage then no coating at all [3].

3.1. Cleaning. Advance soiling is not simply an aesthetic problem because of air pollution. Cleaning with water is one of the not preferred methods. Too much water is required and this may cause other some negative side effects on buildings materials. Cleaning with laser is very expensive and mechanical ways always unsuccessful. Using preservatives chemicals is other important point for the treatment which must be done after cleaning.

Being a slightly soluble compound, gypsum, along with pollution, is washed away by rain, leaving the surface of the stone , clean but slightly eroded and vulnerable to further decay. In sheltered areas the sulphuric acid condenses on the stone, reacts with it, and binds in particulate pollutants, form in a black crust on the surface. Some durable limestone can retain this unsightly gypsum crust indefinitely, while less durable ones from blisters which eventually burst leaving powdery decayed stone behind. In time the decayed stone falls off leaving fresh limestone exposed to further decay [1].

3.2. Plaque disease. Restorers and conservationists created the name in Europe "plaque disease" which is called "calsin" by French and Belgian professionals to refer to the results of weathering [6] and air pollution. While plaque disease is a most dangerous illness for some stone, it can be determined as an auto control element for sandstone.

Observation of the process of disease in the sandstone which is containing calcareous is like this:

Acidic reactions are experienced following concequitif rainfalls. On the other hand, rain water which contains pollution materials penetrates into stone building material and dissolves the calcareous carbonate. As water evaporates back to the air it leaves dissolved materials behind. This causes an accumulation of these materials on the surface. A few millimetres of this layer hardens in time and changes the structure and colour of material(see figure), or the dissolved materials in stone may recrystallize on the face . This can be seen on the places where rain cannot wash of the building [5].

As a result of this ongoing process the internal dissolving in stone increases and the section next to the crust becomes a sandy section. Because of this weak section, the removal of the plaque can increase erosion for the stone and further the process of deterioration. Therefore, for some calcareous materials, calsin not only has a preservative role for the conservation of historic facades but also it can be the preliminary indication of a dangerous disease *[see Figure]*

Finally, removing of plaque can be named as a faulty restoration. Usually it starts a dangerous erosion process. Unfortunately these kind of faulty restoration studies can easily be seen in developing countries where some building materials has been cleaned unknowingly by mechanical and chemical ways still today. For example in Ýstanbul, in Haydarpasha Gar Station was lost original details because of the sandy section erosion of sandstone material after the mechanical cleaning in 1981.

4. Conclusion And Proposals

In some historic areas in Europe the pollution levels are so high today and will be worst tomorrow. For example; as a result of high level air pollution, historic and archaeological buildings in Rome are in danger; However the traffic is banned in some important areas as a part of the Rome rehabilitation project that can be one of the weak solution against to the affect of air pollution.

1.Transmission of global statistical data on regional level for the control of S, C, and N as an interdisciplinary approach must be develop.

2.The periodical inspection system on historic monuments due to the air pollution as a joint multilateral scientific research project must me produce with combined efforts from specialists from several different fields.

The importance of historical monuments and sites shouldn't be forgotten at all. Also affect of air pollution on historic heritage must be considered in large scale computation in air pollution modelling.

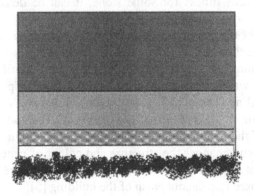

Figure 1: Plaque Disease on Sandstone [6], the upper layer is soft stone, the next layers are: a weak section, sandy section , inner crust and polluted crust resp.

References

1. Ashurst, J. and Dimes, F. G. (1990), *Conservation of Building and Decorative Stone*, Vols; I and II, London.
2. Atkins, P. W. (1989), *General Chemistry, Scientific American Books*, New York.
3. Cezar, T. M. (1998), Calcium Oxalate; A Surface Treatment for Limestone, *Journal*, *4*, London.
4. Guari, K. L. and Gwinn, J. A. (1983), Deterioration of Marble in Air Containing 5-10 ppm. Soub>2 and Noub>2, *Durability of Building Materials*, pp.241-254.
5. Guari, K. L., Popl, Ý, R and Sarma, A.C. (1982), Effect of The Relative Humidity and Grain Size on the Reaction Rates of Marble at High Concentration of SO_4 , *Durability of Building Materials*, pp. 209-216.
6. Kieslinger, A. (1967), Principal Factors in Weathering of Natural Building Stones, *ICOMOS Meeting on Weathering of Stones*, Paris.
7. Skoullikidis, T. (1985), *The Effects of Air Pollution on Historic Buildings And Monuments*, Brussels.
8. Webb, A. H., Bawden, R. J., Busby, A. K. and Hopkins, J. N. (1990), On the Effects of Air Pollution on Limestone Degradation in Great Britain, *International Conference on Acidic Deposition: its Nature and Impacts*, Vol. 26B, pp. 165-181, Glasgow.

MODELLING OF THE LONG-TERM ATMOSPHERIC TRANSPORT OF HEAVY METALS OVER POLAND

A. MAZUR
Meteorology Centre
Institute of Meteorology and Water Management - Warsaw Division
61 Podlesna str., PL-01-864 Warsaw, Poland

Abstract

A regional model for atmospheric transport of four heavy metals: arsenic, cadmium, lead and zinc over the Polish territory is described in this paper. It represents an Eulerian, three-dimensional approach to the transport problem. Heavy metal emissions from Polish sources were collected and compiled to create the emission data base for Poland. This inventory was then used to distribute the emissions in the model grid squares. Calculated concentration/deposition maps for Poland for 1995 are presented in the paper. The model version presented here is compact and fast and can be easily implemented, even on a personal computer.

1. Introduction

Poland is one of the biggest emitters of various heavy metals, e.g. arsenic, cadmium, lead and zinc, in Europe. This leads to enormous contamination of the natural environment. Recently, many studies on this subject were launched. This particular research was focused on achieving of two main goals:

1. Developing 3-D Eulerian model of atmospheric transport of heavy metals. REMOTA is a simulation model that allows describing the dispersion of multiple pollutants. The processes as horizontal advection and vertical diffusion, dry deposition and wet removal are accounted for in the model. The governing equations are solved in terrain-following co-ordinates. The numerical solution is

Z. Zlatev et al. (eds.), Large-Scale Computations in Air Pollution Modelling, 221–234.
© *1999 Kluwer Academic Publishers. Printed in the Netherlands.*

based on discretization applied on a staggered grids. Conservative properties are fully preserved within the discrete model equations. Advective terms in horizontal directions are treated with the Area Flux Preserving method (the Bott's type scheme) with boundary conditions assumed zero at income and "open" at outcome flows; vertical turbulent diffusion - with the implicit scheme combined with Gaussian elimination method. The bottom boundary condition is the dry deposition flux, the top boundary condition is an "open" one. Dry deposition velocity is postulated as a function of terrain roughness, friction velocity and diameter of a particle according to Sehmel's model, while washout ratio is assumed constant (depending on pollutant only).

2. Gathering and processing of available data on heavy metals emissions. The emissions inventory was created strictly for the needs of the research project. It includes emissions from: power plants (both commercial and industrial), heating power plants (communal and industrial), steel mills and non-ferrous metals industry, cement production, chemical fertilizer production and from mobile sources (emissions from traffic).

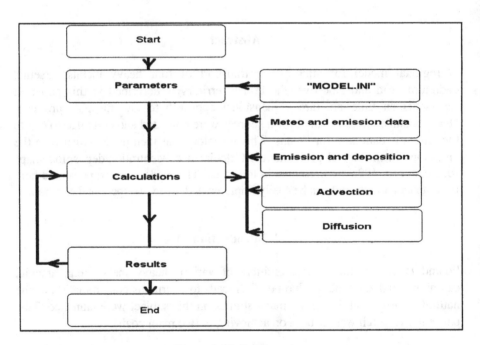

Figure 1. Model flow chart

2. Model description

It is three-dimensional, Eulerian type model based on similar model for Europe [2, 3]. The numerical grid system for this particular research consists of 47x27 cells of the size 27.5x17.5 km. The model is composed of three main modules; the first sets parameters of the model (i.e. local deposition coefficient, washout ratio, dry deposition velocity), the second solves the advection-diffusion equation, and the last prepares output files for graphical presentation and statistical analysis. Flow chart for the model is shown in the Figure 1.

During a simulation the program gets user-defined parameters from file "MODEL.INI", then it begins its main loop of calculations: reads all necessary input data (emissions, meteorological fields etc.), performs computations (advection, diffusion, deposition) and stores the results (if required).

The main equation of the model is three-dimensional advection-diffusion equation of the following form:

$$\frac{\partial c_i}{\partial t} + \frac{\partial c_i u_j}{\partial x_j} = \frac{\partial}{\partial z}(K\frac{\partial c}{\partial z}) - \left(\frac{v_d}{h} + \frac{wP}{h}\right)c_i + (1 - \alpha_i)e_i \quad (1)$$

where: c_j, i=1..4 - mass of As, Cd, Pb, Zn, respectively, e_i, i=1..4 - emission of particular metal, a_i, i=1..4 - local deposition coefficient, u_j, j=1,2 - horizontal wind field components [m/s], x_j, j=1,2 - horizontal coordinates, h - mixing height [m], t - time, P - precipitation amount [mm/6h], w - washout ratio, v_d - dry deposition velocity [m/s], K - eddy diffusivity coefficient (z-dependent).

The left side of equation is a substantial derivative of each metal mass. The first term on the right side describes the loss of mass due to dry and wet deposition, the second one - production of mass due to emission in the given grid. The equation is solved using the Area Flux Preserving method (advection; [4]) and a Crank-Nicholson method (diffusion).

The decay of mass in the vicinity of each emitter is included in the equation in the form of a constant coefficient denoted as the local deposition value a, depending on the meteorological conditions, height of emission source etc.

In each time step, total deposition D_{tot} (sum of dry and wet) is computed first, then dry D_d and wet D_w depositions are calculated. Denoting total deposition coefficient as k_{tot} and dry and wet deposition coefficients as k_d and k_w, respectively, one can get:

$$D_{dry} = D_{tot} \cdot \frac{k_{dry}}{k_{tot}}, D_{wet} = D_{tot} \frac{k_{wet}}{k_{tot}}$$

$$k_{dry} = \frac{v_d}{h}, k_{wet} = \frac{wP}{h}, k_{tot} = k_{dry} + k_{wet} \tag{2}$$

$$D_{tot} = c_i \cdot \left[1 - \exp(-\frac{v_d}{h} \cdot \Delta t) \cdot \exp(-\frac{wP}{h} \cdot \Delta t) \right]$$

with Δt - time step. The dry deposition velocity is a function of roughness height, friction velocity and particle diameter [5]. In turn, the washout ratio in this model is also assumed constant (and equal to $5 \cdot 10^5$).

3. Preliminary inventory of arsenic, cadmium, lead and zinc emissions from polish sources.

The basis for the data base were inquiries (direct communication with selected factories and power plants), information prepared by governmental agencies, private enterprises ([6], [7], [8]) and data measured at several power plants and factories (e.g. [9], [10], [11]). In several cases the emission factors prepared by Pacyna ([12], [13], [14], [15]) were used.

Data base contains the following information:
- emission - annual amounts of emission of As, Cd, Pb and Zn from particular sources
- emission parameters
    - height and diameter of a stack
    - temperature and velocity of outflow gases
- auxiliary data
    - energy production - commercial branch
        - annual emission of dust
        - type of furnace
    - energy production - industrial branch
        - installed power
    - mobile sources
        - total mileage of roads
        - traffic intensity

Emission from the coal combustion processes was calculated for each source using one of the following formulas:

$$E = C\, F_s\, (100\text{-}\eta) \tag{3}$$

where C - metal concentration in coal (Table 1), F_s - source-dependent factor, h - effectiveness of dust collectors, or

$$E = C_p\, W_p \tag{4}$$

where C_p - metal concentration in a stack dust (see Table 2), W_p - dust emission factor.

Table 1. Averaged concentration of As,Cd,Pb and Zn in coal (ppm)

Metal	Pit coal		Brown coal	
	European	Polish	European	Polish
As	-	5.9	-	32.5
Cd.	0.47	1.9	0.61	0.65
Pb	4.9	27.0	4.7	6.0
Zn	35.0	125.0	28.0	46.3

Table 2. Averaged concentration of As,Cd,Pb and Zn in stack dust (ppm)

Metal	estimated	pit-coal (measured)	brown coal (measured,)
As	75.0	55.4	21.5
Cd	24.3	6.6	11.3
Pb	300.1	213.0	175.0
Zn	435.0	537.0	355.0

The effective emission height for every emission source was computed using Carson-Moses' formula:

$$H_p = H + 5.32\frac{\sqrt{Q}}{u_h} \tag{5}$$

where Q is heat emission, H - height of a stack, u_p - wind velocity at H.

These data have been applied to the model. The histogram of the effective emission heights is shown in Figure 2.

The locations of main (point) emission sources are shown in Figures 3a-3d.

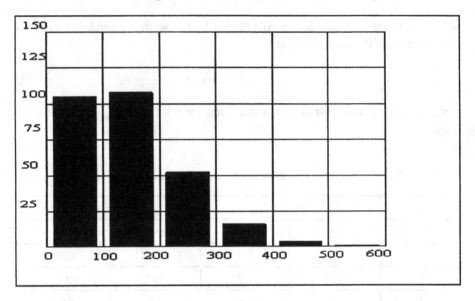

Figure 2. Histogram (no. of cases) of effective emission heights (meters)

4. Results and discussion.

The first model run resulted in concentration and deposition patterns for every metal. These patterns are shown in figures 4 - 7.

The results show that the southern part of Poland (Upper Silesia, Sudety Mountains and Cracow region) is highly affected by contamination. Local maximum of deposition can be located in the vicinity of the biggest point source(s) in industrial regions. Emissions from traffic (leaded gasoline) give the main contribution to the deposition of lead especially close to the main roads and to highways. "The cleanest" part of Poland seems to be the north-eastern one (Bialowieza Primeval Forest and Suwalki Voivodship).

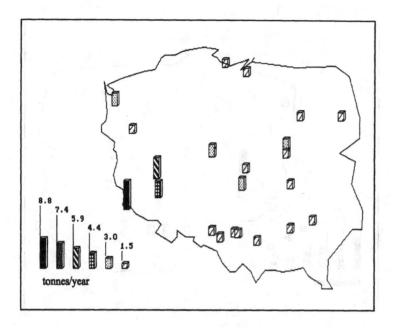

Figure 3a. Main emission sources of As

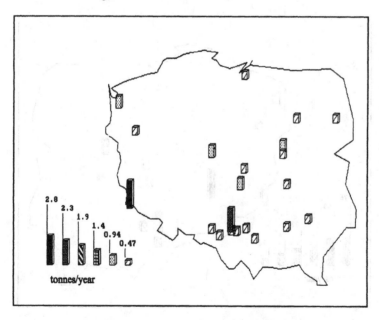

Figure 3b. Main emission sources of Cd

228

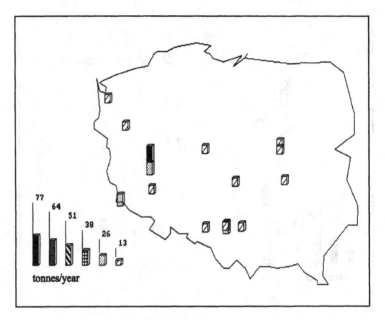

Figure 3c. Main emission sources of Pb

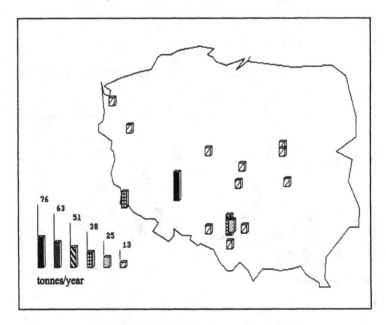

Figure 3d. Main emission sources of Zn

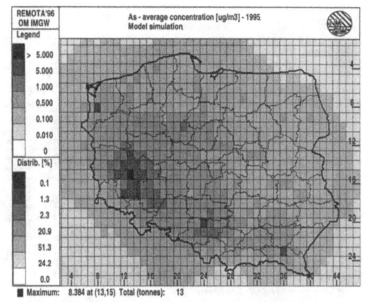

Figure 4a. Model results - average concentration of As

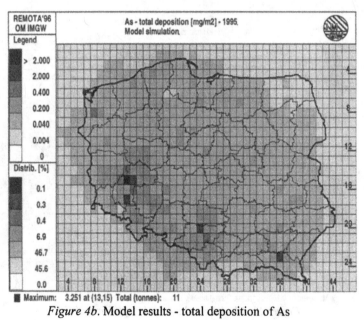

Figure 4b. Model results - total deposition of As

230

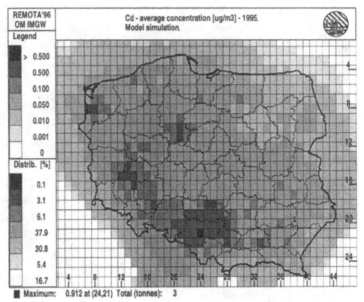

Figure 5a. Model results - average concentration of Cd

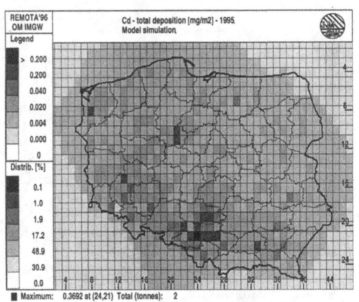

Figure 5b. Model results - total deposition of Cd

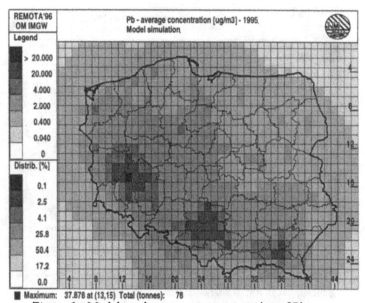

Figure 6a. Model results - average concentration of Pb

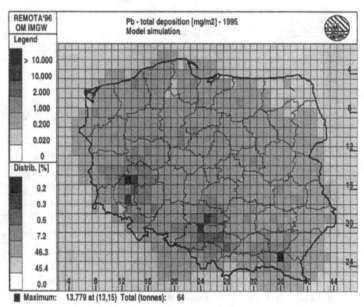

Figure 6b. Model results - total deposition of Pb

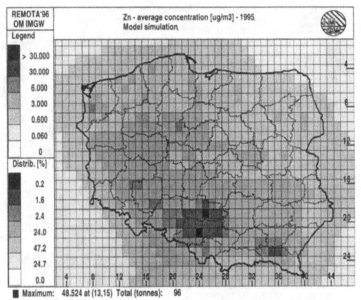

Figure 7a. Model results - average concentration of Zn

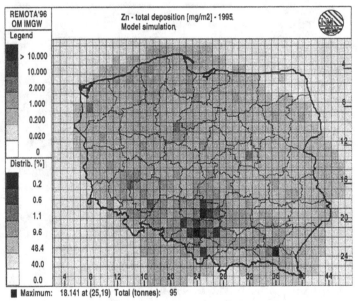

Figure 7b. Model results - total deposition of Zn

5. Summary and conclusions

The main results of the research work are:
- creation of unique heavy metals emission data base for Poland;
- developing the operational version of model of atmospheric transport of heavy metal for Poland.

Results gained during research point out, first, that natural environment of Poland is greatly contaminated, second, that emissions are dangerously high, and, moreover, about 60% of emitted matter is dispersed outside of Poland.

Access to the data, especially the emission ones, is a separate question. It obviously is a necessary condition to carry out this kind of research. Contrary to the single point source case, neglecting one source (or a group of sources) can change the overall results dramatically. That is why so much attention has been paid to the problem in this study.

Acknowledgment

This research was supported by Polish Ministry of Environment Protection, Natural Resources and Forestry.

References

1. Bartnicki, J. (1986), An efficient positive definite method for the numerical solution of the advection equation. IIASA Working Paper WP-86-35, Institute for Applied Systems Analysis, Laxenburg, Austria.
2. Bartnicki, J . (1991), Long range transport of heavy metals from Poland computed by an Eulerian model, in: van Dop H., Steyn D.G. (Eds.) *Air pollution modelling and its application* VIII. Plenum Press, pp. 339-348.
3. Bartnicki, J., Modzelewski, H., Bartnicka-Szewczyk, H., Saltbones, J., Berge, E. and Bott, A. (1993), *An Eulerian model for atmospheric transport of heavy metals over Europe: Model development and testing.* Techn. Report 117, Norwegian Meteorological Institute, Oslo, Norway.
4. Bott, A. (1989), A positive definite advection scheme obtained by nonlinear renormalization of the advective fluxes. *Mon. Wea. Rev.* 117, 1006-1015.
5. Sehmel, G. A. (1980), Particle and gas dry deposition: A review. *Atmos.Environ.* 14, 983-1011.
6. CIE (1991), *The statistics of Polish energetics - 1990.* Warsaw (in Polish).
7. CIE (1992), *Energetic balance in Poland by means of OECD, EUROSTAT and UN statistics.* Warsaw (in Polish).

8. Opoczynski, K. and Rosinska, M. (1992), *Traffics in 1990*. TRANSPROJEKT, Warsaw (in Polish).
9. Zareba , J. (1991), Emission of heavy metals due to cement production. *Cement, Wapno, Gips.* **11** (in Polish).
10. Konieczynski, J. (1990), Emission factors for metals and trace elements for the coal combustion processes. *Air Protection* **4**, 80-85 (in Polish).
11. Konieczynski, J., Szeliga, J. and Pason, A. (1992), Emissions of PAH in the communal sector. *Air Protection* **6**, 133-135 (in Polish).
12. Pacyna, J. M. (1984), Estimation of the atmospheric emission of trace elements from anthropogenic sources in Europe. *Atmos. Environ.* **18**, 41-50.
13. Pacyna, J. M. (1986), Emission factors of atmospheric elements, in Nriagu J.O., Davidson C.I. (Eds), *Toxic metals in the atmosphere*. J. Wiley & Sons, Inc., New York
14. Pacyna, J. M. (1989) ,Technological parameters affecting emissions of trace elements from major anthropogenic sources, in J. M. Pacyna and B. Ottar (eds.), *Control and fate of atmospheric trace metals*.
15. Pacyna, J. M. (1991), Emission factors of atmospheric Cd, Pb an Zn for major source categories in Europe in 1950 through 1985. NILU-Report 30/91, NILU, Lillestroem, Norway.
16. Hrehoruk, J., Bartnicki, J., Mazur, A., Grzybowska, A., Modzelewski, H., Frydzinska, B. and Bartnicka, H. (1992), Modelling the Transport of Air Pollution. IMWM Tech. Report P-7 (in Polish).
17. Hrehoruk, J., Grzybowska, A., Mazur, A., Frydzinska, B., Bartnicki, J., Modzelewski, H. and Bartnicka, H. (1993), Regional Model for Atmospheric Transport of Heavy Metals over Poland. IMWM News, **3**.
18. Hutton, M. and Symon, C. (1986), The quantities of Camium, Lead, Mercury and Arsenic entering the U.K. environment from human activities. *Sci. Total Envir.* **49**, 409-419
19. Mejstrík, V. and Svacha, J. (1988), The fallout of particles in the vicinity of coal fired power plants in Czechoslovakia. *Science Tot. Environ.* **72**, 43-55
20. Sato, K. and Sada, K. (1992), Effects of emissions from a coal-fired power plant on surface soil trace element concentrations. *Atmos. Environ.* **26A**, 325-331.
21. Starkova, B. (1991), Heavy metals in emmisions from Czechoslovak combustion processes. Paper presented at The ECE Task Force Group on Heavy Metals Emissions, Prague.
22. Wesely, M. L. and Lesht, B. M. (1989), Comparison of RADM dry deposition algorithms with a site-specific method for inferring dry deposition. *Water, Air and Soil Pollut.* **44**, 273-293.

POLLUTION TRANSMISSION IN THE AIR

CS. MÉSZÁROS, T. RAPCSÁK and Z. SÁGI
Department of Operations Research and Decision Systems,
Computer and Automation Research Institute,
Hungarian Academy of Sciences
H-1518 Budapest, Kende u. 13-17., POB 63, Hungary

1. Abstract

In the paper, an overview is provided on the application of air pollution transmission modelling based on two case studies. Besides a mathematical model, some computational experiments and the 2D-visualization of the results by maps are presented.

Keywords: pollution transmission in the air, parabolic partial differential equation, visualization by maps.

2. Introduction

In Hungary, a growing interest is manifested in studying the state of the environment. Before starting any installation, construction or investment, an environmental impact assessment is obligatory in order to obtain the permit of the authorities from the point of view of protecting the environment. The main requirements of an environmental impact assessment are to take into account the environmental effects, the probable changes of those to be affected by the corresponding effects and to elaborate suggestions for avoiding essential and unreversible damage for the environment.

From among ten real-life applications, two were related mainly to the possible danger of air pollution. A part of the environmental impact assessment for a planned battery recycling center and a part of the environmental monitoring system for an aluminium recycling foundry in the same region was to investigate and show what happens if a certain quantity of pollution appears in the air from a chimney. In an environmental impact assessment, by law, the Hungarian National Standards (HNS) should have to be followed which contain instructions and regulations for the necessary data and formulae of the different situations, but not the mathematical model with the initial and bound-

235

Z. Zlatev et al. (eds.), Large-Scale Computations in Air Pollution Modelling, 235–247.
© 1999 *Kluwer Academic Publishers. Printed in the Netherlands.*

ary conditions, neither the assumptions for the parameters. Thus, the model validation in the concrete cases seemed to be necessary.

In the paper, an overview is provided on the application of air pollution transmission modelling in these two recent projects, completed by our Department. In the second section, some details of the projects are given. In Section 3, the mathematical background of some air transmission models is reviewed, performing the model validation, whilst in the fourth section, the Hungarian National Standards that prescribe the formulae to be used for modelling purposes are presented. The fifth section contains considerations with respect to the numerical procedures used, and representative results.

3. Two real-life applications

3.1. ENVIRONMENTAL IMPACT ASSESSMENT FOR A PLANNED BATTERY RECYCLING CENTER

Approximately 60% of the chemical energy source prodeuction consists of acid lead accumulators. These batteries, after having become waste, bear considerable environmental risk due to their toxic components, some of which (primarily, lead) are precious industrial raw materials. Therefore, in industrially developed countries, over 90% of the waste batteries is collected and recycled.

No recycling of waste batteries has not been carried out in an organized way in Hungary for about 10 years. This fact involves negative economic, ecological and political consequences. This way, the Hungarian economy loses the possible profit deriving from recycling, furthermore, lead must be imported. Without recycling, the collecting system cannot work properly: the amount of waste batteries is estimated between 25000 and 30000 tons per year (containing 13000 to 15600 tons of recyclable lead), while the amount of collected waste accumulators is about half of that. The large number of waste batteries not collected implies a considerable environmental risk. The collected accumulators could be exported to countries with free recycling capacity, but export is impeded by the Basel Convention and the stricter OECD new regulations.

After several efforts with no success, though the sphere of private capital was involved, a consortium was founded for having a lead battery recycling center built and running it. The operation of the planned factory will imply environmental effects, out of which, emission of air contaminants is the critical factor. The location of the settlement is an old industiral establishment, already heavily affected by metallurgical activities of more than 50 years. There are smaller industrial units in the establishment still in operation, the contributory effects of which have to be considered as well. Based on licencing, our Department as the prime contractor, completed a detailed environmental impact assessment ([3]), by involving a subcontractor and about 15 experts. The so-called environmental protection permit necessary for the investment was released by the authorities.

3.2. Environmental monitoring system for an aluminium recycling foundry

In industrialized countries, it is important to continuously check the pollutant emission of factories with environmental load. In case of approaching the tolerance levels, the technological process can be altered on the signal of the monitoring system (or the monitoring system itself launches the required intervention). Moreover, sometimes also the state of the environmental elements can be continously checked by measuring contaminant concentrations in some critical locations.

In Hungary, the contaminant emission values are reported to the competent authorities on a yearly basis. These reports are not required to be backed by measuring or computation, estimations are accepted as well. The state of the environment is checked by infrequent, random testings. The loose of control leads to abuse in some cases. Not surprisingly, without the possibility to check the emission values and environmental state properly, people lost their trust: operating factories have to face with the protests of local dwellers, and setting up a new plant is often hampered by the citizens' resistance. This happened in the case of the planned battery recycling center. Recently, besides the traditional criteria of locating the industrial investments, such as the location of raw materials, market size, etc., one of the most important factors is where the inhabitants allow to set up the factory.

A possible solution may be the introduction of a modern monitoring system that provides reliable data on the environmental effects of factories. This way, no emission above the tolerance level or deterioration of environmental state could occur, on the other hand, factories with environment-friendly technology can bear the support of the citizens. Recognizing these facts, the owner of an aluminium recycling foundry decided to install a complete environmental monitoring system at the factory, since the production is planned to be increased from tripling to quadrapling. The task of our Department was to elaborate the methodology of environmental monitoring, to develop and install the corresponding software system (for further details, see [4]). It is worthwhile mentioning that this foundry is located in the same establishment where the battery recycling center is planned to be set up.

4. Mathematical model for the air pollution transmission and the solution

In the applications, a relatively small area is affected, therefore, following the terminology used in [2], a Lagrangian type transmission model is best suited for the problem.

4.1. TRANSMISSION EQUATION WITH BASIC ASSUMPTIONS

Transmission equation ([1]) related to different species that should be solved with the proper assumptions, initial and boundary conditions in the domain of $-\infty < x, y, z < \infty$, is as follows:

$$\frac{\partial C}{\partial t} = \frac{\partial}{\partial x}\left(D_x \frac{\partial C}{\partial x} - uC\right) + \frac{\partial}{\partial y}D_y \frac{\partial C}{\partial y} + \frac{\partial}{\partial z}\left(D_z \frac{\partial C}{\partial z} + wC\right) - \lambda C + F,$$
$$(x, y, z, t) \in \Re^3 \times \Re^+,$$

$$(1)$$

where the function $C : \Re^3 \times \Re^+ \to \Re$ shows the concentration of the contaminant at every point $(x, y, z) \in \Re^3$ and at the moment $t \in \Re^+$; $u : \Re^3 \times \Re^+ \to \Re$ is the wind speed, D_x, D_y and D_z are the diffusion coefficients, $\Re^3 \times \Re^+ \to \Re$, in the corresponding directions, respectively, w is the speed of sedimentation, λ is the rate of escape from the atmosphere, and $F : \Re^3 \times \Re^+ \to \Re$ is the pollution source function.

The coordinate system is choosen so that the directon of wind flow (assumed to be horizontal) be parallel to the axis x, i.e., the component of the wind flow in the direction of y be zero. The axis z is vertical. The following basic assumptions hold in the model:

1. λ is constant;

2. w is constant;

3. The wind speed, $u : \Re^3 \times \Re^+ \to \Re$, is independent of the variable t; its dependence on the altitude is determined by the relation

$$u(t, x, y, z) = \tilde{u}(x, y) \left(\frac{z}{z_0}\right)^p, \quad (x, y, z, t) \in \Re^4,$$

where z_0 denotes the altitude of the speedometer for the wind, $\tilde{u} : \Re^2 \to \Re$ is the wind speed (detected at the height of the speedometer), and p denotes the exponent of the wind profile. This empirical formula is valid above the surface only, i.e., in the open half-space $z > 0$.

4. The source function $F : \Re^3 \times \Re^+ \to \Re$ is independent of time and is point-like, i.e.,

$$F(t, x, y, z) = \delta(x)\delta(y)\delta(z - H)Q, \quad (x, y, z, t) \in \Re^3 \times \Re^+,$$

where δ denotes the Dirac-function.

The solutions of equation (1) subject to the given bounded initial condition

$$C(0, x, y, z) = C_0(x, y, z), \quad (x, y, z) \in \Re^3, \tag{2}$$

and various conditions for the coefficients, are examined which are *bounded* $(C \leq K < \infty, \quad K$ constant) in the entire space except for the point $(0, 0, H)$, where the coordinates of the base point of the chimney is $(0, 0, 0)$ and H denotes the effective height of the emission.

Considering stationary phenomena, it is important to determine the function

$$\lim_{t \to \infty} C(t, x, y, z) = c(x, y, z), \quad (x, y, z, t) \in \Re^3 \times \Re^+, \tag{3}$$

if any. It will be shown that function c in (3) satisfies the equation

$$\frac{\partial}{\partial x}\left(D_x \frac{\partial c}{\partial x} - uc\right) + \frac{\partial}{\partial y}D_y \frac{\partial c}{\partial y} + \frac{\partial}{\partial z}\left(D_z \frac{\partial c}{\partial z} + wc\right) - \lambda c + F = 0,$$
$$(x, y, z) \in \Re^3, \tag{4}$$

and is bounded in the entire space except for $(0, 0, H)$; furthermore, it is the only function with these properties. It turns out that in the cases discussed, the values of function C_0 are indifferent for both the existence and value of the function c. Nevertheless, the difference $C(t, x, y, z) - c(x, y, z)$, considered at a fix point $(x, y, z) \in \Re^3$ as the function of t describes the stabilization process.

Remark. Equation (4) can be reached in another way. It could be examined whether problem (1) has a stationary solution, i.e., a solution independent of time. In the affirmative, the solution necessarily satisfies (4) as well. However, this means that the function $C_0 : \Re^3 \to \Re$ is a solution of (4) which is generally not true.

4.2. PASQUILL-TYPE MODEL

By considering stationary problems only, it is assumed that the coefficients of equation (4) fulfil the following relations:

1. $\lambda = 0$;

2. $w = 0$;

3. $\hat{u}(x, y) \equiv u$, $p = 0$, i.e., $u(t, x, y, z) \equiv u$, (u constant);

4. $D_x = 0$;

5. $D_y > 0$, $D_z > 0$.

Then, equation (4) becomes the following parabolic one:

$$u\frac{\partial c}{\partial x} = D_y\frac{\partial^2 c}{\partial y^2} + D_z\frac{\partial^2 c}{\partial z^2} + F, \quad (x, y, z) \in \Re^3, \tag{5}$$

By integrating equation (5) over the whole space, the continuity law is obtained. In the space of bounded functions defined on the punctured whole space, solution of (5) is of form

$$c(x, y, z) = \frac{Q}{4\pi\sqrt{D_y D_z}x}\exp\left\{-\frac{u}{4x}\left[\frac{y^2}{D_y} + \frac{(z-H)^2}{D_z}\right]\right\},$$
$$(x, y, z) \in \Re^3 \setminus \{(0, 0, H)\}. \tag{6}$$

By using the notations $\sigma_y^2(x; u) = \frac{2D_y x}{u}$, $\sigma_z^2(x; u) = \frac{2D_z x}{u}$, the form of the solution is

$$c(x, y, z) = \frac{Q}{2\pi\sigma_y(x;u)\sigma_z(x;u)u}\exp\left\{-u\left[\frac{y^2}{2\sigma_y^2(x;u)} + \frac{(z-H)^2}{2\sigma_z^2(x;u)}\right]\right\},$$
$$(x, y, z) \in \Re^3 \setminus \{(0, 0, H)\}. \tag{7}$$

The solution in this form appeared in [6] for the first.

Based on the new notations, a different form of equation (5) is

$$\frac{\partial c}{\partial x} = \frac{\sigma_y^2(x; u)}{2x}\frac{\partial^2 c}{\partial y^2} + \frac{\sigma_z^2(x; u)}{2x}\frac{\partial^2 c}{\partial z^2} + \frac{F}{u}, \quad (x, y, z) \in \Re^3 \setminus \{(0, 0, H)\},$$
$$\tag{8}$$

where the solution (7) was determined under the conditions that $\sigma_y^2(x; u)$, $\sigma_z^2(x; u)$, $(x, u) \in \Re^2$, are linear in the variables x and $\frac{1}{u}$, respectively. If these conditions do not hold, then in general, the solution of (8) cannot be given in the form of (7).

5. The Hungarian National Standards

The HNS, with respect to air ([5]), prescribe formulae for modelling the features of the pollution transmission in the air, however, behind the Standards there are no exactly built mathematical models that the computation formulae could be derived from. The model discussed in the preceeding part contains some assumptions that generally do not hold, and considerably affect the transmission process, but it makes the basic equation with the initial and boundary conditions clearer similarly to the assumptions for the parameters. In the formulae of the HNS, the speed of sedimentation is not assumed to be zero, and the escape from the atmosphere is also taken into account. Furthermore, the concentration function is defined above the surface only $(z > 0)$, which implies that the effect of the surface has to be considered: a fraction of the contamination gets deposed on the surface, the remaning part is reflected.

In case the considered contaminant is gaseous, the corresponding formula involves a complete reflection:

$$c(x,y,z) = \frac{Q}{2\pi\sigma_y\sigma_z u}\exp\left[-\frac{1}{2}\left(\frac{y}{\sigma_y}\right)^2\right]\left\{\exp\left[-\frac{1}{2}\left(\frac{z-H}{\sigma_z}\right)^2\right] + \left[-\frac{1}{2}\left(\frac{z+H}{\sigma_z}\right)^2\right]\right\}$$

$$\exp\left(-\frac{(ln2)x}{uT_{\frac{1}{2}}^{SZ}}\right)\exp\left(-\frac{(ln2)x}{uT_{\frac{1}{2}}^{N}}\right)\exp\left(-\frac{(ln2)x}{uT_{\frac{1}{2}}^{A}}\right),$$

$$(x,y,z) \in \Re^2 \times \Re_+,$$

$$\tag{9}$$

where $T_{\frac{1}{2}}^{SZ}, T_{\frac{1}{2}}^{N}$ and $T_{\frac{1}{2}}^{A}$ denote the half-peirod for dry and wet deposition and chemical transformation, respectively.

In the case of solid pollutants, the following formula is used:

$$c(x,y,z) = \frac{Q}{2\pi\sigma_y\sigma_z u}\exp\left[-\frac{1}{2}\left(\frac{y}{\sigma_y}\right)^2\right]$$

$$\left\{\exp\left[-\frac{1}{2}\left(\frac{H-\frac{wx}{u}-z}{\sigma_z}\right)^2\right] + g\left[-\frac{1}{2}\left(\frac{H-\frac{wx}{u}+z}{\sigma_z}\right)^2\right]\right\}\exp\left(-\frac{(ln2)x}{uT_{\frac{1}{2}}^{N}}\right),$$

$$(x,y,z) \in \Re^2 \times \Re_+,$$

$$\tag{10}$$

where g denotes the reflection coefficient, i.e., a portion g of the contaminant particles reaching the surface is reflected, the rest deposes on the surface. $T_{\frac{1}{2}}^{N}$ denotes the helf-period for wet deposition.

6. Computational issues and visualization

In our environmental impact assessment project, a complex model was built and evaluated for the transmission of different pollutions. By using this model, the transmission equation deduced from HNS can be solved and the results visualized.

To solve the transmission equation three steps are necessary which are as follows:

1. the determination of Pasquill stability categorie
   input data: wind speed and either the surface radiation balance or the measure of cloud covering

output: Pasquill stability category and the exponent of the wind profile

2. the calculation of the effective height of the emission
   input data: wind speed, exponent of the wind profile, temperature of the air, height of the chimney, diameter of the chimney, flow velocity of the outgoing gas, temperature of the outgoing gas
   output: effective height of the emission

3. the calculation of the different pollutants at fixed points by formulae (9) and (10).

Our tasks in the preparation of the environmental impact assessment included the modelling of the transmission of different contaminations in the air, soil, surface and ground water. In such a real-life application, the presentation of the results is of great importance, therefore the results were presented by digitized maps (2D-visualization).

First, an area approximately of 5 km × 8 km was selected, having the planned factory nearly in its center. For our base map the corresponding part of the official geological map on the scale of 1:30000 was taken. After digitizing different data from this area, such as the height contours, places of different sample points, roads and different objects having importance in our investigations, putting in a database and connecting to the base map, the relief map of the area was obtained (Figure 1) which served as information source for our computations.

For visualizing the results of air pollution, the examined area was covered by an about 100 × 160 grid. In the points of this grid, the air pollution was computed corresponding to HNS for certain pollutants by a program. The estimations of the concentration in the rest of the points on the maps were achived by statistical interpolation functions. Both results were stored in a database which was exported to the software package "Surfer" in order to create the final presentations. By developing the environmental monitoring software system in the same area, both the mathematical models and the software were refined, thus the present version does not use the "Surfer" and is linked to the monitoring program written in C++ and JAVA.

In the environmental impact assessment the pollution transmission problem in the air was solved for the species lead and sulphur dioxid. The surface immission was focused on, as those affected by the air pollution are located on the surface (i.e., humans, vegetation, etc.). Since the atmospherical parameters may differ considerably daytime and night as well as during the seasons, four scenarios were studied accoringly: summer half-year daytime, summer half-year night, winter half-year daytime and winter half-year night. In each case, averages of 30 years with respect to the temperature and the measure of cloud covering were used (provided by the National Meteorological Service).

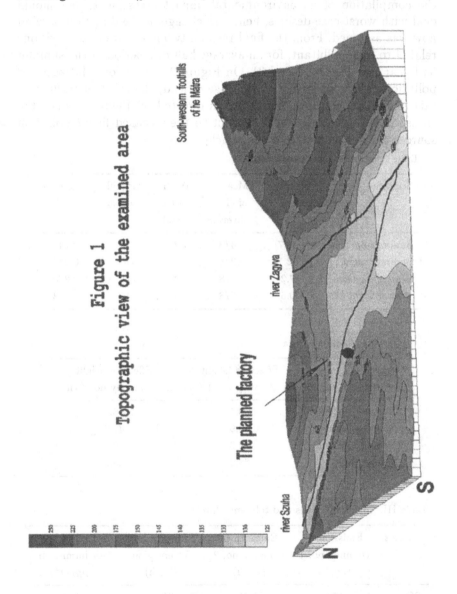

Figure 1
Topographic view of the examined area

The meteorological input data with the corresponding Pasquill stability category and exponent of wind profile can be found in Table 1. The effective heights of the emission are shown in Table 2 for all the four scenarios and for both chimneys. Table 3 contains the expected emission values (determined from the details of the technology) and the tolerance with respect to the emission and the immission for 24 hours. During the compilation of an environmental impact assessment, one should deal with worst-case-designs, hence no change in the direction of wind flow was assumed. From the final results, two maps are presented, one related to each pollutant, for an average half-day period in the summer half-year (see Figures 2 and 3). On Figure 2, the regional background pollution can be observed. In the case of lead, the effect of traffic was taken into account, but this factor is omitted on Figure 3, since the transmission calculation with respect to the so-called linear pollution sources are not discussed in the article.

Table I. Meteorological parameters

Scenario	Tempera-ture ($^\circ C$)	Measure of cloud covering	Wind speed (m/s)	Pasquill category	Exponent
Summer daytime	22.7	4/8	2	B	0.143
Summer night	12.1	3/8	2	F	0.440
Winter daytime	7.2	5/8	2	D	0.270
Winter night	0.5	5/8	2	E	0.363

Table II. Chimney data

Scenario	Effective height for chimney no. 1 (m)	Effective height for chimney no. 2 (m)
Summer daytime	90.7	56.7
Summer night	60.9	42.3
Winter daytime	76.8	50.1
Winter night	68.2	66.2

Table III. Emission values and tolerance levels

Species	Emission of chimney no. 1 (kg/h)	Emission of chimney no. 2 (kg/h)	Tolerance level for emission (kg/h)	Tolerance level for immission ($\mu g/m^3$)
SO_2	17.2	0.01	26.25	150
Pb	0.021	0.025	0.0525	0.3

Figure 2

Forecasted distribution of sulphur dioxid pollution
Summer half-year, daytime

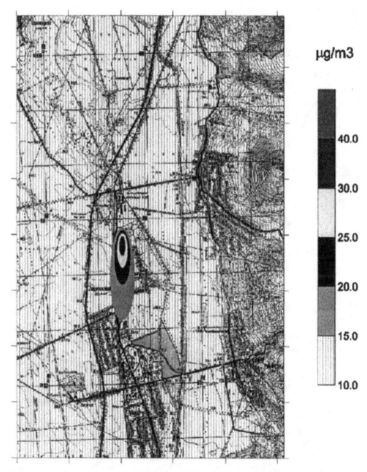

It is emphasized that in our models, closed formulae are used, without any computational difficulties, thus the computed concentrations do not depend on t he density of the grid. It is to be noted, however, that the uncertainty in the parameters may cause a relatively high uncertainty in the results, e.g., in the case of atm ospherical parameters. For this reason, the pollution was modelled also in extreme cases, including hav aria.

Figure 3

Forecasted distribution of lead pollution
Summer half-year, daytime

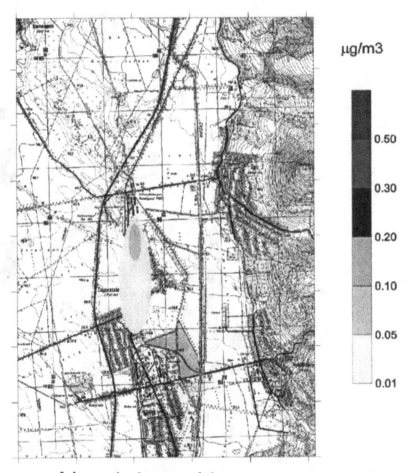

In course of the monitoring, one of the main issues was portability, i.e., the software developed can be used at other locations without any change in the code, the parameters depending on the current task take the role of installation parameters. Sulphur dioxid, carbon monoxid, nitrogen oxides, non-toxic solid components, carbon black, aluminium, chrome, copper, iron, magnesium, manganese, lead and zinc are to be traced in the environment of the aluminium foundry.

It was equally important to automatize the monitoring system as much as possible. By taking the surface radiation balance as an input parameter instead of the measure of cloud covering, the air transmission modelling process could be fully automatized, since this parameter can be automatically measured.

References

1. Bede, G. and Gács, I. (1980), *Transmission on pollutants in the air*, Technical University of Budapest, Budapest. (in Hungarian)
2. Brandt, J., Mikkelsen, T., Thykier-Nielsen, S. and Zlatev, Z. (1996), Using a Combination of Two Models in Tracer Simulations, *Mathematical Computational Modeling*, **23**, pp. 99–115.
3. CAI (1996), *Detailed environmental impact assessment study for setting up a waste accumulator recycling center in the former 'Qualital' establishment, Apc*, in Hungarian.
4. CAI (1998), *Software environmental monitoring system and its implementation at the company seat of B. T. Holding Foundry Ltd.*, in Hungarian.
5. *Hungarian National Standards.*, no. 21457 and no. 21459
6. Pasquill, F. (1962), *Atmospheric Diffusion*, Van Nostrand, London.
7. Tikhonov, A. N. and Samarskii, A. A. (1953), *Differential equations of mathematical physics*, Publisher of Technical-Scientific Literature, Moscow, in Russian.
8. Vladimirov, V. S. (1967), *Differential equations of mathematical physics*, Nauka, Moscow, in Russian.

BAYESIAN HEURISTIC APPROACH (BHA) AND APPLICATIONS TO OPTIMIZATION OF LARGE SCALE DISCRETE AND CONTINUOUS MODELS

J. MOCKUS

Institute of Mathematics and Informatics
Kaunas Technological University, Vytautas Magnus University
jonas@optimum.mii.lt

1. Introduction

The traditional numerical analysis considers optimization algorithms which guarantee some accuracy for all functions to be optimized. This includes the exact algorithms (that is the worst case analysis). Limiting the maximal error requires a computational effort that in many cases increases exponentially with the size of the problem. The alternative is average case analysis where the average error is made as small as possible. The average is taken over a set of functions to be optimized. The average case analysis is called the Bayesian Approach (BA) [4, 12].

There are several ways of applying the BA in optimization. The Direct Bayesian Approach (DBA) is defined by fixing a prior distribution P on a set of functions $f(x)$ and by minimizing the Bayesian risk function $R(x)$ [3, 12]. The risk function describes the average deviation from the global minimum. The distribution P is regarded as a stochastic model of $f(x)$, $x \in R^m$ where $f(x)$ might be a deterministic or a stochastic function. In the Gaussian case assuming (see [12]) that the $(n+1)$th observation is the last one

$$R(x) = \frac{1}{\sqrt{2\pi} s_n(x)} \times$$
$$\int_{-\infty}^{+\infty} \min(c_n, z) e^{-\frac{1}{2}\left(\frac{y - m_n(x)}{s_n(x)}\right)^2} dz, \tag{1}$$

Here $c_n = \min_i z_i - \epsilon$, $z_i = f(x_i)$, $m_n(x)$ is the conditional expectation given the values of z_i, $i = 1, ..., n$, $d_n(x)$ is the conditional variance, and $\epsilon > 0$ is a correction parameter.

The objective of DBA (used mainly in continuous cases) is to provide as small average error as possible while keeping the convergence conditions.

Z. Zlatev et al. (eds.), Large-Scale Computations in Air Pollution Modelling, 249–260.
© *1999 Kluwer Academic Publishers. Printed in the Netherlands.*

The Bayesian Heuristic Approach (BHA) means fixing a prior distribution P on a set of functions $f_K(x)$ defining the best values obtained using K times some heuristic $h(x)$ to optimize a function $v(y)$ of variables $y \in R^n$ [13]. As usual the components of y are discrete variables. The heuristic $h(x)$ defines an expert opinion about the decision priorities. It is assumed that the heuristics or their "mixture" depend on some continuous parameters $x \in R^m$, where $m < n$.

2. Direct Bayesian Approach (DBA)

Now all these ways will be considered in detail starting from the DBA. The Wiener process is common [9, 15, 17] as a stochastic model applying the DBA in the one-dimensional case $m = 1$.
Extending the Wiener process to $m > 1$ the Markovian property disappears.

Replacing the regular consistency conditions by some weaker assumptions the following simple expression of $R(x)$ is obtained using the results of [12].

$$R(x) = \min_{1 \leq i \leq n} z_i - \min_{1 \leq i \leq n} \frac{\|x - x_i\|^2}{z_i - c_n}.$$

The aim of the DBA is to minimize the expected deviation. In addition, DBA has some good asymptotic properties, too. It is shown in [12] that

$$d^*/d_a = \left(\frac{f_a - f^* + \epsilon}{\epsilon} \right)^{1/2}, \ n \to \infty$$

where d^* is density of x_i around the global optimum f^*, d_a and f_a are average density of x_i and average value of $f(x)$, and ϵ is the correction parameter in expression (1). That means that DBA provides convergence to the global minimum for any continuous $f(x)$ and greater density of observations x_i around the global optimum if n is large. Note that the correction parameter ϵ has a similar influence as the temperature in simulated annealing. However, that is a superficial similarity since, using DBA the good asymptotic behavior should be regarded just as an interesting "by-product". The reason is that Bayesian decisions are applied for the small size samples where asymptotic properties are not noticeable.

Choosing the optimal point x_{n+1} for the next iteration using DBA one solves a complicated auxiliary optimization problem minimizing the expected deviation $R(x)$ from the global optimum . That makes the DBA useful mainly for the computationally expensive functions of a few ($m < 20$) continuous variables.

3. Bayesian Heuristic Approach

Using DBA the expert knowledge is included by defining the prior distribution. In BHA the expert knowledge is involved by defining the heuristics and optimizing their parameters using DBA.

If the number of variables is large the Bayesian Heuristic Approach (BHA) is preferable. That is the case in many discrete optimization problems. As usual these problems are solved using heuristics based on an expert opinion. Heuristics often involve randomization procedures depending on some empirically defined parameters. The examples of such parameters are the initial temperature if the simulated annealing is applied or the probabilities of different randomization algorithms if their mixture is used. In these problems the DBA is a convenient tool for optimization of the continuous parameters of various heuristic techniques. That is called the Bayesian Heuristic Approach (BHA) [13].

The example of knapsack problem illustrates the basic principles of BHA in discrete optimization. Given a set of objects $j = 1, ..., n$ with values c_j and weights g_j, find the most valuable collection of limited weight.

$$\max_y v(y), \; v(y) = \sum_{j=1}^{n} c_j y_j, \; \sum_{j=1}^{n} g_j y_j \leq g.$$

Here the objective function $v(y)$ depends on n Boolean variables $y = (y_1, ..., y_n)$, where $y_j = 1$ if object j is in the collection, and $y_j = 0$ otherwise. The well known greedy heuristics $h_j = c_j/g_j$ is the specific value of object j. The greedy heuristic algorithm: "take the greatest feasible h_j", is very fast but it may get stuck in some non-optimal decision.

A way to force the heuristic algorithm out of such non-optimal decisions is by taking decision j with probability $r_j = \rho_x(h_j)$, where $\rho_x(h_j)$ is an increasing function of h_j and $x = (x_1, ..., x_m)$ is a parameter vector. The DBA is used to optimize the parameters x by minimizing the best result $f_K(x)$ obtained applying K times the randomized heuristic algorithm $\rho_x(h_j)$. That is the most expensive operation of BHA therefore the parallel computation of $f_K(x)$ should be used when possible reducing the computing time in proportion to a number of parallel processors.

Optimization of x adapts the heuristic algorithm $\rho_x(h_j)$ to a given problem. Let us illustrate the parameterization of $\rho_x(h_j)$ using three randomization functions: $r_i^l = h_i^l / \sum_j h_j^l$, $l = 0, 1, \infty$ Here the upper index $l = 0$ denotes the uniformly distributed component and $l = 1$ defines the linear component of randomization. The index ∞ denotes the pure heuristics with no randomization where $r_i^\infty = 1$ if $h_i = \max_j h_j$ and $r_i^\infty = 0$, otherwise. In this case parameter $x = (x_0, x_1, x_\infty)$ defines the probabilities of using randomizations $l = 0, 1, \infty$ correspondingly. The optimal x may be applied solving different but related problems, too [13]. That is very important in the "on-line" optimization adapting the BHA algorithms to some unpredicted changes.

Another simple example of BHA application is by trying different permutations of some feasible solution y^0. In this case heuristics are defined as the difference $h_i = v(y^i) - v(y^0)$ between the permuted solution y^i and the original one y^0. The well known simulated annealing algorithm illustrates the parameterization of $\rho_x(h_j)$ depending on a single parameter x. Here the probability of accepting a worse solution is equal to $e^{-h_i/x}$, where x is the "annealing temperature".

3.1. APPLICATION EXAMPLES

The comparison of BHA with exact $B\&B$ algorithms solving a set of the flow-show problems shows the table from [13].

Table I. Flow-Shop Problem, Simulation Results

$R = 100, K = 1, J = 10, S = 10, O = 10$					
Technique	f_B	d_B	x_0	x_1	x_∞
BHA	6.18	0.13	0.28	0.45	0.26
CPLEX	12.23	0.00	—	—	—

where S is the number of tools, J is the number of jobs, O is the number of operations, f_B, x_0, x_1, x_∞ are the mean results, d_B is the variance, and "CPLEX" denotes the standard MILP technique truncated after 5000 iterations. The table shows that in the randomly generated flow-shop problems the average make-span obtained by BHA was almost twice less that obtained by the exact $B\&B$ procedure truncated at the same time as BHA . The important conclusion is that stopping the exact methods before they reach the exact solution is not a good way to obtain the approximate solution.

The main objective of BHA is improving any given heuristic by defining the best parameters and/or the best "mixtures" of different heuristics. The heuristic decision rules mixed and adapted by BHA often outperform (in terms of speed) even the best individual heuristics as judged by the considered examples. In addition, BHA provides almost sure convergence. However, the final results of BHA depend on the quality of the specific heuristics including the expert knowledge. That means the BHA should be regarded as a tool for enhancing the heuristics but not for replacing them.

Many well known optimization algorithms such as Genetic Algorithms (GA) [7], GRASP [10], and Tabu Search (TS) [6], may be regarded as generalized heuristics that can be improved using BHA. There are a number of heuristics tailored to fit specific problems. For example, the Gupta heuristic was the best one while applying BHA to the flow-shop problem [13].

Genetic Algorithms [7] is an important "source" of interesting and useful stochastic search heuristics. It is well known [2] that the results of the genetic algorithms depend on the mutation and cross-over parameters. The Bayesian Heuristic Approach could be used in optimizing those parameters.

In the GRASP system [10] the heuristic is repeated many times. During each iteration a greedy randomized solution is constructed and the neighborhood around that solution is searched for a local optimum. The "greedy" component constructs a solution, one element at a time until a solution is constructed. A possible application of the BHA in GRASP is in optimizing a random selection of a candidate to be in the solution because different random selection rules could be used and their best parameters should be defined. BHA might be useful as a local component, too, by randomizing the local decisions and optimizing the corresponding parameters.

In tabu search the issues of identifying best combinations of short and long term memory and best balances of intensification and diversification strategies may be obtained using BHA.

4. Optimization of Time Series Model

Modeling ecological, economic and financial time series using the autoregressive moving average (ARMA) method has attracted the attention of many researchers and practitioners in recent years.

Define the ARMA model for prediction of variable y_t as

$$y_t = \sum_{i=1}^{p} a_i y_{t-i} + \sum_{i=1}^{q} b_i \epsilon_{t-i} + \epsilon_t \tag{2}$$

Assume that

$$y_{t-i} = 0, \ \epsilon_{t-i} = 0, \ if \ t \leq i. \tag{3}$$

Here parameters a_i, $i = 1, ..., P$ and b_j, $j = 1, ..., Q$ show how y_t depends on the previous values y_{t-i} and previous residuals ϵ_{t-j} defining the differences between predicted and observed values.

4.1. STRUCTURAL OPTIMIZATION

The objective of the traditional time series models is to minimize the deviation from the available data. One may call these as the best fit models. The models that fit best to the past data will predict the future data well if no changes happen in the system. Otherwise, the best fit to the past data can be irrelevant or even harmful. Thus a model of stable structure is needed which is insensitive to the system changes and thus may predict the uncertain future better by eliminating the "nuisance" parts from the structure of the model.

In traditional ARMA models it is assumed that parameters a and b remain approximately constant . Now we shell consider the case were values and numbers of parameters a, b change.

Regarding this problem the available data $Y = (y_t, t = 1, ..., T)$ is divided into two parts $Y_0 = (y_t, t = 1, ...T_0)$ and $Y_1 = (y_t, t = T_0 + 1, ..., T)$. The goodness of fit is described by continuous variables $C = (a_i, i = 1, ..., P, b_j, j = 1, ..., Q)$. The model structure is determined by discrete structural variables S including two integer variables P and Q and $P+Q$ Boolean variables $S = (P, Q, s_i^a, i = 1, ..., P, s_j^b, j = 1, ..., Q)$. A Boolean variable is equal to unit if the corresponding component of model is included and is equal to zero, otherwise. For example, $s_i^a = 1$, if the parameter a_i is included , and $s_i^a = 0$ if not. Denote by $R_t(S, C, Y)$ the predicted value of the ARMA model at fixed parameters (S, C) using the data $(y_1, ..., y_{t-1}) \subset Y$.

The difference between the ARMA prediction and the actual data y_t is denoted by $\epsilon_t(S, C, Y) = y_t - R_t(S, C, Y)$. Denote by $C_0(S)$ the fitting parameters C which minimize the sum of squared deviations $\Delta_{0,0}(C, S)$ using the first data set Y_0 at fixed structure parameters S.

$$\Delta_{0,0}(C, S) = \sum_{t=1}^{T_0} \epsilon_t^2(S, C, Y), \tag{4}$$

$$C_0(S) = \arg\min_C \Delta_{0,0}(C, S). \tag{5}$$

We stabilize the structure S by minimizing the sum of squared deviations $\Delta_{1,0}(S)$ using the second data set Y_1 and the fitting parameters $C_0(S)$ that were obtained from the first data set

$$\Delta_{1,0}(S) = \sum_{t=T_0+1}^{T} \epsilon_t^2(S, C_0(S), Y), \tag{6}$$

$$S_1 = \arg \min_S \Delta_{1,0}(S). \qquad (7)$$

The usefulness of the stabilization follows from the observation that any optimal estimate of time series parameters using a part Y_0 of the data $Y = Y_0 \cup Y_1$ is optimal for another part Y_1 only if all the parameters remain the same. Otherwise one may obtain a better estimate eliminating the changing parameters from the model. For example, in the case of changing parameters $(a_i, \; i = 2, ..., P, \; b_j, \; j = 1, ..., Q)$ of the ARMA model the best prediction may be obtained by elimination of all the parameters except the first one (see the first and the third row in Table 5.1).

The simplest way of structure stabilization of ARMA model is if only two structural parameters, namely P and Q are optimized by choosing the best pair (P, Q) where $P_{min} \leq P \leq P_{max}, \; Q_{min} \leq Q \leq Q_{max}$. If the ranges $P_{max} - P_{min}$ and $Q_{max} - Q_{min}$ are large then some heuristics should be used with parameters optimized by BHA. The software was developed using exhaustive search for the best pair (P, Q) if the ranges $P_{max} - P_{min}$ and $Q_{max} - Q_{min}$ are not large. Otherwise the simulated annealing with "annealing temperature" optimized by BHA is applied.

4.2. EVALUATION OF ARMA PREDICTION ERRORS

We compare the "next-day" ARMA prediction results and the simple "Random Walk (RW)" model, where $y_{t+1} = y_t$ for a given t . Table 5.1 shows the difference between the mean square deviations of ARMA and RW models defined as $\Delta = 100(\delta(ARMA) - \delta(RW))/\delta(RW)$. Here $\delta(ARMA)$ and $\delta(RW)$ are average deviations of ARMA and RW models from actual data in the estimating period of time, from the end of testing set until the end of complete set. In the Table 5.1 Δ is called as the "Optimal Deviation", the part of data used for parameter optimization is defined as the "Learning Set". The "Testing Set" means the data used for structural optimization. The table 5.1 is prepared using Intel Co. stocks closing rates (the column "Intel Stocks") and DM/$ exchange rates (the column "DM rates") for the period of $T = 450$ days and daily call rates of some call center for 378 days (the column "Call Center").

The DM/$ exchange rates are predicted using not only previous DM/$ rates but the $/£, yen/$ an French frank/$ rates, too, as some external factors. The results for "DM Rates" of Table 5.1 are very close to the Wiener case where the "Optimal" a[0]=a[1]=a[2]=0[1] and a[3]=1 because only the fourth component a[3] of 4-dimensional ARMA model relates to DM/$ exchange rates, the rest components relates to other

---

[1] In the table 5.1 we number parameters from zero, as in C programs.

currencies. Similar results are obtained for the one-dimensional ARMA model using Intel stock rates.

As a contrast, predicting the call rates the ARMA model works very well using a lot of previous data (the optimal P is 21 days).

In almost all the cases the structural optimization improves the results considerably. That suggests that the ARMA parameters a and b cannot be regarded as constants.

5. Networks Optimization

Optimal resource allocation problems often are reduced to network optimization that is convenient in formulation and visualization.

An application of Global Line-Search (GCS) to the optimization of networks is considered in this section as an example. Advantages and disadvantages are discussed. It is shown (see [11]) that GCS provides the global minimum after a finite number of steps in two cases of piecewise linear cost functions of arcs. The first case is, where all cost functions are convex. The second case is, where all costs are equal to zero at zero flow, and equal to some constant at non-zero flow. In other cases the global line-search approaches the global minimum with a small average error. Therefore this algorithm is regarded as a good heuristics to be enhanced using the BHA.

5.1. GLOBAL COORDINATE-SEARCH (GCS)

Suppose that the objective function $v(y), y = (y_j, j = 1, ..., J)$ is approximately expressed as a sum of components depending on only one "coordinate" $y_j \in A_j$ where A_j is a continuous or discrete subset of R^S.

$$v(y) = \sum_{j=1}^{J} v_j(y_j). \tag{8}$$

Then the original (SJ)-dimensional optimization problem can be reduced to a sequence of S-dimensional optimization problems. If decomposition (8) is exact, then we obtain the global optimum after J steps of optimization. If sum (8) represents $f(x)$ approximately, then we obtain some approximation of global optimum. In this case the results depend on the starting point x^0 which is defined using expert opinion. Random definition of x^0 is justified only if no expert opinion is available.

Table II. Average Prediction Errors

Parameter	Intel Stocks	Call Center	DM Rates
Number of Factors	1	1	4
End of Learning Set	150	176	600
End of Testing Set	300	352	1200
End of Complete Set	450	378	1800
Datafile0	intclosing	callsdaily	DMrates
Datafile1			LBrates
Datafile2			YENrates
Datafile3			FRrates
Pmax Qmax	10 1	41 4	20 5
Optimal Po Qo	10 1	21 1	4 0
Optimal Deviation	-8.309595e-01	-6.074035e+01	2.602389e+00
Optimal bo[0] =	-5.291304e-02	2.000000e-01	
Optimal : ao[0]=	1.005026e+00	7.070145e-01	1.874834e-03
Optimal : ao[1]=	-1.603190e-02	2.942113e-02	4.943231e-04
Optimal : ao[2]=	9.327068e-03	-2.283435e-02	1.901402e-03
Optimal : ao[3]=	-1.962328e-02	1.535324e-01	9.603935e-01
Optimal : ao[4]=	2.746916e-02	-7.792136e-02	
Optimal : ao[5]=	1.342659e-02	5.200528e-01	
Optimal : ao[6]=	-1.830024e-02	-3.050167e-01	
Optimal : ao[7]=	-2.024879e-02	-8.005017e-02	
Optimal : ao[8]=	2.393244e-02	5.328229e-03	
Optimal : ao[9]=	-5.652583e-03	3.985322e-02	
Optimal : ao[10]=		4.217983e-02	
Optimal : ao[11]=		2.531133e-01	
Optimal : ao[12]=		-1.566629e-01	
Optimal : ao[13]=		-5.556350e-02	
Optimal : ao[14]=		4.687828e-04	
Optimal : ao[15]=		-2.051560e-01	
Optimal : ao[16]=		4.439837e-03	
Optimal : ao[17]=		1.718254e-01	
Optimal : ao[18]=		-1.900502e-01	
Optimal : ao[19]=		1.082512e-01	
Optimal : ao[20]=		5.682016e-02	

The result of step i is regarded as the initial point for the $i+1$-th step of optimization. The optimization stops, if no change occurs during J

steps. The difference from the traditional version of coordinate-search
is that the search is not local but global. As usual, it helps to approach the global minimum closer. There are interesting cases when
the global coordinate-search reaches the global minimum, and the local
coordinate-search does not (see [11]).

In those cases when GCS does not reach the global minimum the
BHA should be used. In the terms of BHA the described GCS procedure
choosing the minimizer of $v(y)$ for each coordinate y_j is regarded as the
"Pure Permutation Heuristics". The "Monte Carlo Heuristics" would
be a random choice of some point in A_j regardless of $v(y)$. The "Linear
Randomization Heuristics" means that at the current iteration n one
chooses a point in A_j with probability proportional to the heuristics
h_k defined as $h_j = v_{max}^n - v(y^k) + \epsilon$. Here v_{max}^n is the maximal value of
the function $v(y)$ achieved until the n-th iteration and ϵ is a positive
constant needed to keep the heuristic h_k positive.

In this example the objective of BHA would be to define such a
"mixture" x_0, x_1, x_2, $x_i \geq 0$, $x_0 + x_1 + x_2 = 1$ of these heuristics
that provides the best results v^N after N repetitions. Here x_0, x_1, x_2
denotes the probabilities of using the Monte Carlo, the linear the pure
heuristics.

During late sixties and early seventies a similar algorithm was used
for optimal extension of large power system in designing new power
transmission lines of 110 KV, 220 KV and 330 KV. The starting point
was defined by expert designers of power systems. The results indicate
that the global line-search is a good heuristic that may be improved
by randomization and parameter optimization in the framework of the
BHA.

Average deviation was about $2 - 3\%$ when the starting point is
defined by expert opinion. Application of BHA improves the average
deviation by the factor of 10. This is achieved at the price of extensive
initial computing effort obtaining the best mixture of the Monte Carlo,
linear and pure heuristics. Since this mixture is defined it is applied to
related network optimization problems skipping the additional costly
updating.

Pardalos and Rosen [14] present an approximation technique based
on piecewise linear underestimation of concave cost functions $f_j(x_j)$.
The resulting model is a linear, zero-one, mixed integer problem. A
direct comparison of this approach and global line-search techniques
is an interesting problem of future research. For a review of results in
network optimization see [8].

6. Software for Global Optimization

The global optimization software is in four versions: portable Fortran Library, interactive software for Turbo C , interactive software for C++ compiler and X-Window, interactive software for Java . One may notice a cycle of portability in this sequence of software versions. The sequence is started from by the portable Fortran library and is concluded by Java language. The two systems in between are more difficult to port. The Turbo C system is for DOS-compatible operating systems and C++ is for the UNIX environment. Fortran, Turbo C and C++ versions are described in [13].

References

1. S. Andradottir, S. (1996), A global search method for discrete stochastic optimization, *SIAM J. Optimization*, **6**, pp. 513–530.
2. Androulakis, I. P. and Venkatasubramanian, V. (1991), A genetic algorithmic: Framework for process design and optimization, *Comput. Chem. Engng.*, **15**, pp. 217–228.
3. DeGroot, M. (1988), *Optimal Statistical Decisions*, McGraw-Hill, New York.
4. Diaconis, P. (1988), Bayesian numerical analysis, in *Statistical Decision Theory and Related Topics*, Springer, Berlin, pp. 163–175.
5. Dzemyda, G. and Senkiene, E. (1990), Simulated annealing for parameter grouping, in *Transactions on Information Theory Statistical Decision Theory Random Processes*, Praque, pp. 373–383.
6. Glover, F. (1994), Tabu search: improved solution alternatives, in *Mathematical Programming. State of the Art, 1994*, University of Michigan, pp. 64-92.
7. Goldberg, D. E. (1989), *Genetic Algorithms in Search, Optimization, and Machine Learning*, Addison-Wesley, Reading, MA.
8. Guisewite, G. M. and Pardalos, P. M. (1990), Minimum concave-cost network flow problems: Applications, complexity, and algorithms. *Ann. of Operations Research*, **25**, pp. 75–100.
9. Kushner, H. J. (1964), A new method of locating the maximum point of an arbitrary multi-peak curve in the presence of noise, *J. of Basic Engineering*, **86**, pp. 97–100.
10. Mavridou, T, Pardalos, P. M., Pitsoulis, L. S. and Resende, M. G. C. (1998), A GRASP for the biquadratic assignment problem, *European Journal of Operations Research*, **105**, pp. 613–621.
11. Mockus, J. (1967), *Multi-modal Problems in Engineering Design*, Nauka, Moscow, in Russian.
12. Mockus, J. (1989), *Bayesian approach to global optimization*, Kluwer Academic Publishers, Dordrecht-London-Boston.
13. Mockus, J., Eddy, W., Mockus, A., Mockus, L. and Reklaitis, G. (1997), *Bayesian Heuristic Approach to Discrete and Global Optimization*, Kluwer Academic Publishers, Dordrecht-London-Boston.
14. Pardalos, P. M. and Rosen, J. B. (1987), *Constrained global optimization: Algorithms and applications*, Springer, Berlin.

15. Saltenis, V. R. (1971), On a method of multi-extremal optimization, *Automatics and Computers (Avtomatika i Vychislitelnayya Tekchnika)*, **3**, pp. 33–38, in Russian.

16. Strongin, R. G. (1978), *Numerical methods in multi-extremal problems*, Nauka, Moscow, in Russian.

17. Torn, A. and Zilinskas, A. (1989), *Global optimization*, Springer, Berlin.

18. Van Laarhoven,P. J. M., Boender, C. G., E., Aarts, E. H. L. and A.H.D. RinnooyKan, A. H. D. (1989), A bayesian approach to simulated annealing, *Probability in the Engineering and Information Sciences*, **3**, pp. 453–475.

NONLINEAR ASSIGNMENT PROBLEMS

P. M. PARDALOS and L. S. PITSOULIS
Center for Applied Optimization and
Department of Industrial and Systems Engineering
University of Florida, Gainesville, FL 32611-6595, USA

T. D. MAVRIDOU
Technical University of Crete, Chania, Crete, Greece

Abstract

For all practical purposes three types of nonlinear assignment problems
(NAPs) have emerged in the literature, the Quadratic, the Cubic and
the Biquadratic assignment problems. In this paper we describe the
importance of NAPs in the context of location theory, and we present
a brief survey of recent developments in the design and implementation
of efficient exact and heuristic algorithms for solving NAPs. All three
types of NAPs belong to the NP-hard complexity class, and hence there
is a need for good heuristics. A Greedy Randomized Adaptive Search
Procedure (GRASP) for solving QAP instances is presented together
with preliminary computational results.

1. Introduction

The assignment of industrial facilities to a set of geographical locations,
can play an important role in the modeling of air pollution problems.
The cost resulting from the assignment of facilities to given locations
is often a function of the *flow* of material between facilities and the
distance between locations. This problem is called the Quadratic Assignment Problem (QAP), and is a core problem in facility location
theory. The QAP was first introduced by Koopmans and Beckmann in
1957 as a mathematical model for the location of a set of indivisible
economical activities [12]. Specifically, we are given three real input
matrices, $F = (f_{ij})$, $D = (d_{kl})$ and $B = (b_{ij})$, where f_{ij} is the flow
between the facility i and facility j, d_{kl} is the distance between the
location k and location l, and b_{ij} is the cost of placing facility i at

Z. Zlatev et al. (eds.), Large-Scale Computations in Air Pollution Modelling, 261–273.
© *1999 Kluwer Academic Publishers. Printed in the Netherlands.*

location j. The QAP is formulated as follows:

$$\min_{p \in \Pi_N} \sum_i^n \sum_j^n f_{ij} d_{p(i)p(j)} + \sum_i^n b_{ip(i)} \tag{1}$$

where Π_N is the set of all permutations $p : N \to N$ where $N = \{1, 2, \ldots, n\}$, and n is the number of facilities and locations. Each individual product $f_{ij} d_{p(i)p(j)}$ is the cost of assigning facility i to location $p(i)$ and facility j to location $p(j)$. In the context of facility location the matrices F and D are symmetric with zeros in the diagonal, and all the matrices are nonnegative.

In a similar fashion we define all the non-linear assignment problems (NAPs). Note that in all these problems we have the same feasible domain, that is the set of permutations Π_N. In general the problem is non-convex and we want to find the global solution or an approximate solution. From the complexity point of view non-linear assignment problems are NP-hard.

2. Algorithmic Developments

In this section we will briefly discuss several exact and heuristic algorithms that have been implemented for solving the quadratic, the cubic (three-index) and the biquadratic assignment problem.

2.1. EXACT ALGORITHMS

An exact algorithm for a combinatorial optimization problem provides the global optimal solution to the problem. The exact algorithms that have been used for the QAP are *cutting plane algorithms, dynamic programming*, and *branch and bound methods*.

Cutting plane algorithms have been developed for the *concave minimization formulation* of the QAP. Although theoretically the problem structure favored the cutting plane procedure, in practice the number of cuts required for termination was practically impossible. However the heuristics derived from the cutting plane procedures produced good quality solutions.

Branch and bound algorithms have been applied successfully to many combinatorial optimization problems, and they appear to be the most efficient exact algorithms for solving the QAP. Although there are different branch and bound techniques we will try to describe the general concept. At the beginning we obtain a good feasible solution to the problem by means of a heuristic, and we treat this solution as an upper bound. The problem is then decomposed into a finite number of subproblems and a lower bound is established for each. Then each of these subproblems is decomposed in the same fashion, and by doing so we construct a search tree. However, many subproblems are ignored using a lower bound, a procedure called pruning, so we avoid complete enumeration of the whole feasible set. For the QAP there are three types of branch and bound algorithms and namely:

- *Single assignment algorithms*

- *Pair assignment algorithms*

- *Relative positioning algorithm*

All of the above algorithms work by iteratively constructing an optimal permutation starting from an empty permutation. The single assignment algorithms seem to be the most efficient and the pair assignment algorithms do not have favorable computational results. For a complete survey of exact algorithms and lower bounds for NAPs look [4, 6, 23].

2.2. HEURISTICS

Although substantial improvement has been done in the development of exact algorithms for the QAP, problems of dimension $n > 20$ are still not practical to solve in terms of computational time and this gave rise to the need for suboptimal algorithms or heuristics, which will provide good quality solutions in a reasonable time. Much research has been devoted to the development of such heuristics, and there are currently five types of heuristic algorithms for the QAP:

- Construction methods.

- Limited enumeration methods.

- Improvement methods.

- Simulated annealing.

- Genetic algorithms.

2.2.1. *Construction Methods*

Construction methods were introduced by Gilmore [8], and work as follows. Let the permutations $p_0, p_1, \ldots, p_{n-1}$ be partial permutations and let heur(i, j) be some heuristic procedure that assigns facility i to location j. Denote the set Γ as the set of already assigned pairs of facilities to locations. The procedure update will make a permutation p_i by adding the assignments (i, j) to p_{i-1}. The algorithm then works as

```
procedure construction(p₀, Γ)
1  p = {};
2  do i = 1,...,n − 1 →
3    if (i, j) ∉ Γ →
4      heur(i, j);
5      update(pᵢ, (i, j));
6      Γ = Γ ∪ (i, j);
7    fi;
8    p = pᵢ;
9  od;
10 return(p)
end construction;
```

Figure 1. Pseudo-code for a construction method

shown in Figure 1. The heuristic procedure $heur(i, j)$ could be any valid heuristic such as to assign $(i, j) \notin \Gamma$ in a greedy fashion or by employing local search. Based on the construction procedure the CRAFT heuristic which has been developed by Armour and Buffa [2] is one of the oldest heuristics used by the industry.

2.2.2. *Limited Enumeration Methods*
The limited enumeration methods suggest that since the optimum solution need not to be found, a good suboptimal solution can be obtained in the early stages of the enumerative procedure. In other words, instead of searching the entire feasible space with an exact algorithm, impose a limit of some type in the enumeration process. The limit could be a time limit, or an iteration limit beyond which the algorithm will stop, reporting the best solution found. Another option would be to manipulate the upper bound, that is to decrease it if there is no improvement in the solution, which will result in deeper cuts in the search tree and will speed up the process. This method can be used in conjunction with exact algorithms, and with certain heuristics that perform elaborate searches in the feasible space.

2.2.3. *Improvement Methods*
In this area there are two popular heuristic procedures, the local search and tabu search procedures, which have been used extensively as efficient procedures in many combinatorial optimization problems. Improvement methods consist of an iterative process which starts with a feasible solution, and improves it until some stopping criterion is

reached. Maybe one of the most important obstacles in these methods, as in most of the heuristics, is to overcome local optimality which the algorithm might get trapped into.

In local search algorithms it is desired to start with a good solution, and then the algorithm searches in the neighborhood of the current solution for a better solution and it replaces the current one with the best found. When no further improvement can be made, the algorithm stops. More specifically, if we denote the neighborhood of a feasible solution s to the problem P by $N(s)$, then we can describe the local search in pseudocode (see Figure 2. For an efficient local search algorithm

```
procedure local(P, s)
1  do s is not local minimum →
2      Find a better solution t ∈ N(s);
3      Assign s = t;
4  od;
5  return(s as local optimal for P)
end local;
```

Figure 2. Pseudo-code for local search

one should try to choose a good neighborhood structure, efficient procedures for searching the neighborhood, and a good starting solution. The use of good data structures has proven to be very significant for the efficiency of these procedures. For the QAP we have the λ-exchange, k-exchange, and N^* neighborhood local searches. Although it seems intuitively sensible to proceed in a sequence of improving solutions, this procedure sometimes gets easily "trapped" at local optima. It has been observed that sometimes it is more favorable to perform local search in a solution that is not the best found in the current neighborhood. This is the main concept under *tabu search* which can be viewed as a modified local search. Given a local optimum solution s_0 with a neighborhood $N(s_0)$, the procedure examines at each iteration all the neighborhood $N(s_0)$. If there are no improving moves, the procedure chooses the move that least degrades the objective function. A *tabu list* is kept to avoid returning to the local optimum from which the procedure has been escaped. A basic element of tabu search is the *aspiration criterion* upon which tabu moves are allowed to be chosen if they seem to be relevant, as exemplified by those that lead to a better solution than the best found so far. A termination criterion for the tabu procedure would be a limit in the consecutive moves at which no improvement occurs.

2.2.4. *Simulated Annealing Methods*

Annealing refers to a process of cooling material slowly until it reaches a stable state. At the early stages of the cooling process the particles of the material might change at high-energy states, but at lower temperatures the particles move to low-energy states until the material is completely frozen. In simulated annealing we can think of each stage of the search, as being carried out under lower temperature than that which occurred at the previous stage. Analogous to the cooling process, at early stages of the search we are flexible enough to move to a worse solution, but at later stages of the search (low temperatures) we are less flexible. Let C_{p_i} denote the objective function cost of a given starting permutation p_i, and let us use neighborhood search. The procedure then selects randomly from the neighborhood of the p_i. A neighbor p_j with corresponding C_{p_j} will become the next neighborhood generation permutation if $C_{p_j} < C_{p_i}$. If $C_{p_j} \geq C_{p_i}$ then the p_j will become the next neighborhood generation permutation with a probability of $\pi_{ij} = \min(1, e^{\frac{-\delta_{ij}}{T(i)}})$ where $\delta_{ij} = C_{p_j} - C_{p_i}$ and $T(i)$ is the temperature at that stage.

2.2.5. *Genetic Algorithms*

Unlike most of the suboptimal algorithms described in the previous sections, genetic algorithms conduct the search through the information of a population consisting a subset of individuals (solutions). Let us treat each feasible solution as an individual, and associate each individual with a fitness value. A population then can be defined as a set of individuals. The population evolves, producing better and better individuals (solutions). Naturally, the key to the success of genetic algorithms, is the selection procedure based on the fitness values of the individuals, which measure how good the individuals are. Usually, genetic algorithms start with a random population, and the fitness value of an individual is simply the corresponding objective function value. After selecting randomly among the best individuals, genetic operators, such as crossover and mutation, are applied toward the selected sample to produce a new improved population. Genetic algorithms are inherently parallel in nature, therefore they are favorable to implement in a parallel environment. Such a parallel application of genetic algorithms is done by Mirchandani and Obata [14].

3. GRASP for Large Sparse Problems

The GRASP is an iterative procedure consisting of two phases at each iteration, a construction phase and a local search phase. At each

GRASP iteration the algorithm founds a solution, displaying the best solution among all iterations as the final one. A GRASP pseudo-code is presented in Figure 3. Line 1 of the pseudo-code is the problem

```
procedure GRASP(RCLSize,MaxIter,RandomSeed)
1 InputInstance();
2 do k = 1,..., MaxIter →
3   ConstructGreedyRandomizedSolution(RCLSize,
>    RandomSeed);
4   LocalSearch(BestSolutionFound);
5   UpdateSolution(BestSolutionFound);
6 od;
7 return BestSolutionFound
end GRASP;
```

Figure 3. Pseudo-code for a generic GRASP

input. Line 2 to 6 correspond to the GRASP iterations. The termination criterion can be, for example, a satisfactory solution value or a specified maximum number of iterations. Line 3 is the construction of a greedy randomized solution, followed by the local search phase (line 4). Line 5 corresponds to the improved solution update. Next we will present a more detailed description of the two phases.

In the construction phase, a greedy randomized solution is iteratively constructed, one element at a time. The procedure is summarized in pseudo-code in Figure 4. Line 1 of the pseudo-code corresponds to the initialization of the solution set. Line 3 shows the construction of a *Restricted Candidate List* (RCL). The RCL contains partially the best candidates, from which an element s will be selected randomly to be added, by the algorithm, in the solution set (line 4). Lets note that the random selection of element s is very important for the algorithm in order to avoid the entrapment in local minima. In the next step the greedy function is evaluated on the remaining elements of the list (excluding s). The loop corresponding to lines 2 to 7 is repeated until a complete solution is constructed.

The second phase follows , where a local search is applied in order to improve the constructed solution. The algorithm starts with and initial solution s of the problem P. The procedure successively replaces the current solution with a better one in the neighborhood $N(s)$ of the current solution and stops when no further improvement is possible. The local search algorithm is described in Figure 2. The key to success

```
procedure ConstructSolution(solution)
1  solution={};
2  do Solution is not complete →
3    MakeRCL(RCL);
4    s=SelectRandomElement(RCL);
5    solution=solution ∪{s};
6    AdaptGreedy(s,RCL);
7  od;
8  return(solution)
end ConstructSolution;
```

Figure 4. GRASP construction phase pseudo-code

for a local search algorithm consists of suitable choice of a neighborhood structure, efficient neighborhood techniques, and the starting solution.

A favorable characteristic of GRASP is the ease with which it can be implemented. First because adaptive greedy functions are known for many problems and second because neighborhood definitions and local search procedures are plentiful. Few parameters need to be set and tuned, i.e. the candidate list size and the number of GRASP iterations. A wide variety of GRASP implementations have appeared in the literature [7], including several optimization problems such as production planning and scheduling, location problems, computer aided process planning, corporate acquisition of flexible manufacturing equipment and other.

4. Computational Experiments

In this section we report computational experience with the implementation of GRASP for solving QAP instances with sparse coefficient matrices. The GRASP is implemented on ANSI standard FORTRAN77 and works on any UNIX platform without modifications.

The GRASP Fortran implementation was tested on a SUN Sparc station with flags 02 -Olimit 800. The set of problems consisted of all problems in QAPLIB [5] of dimension $n \geq 8$ that are pure quadratic assignment problems, and have at least one symmetric distance or flow matrix. Every table contains the following information:

- Problem name: The name under which the specific QAP instance is classified in the literature.

- n: The size of the QAP instance.

- Best known solution value: The best solution value reported in the literature for the specific instance.

- $\max(S_D, S_F)$: The maximum sparsity from the flow matrix F and the distance matrix D.

- solution: The solution found by the heuristic.

- iterations: The number of iterations needed to find the given solution.

- CPU time: The CPU seconds needed to perform the given iterations.

The performance of the heuristic is measured in terms of the solution produced, and the CPU time needed. Naturally, it is desired to have the best known solution, in the least amount of CPU time. As it can be seen from the tables the GRASP heuristic finds the best known value for all the problems in the QAPLIB [5] except for four instances in the chr class and one instance in the nug class. However even in these cases the solution found is within 4% of the best known value in the worst case.

A parallel version of GRASP for solving large scale QAP instances has been implemented and can be found in [20]. Moreover the GRASP has also been successfully implemented to solve general biquadratic instances, of size up to and $n = 34$.

5. Concluding Remarks

In this paper we presented a brief survey of recent developments regarding techniques and software for solving nonlinear assignment problems. In many large problems that arise from location theory, exact global optima are very difficulty to find due to the prohibitive computational requirements. For such problems, heuristics such as the GRASP can obtain acceptable sub-optimal solutions in reasonable CPU time.

Table I. GRASP-sparse for the *esc* problem class

Problem name	n	Best known value	$max(S_D, S_F)$	GRASP-sparse		
				solution	iterations	CPU time
esc16a	16	68	0.64	68	1	5.0-02
esc16b	16	292	0.25	292	1	5.0-02
esc16c	16	160	0.53	160	7	0.33
esc16d	16	16	0.77	16	1	5.0-02
esc16e	16	28	0.77	28	1	3.3-02
esc16f	16	0	0.93	0	1	3.3-02
esc16g	16	26	0.77	26	1	5.0-02
esc16h	16	996	0.25	996	1	6.6-02
esc16i	16	14	0.82	14	1	6.6-02
esc16j	16	8	0.84	8	1	5.0-02
esc32a	32	130	0.82	130	6935	2948
esc32b	32	168	0.75	168	60	29.5
esc32c	32	642	0.80	642	2	0.95
esc32d	32	200	0.85	200	83	31.16
esc32e	32	2	0.85	2	1	0.31
esc32f	32	2	0.87	2	1	0.31
esc32g	32	6	0.91	6	1	0.3
esc32h	32	438	0.90	438	162	67.5
esc64a	64	116	0.95	116	1	8.8
esc128	128	64	0.98	64	22	581.5

Table II. GRASP-sparse for the *nug* problem class

Problem name	n	Best known value	$max(S_D, S_F)$	GRASP-sparse		
				solution	iterations	CPU time
nug12	12	578	0.37	578	45	1.4
nug15	15	1150	0.33	1150	20	1.5
nug20	20	2570	0.30	2570	808	169.9
nug30	30	6124	0.20	6130	20000	11788

Table III. GRASP-sparse for the *chr* problem class

Problem name	n	Best known value	$max(S_D, S_F)$	GRASP-sparse		
				solution	iterations	CPU time
chr12a	12	9552	0.76	9552	42	0.81
chr12b	12	9742	0.76	9742	25	0.53
chr12c	12	11156	0.76	11156	790	15.8
chr15a	15	9896	0.80	9896	1739	74.35
chr15b	15	7990	0.80	7990	1077	44.6
chr15c	15	9504	0.80	9504	1710	75.57
chr18a	18	11098	0.83	11098	17179	1395.6
chr18b	18	1534	0.83	1534	533	39.5
chr20a	20	2129	0.85	2129	9952	1104
chr20b	20	2298	0.85	2362	100000	10964
chr20c	20	14142	0.85	14142	1037	120
chr22a	20	6156	0.87	6176	100000	13798
chr22b	20	6194	0.87	6278	100000	16285
chr25a	20	3796	0.88	3954	13660	3395

Table IV. GRASP-sparse for the *kra* problem class

Problem name	n	Best known value	$max(S_D, S_F)$	GRASP-sparse		
				solution	iterations	CPU time
kra30a	30	88900	0.63	88900	12004	6011

Table V. GRASP-sparse for the *scr* problem class

Problem name	n	Best known value	$max(S_D, S_F)$	GRASP-sparse		
				solution	iterations	CPU time
scr10	10	26992	0.56	26992	9	0.133
scr12	12	31410	0.61	31410	18	0.383
scr15	15	51140	0.62	51140	77	3.783
scr20	20	110030	0.69	110030	8237	1092

Table VI. GRASP-sparse for the *els* problem class

Problem name	n	Best known value	$max(S_D, S_F)$	GRASP-sparse		
				solution	iterations	CPU time
els19	19	17212548	0.68	17212548	51	5.88

References

1. Ananth, G. Y., Kumar, V. and Pardalos, P. M. (1993), Parallel Processing of Discrete Optimization Problems, *Encyclopedia of Microcomputers*, **13**, pp. 129-147, Marcel Dekker Inc., New York.

2. Armour, G. C. and Buffa, E. S., (1963), A heuristic algorithm and simulative approach to relative location of facilities, *Management Science*, 9, pp. 294-309.

3. Balas, E. and Saltzman, M. J. (1991), An Algorithm fir the three-index Assignment Problem, *Operation Research*, **39**, pp. 150-161.

4. Burkard, R. E., Çela, E., Pardalos, P. M. and Pitsoulis, L. (1998), The Quadratic Assignment and Related Problems, in DingZhu Du and P.M. Pardalos (eds.), *Handbook of Combinatorial Optimization*, Vol. 4, Kluwer Academic Publishers, Dordrecht-Boston-London.

5. Burkard, R. E., Karisch, S. and Rendl, F. (1997), QAPLIB – A quadratic assignment problem library. *Journal of Global Optimization*, **10**, pp. 391–403.

6. Çela, E. (1998), *The Quadratic Assignment Problem: Theory and Algorithms*. Kluwer Academic Publishers, Dordrecht-Boston-London.

7. Feo, T. A. and Resende, M. G. C. (1995), Greedy Randomized Adaptive Search Procedures, *Journal of Global Optimization*, **16**, pp. 1-27.

8. Gilmore, P. C. (1962), Optimal and suboptimal algorithms for the quadratic assignment problem, *SIAM Journal on Applied Mathematics*, **10**, pp. 305-313.

9. Glover, F., Taillard, E. and De Werra, D. (1993), A user's guide to tabu search, *Annals of Operation Research*, **41** pp. 3-28.

10. Hall, M. (1976), *Combinatorial Theory*, Blaisdell Company, Waltman, MA.

11. Kirkpatrick, S, Gellat Jr, C. D. and Vecchi, M. P. (1983), Optimization by Simulated Annealing, *Science*, **220**, pp. 671-680.

12. Koopmans, T. C. and Beckmann, M. J. (1957), Assignment problems and the location of economic activities, *Econometrica*, **25**, pp. 53-76.

13. Lawler, E. L. (1963), The quadratic assignment problem, *Management Science*, **9**, pp. 586-599.

14. Mirchandani, P. B. and Obata, T. (1979), *Locational Decisions with Interactions Between Facilities: the Quadratic Assignment Problem a Review*, Working Paper Ps-79-1, Rensselaer Polytechnic Institute, Troy, New York.

15. Nugent, C. E., Vollmann, T. E. and Ruml, J. (1969), An Experimental Comparison of Techniques for the Assignment of Facilities to Location, *Journal of Operations Research*, **16**, pp. 150-173.

16. Pardalos, P. M., Li, Y. and Murthy, K. A. (1992), *Computational Experience with Parallel Algorithms for Solving the Quadratic Assignment Problem*, in O. Balci, R. Sharda, S.A. Zenios (eds.), *Computer Science and Operations Research: New Developments in their Interface*, Pergamon Press, pp. 267-278.

17. Pardalos, P. M. and Guisewite, G. (1993), Parallel Computing in Nonconvex Programming, *Annals of Operations Research* **43**, pp. 87-107.

18. Pardalos, P. M., Murthy, K. A. and Harrison, T. P. (1993), A Computational Comparison of Local Search Heuristics for Solving Quadratic Assignment Problems, *Informatica*, **4** pp. 172-187.

19. Pardalos, P. M., Phillips, A. T. and Rosen, J. B. (1993), *Topics in Parallel Computing in Mathematical Programming*, Science Press, 1993.

20. Pardalos, P. M., Pitsoulis, L. S. and Resende, M. G. C. (1995), A Parallel GRASP Implementation for the Quadratic Assignment Problem, in A. Ferreira and J. Rolim (eds.), *Solving Irregular Problems in Parallel: State of the Art*, Kluwer Academic Publishers, Dordrecht-Boston-London, pp. 111-128.

21. Pardalos, P. M., Pitsoulis, L. S., Resende, M. G. C. and Mavridou, T. D. (1995), Parallel Search for Combinatorial Optimization: Genetic Algorithms, Simulated Annealing, Tabu Search and GRASP, in *Lecture Notes in Computer Science*, Vol(980), Springer, Berlin, pp. 317-331.

22. Pardalos, P. M., Resende, M. G. C. and Ramakrishnan, K. G. (1995), *Parallel Processing of Discrete Optimization Problems*, DIMACS Series, Vol. 20, American Mathematical Society.

23. Pardalos, P. M., Rendl, F. and Wolkowicz, H. (1994), The quadratic assignment problem: A survey and recent developments, in P.M. Pardalos and H. Wolkowicz, (eds.), *Quadratic assignment and related problems*, Vol. 16 of *DIMACS Series on Discrete Mathematics and Theoretical Computer Science*. American Mathematical Society.

SOME EXPERIMENTS IN CONNECTION WITH NEURAL AND FUZZY MODELLING FOR AIR POLLUTION PROBLEMS

G. M. SANDULESCU AND M. BISTRAN

Artificial Intelligence Laboratory - SIA at IPA SA ,
18 Mircea Eliade, Bucharest, Romania,
Phone: +401 23005 91; Fax: +401 230 70 63; e-mail: sand@automation.ipa.ro

Introduction

The possibilities of using procedures based Neural and Fuzzy networks in large scale air pollution models is discussed in this note. Experiments with many such procedures, performed by the authors, indicate that these can be used at least in some parts of existing air pollution models. models.

The authors have tested over 200 different models to see if there are possibilities of applying Neural Modelling in very complicated and time-consuming parts of big analytical models. The conclusion is that Neural Modelling can successively be applied and this is a good alternative way of solving the problems. The success is due to the abilities of these procedures and systems to extract important deterministic patterns and relations, hidden in the air pollution evolution, and then to apply the neural intrinsic interpolation and generalisation features.

Based on the learned similarities, it becomes possible to extract the hidden deterministic relations and rules (as patterns), which describe the future, possible evolution (for forecasting). After that the application of neural multivariate procedures may, potentially, improve the prediction quality and, thus, the use of these methods in some parts of the very complex models or as supplementary sources of comparing prediction operations performed by different algorithms.

Z. Zlatev et al. (eds.), Large-Scale Computations in Air Pollution Modelling, 275–280.
© 1999 Kluwer Academic Publishers. Printed in the Netherlands.

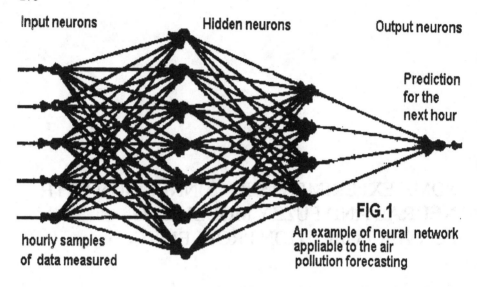

Input neurons   Hidden neurons   Output neurons

Prediction
for the
next hour

FIG.1

hourly samples
of data measured

An example of neural network
appliable to the air
pollution forecasting

Figure 1. An example of Neural Network of the type "feed-forward with BKP training", achieved from 4 levels including 2 hidden levels , which may potentially be used in forecasting of complex evolution or as a part of the very complex systems .

The information related to the specific model is captured and stored, in the course of the training process of the Neural Networks.

Practically, Neural, Fuzzy and Neuro-Fuzzy models are **"games" of the weights**, which are combined, by following complex relations, inside the specific architecture. The wieghts are combined by using the non-linear functions in each neurone.

The output signal from a Neural model is obtained by investigating many combinations of (i) new input signals, (ii) the weights between the nodes and (iii) the non-linear, for instance sigmoidal characteristic, of the neural nodes.

The application of the Neural Nets for modelling , with the view of forecasting of the air pollution evolution is a typical example where success can be expected. This is discussed in [1] and [2]. Another application is given in [3], where the application of the Neural Nets in the evaluation "of the relative contribution of various pollution sources to the pollution of certain ecosystem" is discussed in detail..

Experimental Results

Over 200 neural models are developed and experimented. With the aim is to find suitable modules that can be applied in pollution modelling (and, especially, in Air Pollution Modelling). Both different types of Neural Nets and different types of learning have been tried. The result was a new complex system called **"The Time Machine",** that is an innovative scientific and applicative flexible modelling configuration , based on Artificial Neural and Fuzzy principles (also with the possibilities for both Neuro-Fuzzy and Genetical Algorithm applications). The forecast values, offered by the models in this system dependent on the training conditions. The forecaster values represent future values for the pollution levels for the studied pollutant. The values are either for a given geographical point or interpolated for other geographical points. As a consequence of these good forecasting/interpolation procedures, the results obtained by using the Neural Nets procedures are reliable. This may be used to achieve both simplification of the modelling activities and also reduction of the number of the measurements stations.

The following figures illustrates the modelling possibilities based on the use of the deducted and normalised primary data.

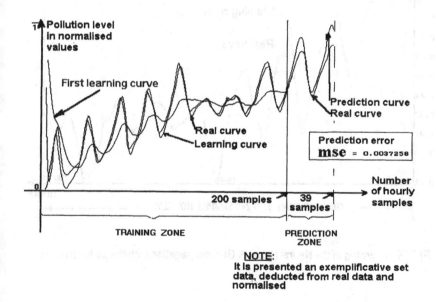

FIG.2 - Forecasting of the pollution level with Self-Learning Neural Network

278

Some other results are given in figures 2-4. These results demonstrate the abilities for modelling and forecasting with Neural Networks from the **"Time Machine"** system. The particular problem studied is the evolution of the NMHC in urban area (based on deducted/exemplificative past evolution). It is also possible to forecast the evolution of SO2 , NO2 , CO etc. From figures 2 and 3 it is seen that the forecaster values obtained by our system are comparable with those given in [1].

3. Conclusions

The Neural Nets are a very interesting alternative of very complex modelling systems. This is especially true for studies at a given geographical point (a given site), but some other applications are also possible.

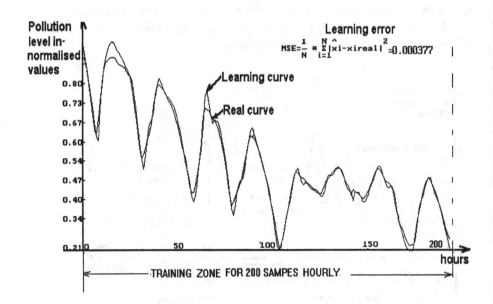

FIG.3- Learning of the Neural Network Backpropagation with the pollution level

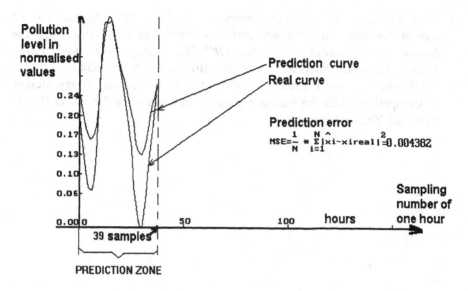

FIG.4 -Prediction of the pollution level with the Neural Network Backpropagation

Today, the application of neural and/or fuzzy modelling procedures is focused, in air pollution, to applications, in the zones of big cities or in some industrial zones. However, it is, in principle, possible to use this technique also in solution of other complex tasks or in some parts of very complex models.

The experiences with the **"The Time Machine"** demonstrate the ability of this system in the field of air pollution modelling. The authors will be happy to collaborate with air pollution specialist in attempts to apply this tool in the solution of different tasks. the pregnant deterministic parts of the big and complex air pollution models .

References

1. Ando, G., Cammarata, A., Fichera, S., Graziani, G. and Pitrone, N. (1995), Neural Networks for the Analysis of the Air Pollution in Urban Areas, EUFIT'95, Aachen .
2. Schuurman, G. and Muller, E. (1994), Back propagation neural networks - recognition vs. prediction capability, *Environs. Toxicol. Chem.*, **13**, pp. 2061-2077

3. Karayiannis, N. B. and Venetsanopulos, A. N. (1990), Applications of Neural Networks to Environmental Protection, in *Proceedings of the International Network Conference (INNC90), July 9-13 1990, Paris*, Kluwer Academic Publishers, Dordrecht-Boston-London, p. 334.
4. Sandulescu, G. and Bistran, M. (1997), The Self - Learning Fuzzy Models in Competition with the Neural Networks, in Proceeding 5th Zittau Fuzzy-Colloqui, Zittau , Germany.

ADVANCED OPERATIONAL AIR QUALITY FORECASTING MODELS FOR URBAN AND REGIONAL ENVIRONMENTS IN EUROPE: MADRID APPLICATION

R. SAN JOSÉ, M.A. RODRIGUEZ, M.A. ARRANZ, I. MORENO
AND R.M. GONZÁLEZ[1].
Environmental Software and Modelling Group, Technical University of Madrid,
Campus de Montegancedo, Boadilla del Monte 28660 (Madrid),
http://artico.lma.fi.upm.es.
(1) Department of Meteorology and Geophysics,
Complutense University of Madrid, Ciudad Universitaria, Madrid (Spain)

1. Introduction

During the last decades an increased interest on Air Quality Models has been found. Historically, air quality models have been developed in the 60' and 70' based on a very simplified solution of the Eulerian transport equation. The simplified differential transport equation had an analytical solution and therefore the results were obtained rapidly. Nowadays, these type of models (Gaussian) are applied with computer times very much reduced in comparison to the 60' and 70' operational models. These models have become quite complex by parameterizing new and more sophisticated problems such as buildings, very complex terrain, etc. However, because of the extraordinary advance of the computer power, during the 80' a new generation of three-dimensional models started to be developed. During the 70' the mathematical basis for solving the Navier-Stokes equation on the air fluid were established by developing sophisticated numerical techniques which conducted to have the first possible approaches for having a full three dimensional solution of the air dynamics. At the end of the 80' decade the first models for mesoscale areas started to appear (MM5 , MEMO, RAMS, etc.). (Pielke [5]; Flassak and Moussiopoulos [1]; Grell et al. [3]) These models offer a detailed diagnostic and prognostic patter of the wind, temperature and humidity -in addition of the pressure, vorticity, turbulent fluxes, etc.- in a three-dimensional mesoscale

Z. Zlatev et al. (eds.), Large-Scale Computations in Air Pollution Modelling, 281–289.
© *1999 Kluwer Academic Publishers. Printed in the Netherlands.*

domain. The Gaussian approach was appropriate for point or lineal sources in a very local domain (5-10 km) but the rigid meteorological conditions and the difficulties of having on-line information to be incorporated to the model made almost impossible to apply these models for regional domains (20-500 km).

These three dimensional mesoscale meteorological models required important computer resources for running the simulations since the fundamental basis continued to be the numerical analysis so that the grid approach required substantial memory demands. First generation of these models were based on the hydrostatic approach but nowadays these models are run under the non-hydrostatic mode which takes into account the full effects of the topography and land use types. The specific humidity is considered as an scalar quantity so that the different gases which are present in the atmosphere are transported following the physical and dynamical laws which are incorporated in the fundamental Navier Stokes partial differential equation system. So that, the atmospheric pollutants are transported in a similar way than the specific humidity.

Parallel to the meteorological development of the mesoscale models, atmospheric chemical models were also developed to take into account the chemical reactions for those pollutants, which were transported by using the advanced three-dimensional models. The chemical reaction schemes were developed for atmospheric gases as a first stage. The CBM-IV is published by Gery et al [2]. This mechanism incorporates more than 200 chemical reaction and more than 80 species. A simplified version is also published with 78 chemical reactions and 38 chemical species. Secondary pollutants started to appear as quite important from the environmental stand point of view such as Ozone, PAN, etc. The strong non-linear relation between ozone concentrations and primary pollutant emission rates started to appear as an important result of applying such a complex modelling systems. The simulations were performed in off-line mode, which means that the chemical models read the meteorological outputs from the non-hydrostatic mesoscale models for a specific episode. The chemical models used to incorporated their own advection and diffusion parts so that a full Eulerian transport model was included in the so-called chemistry model.

A very important module of these diagnostic and prognostic environmental models is the emission model. Emission inventories started to be compiled at the end of 70' in the United States and in the middle of 80' decade European Union lunches structured programs to compile this information by developing European emission factors which appeared to be in some cases quite different from the USA emission factors because of the different composition of fuels, building materials, etc. Also, at the beginning of the 80' decade the first biogenic emissions are carried out for California and by using a quite coarse grid for all the United States. Europe research programs focus on the biogenic emission inventories much later (beginning of 90' decade). Again differences on forest composition, species and land use planning strategies between USA and Europe make difficult to generalise the results from a few different field experiments. In any case, emission inventories are becoming a key issue on environmental models since the uncertainty seems to be quite important when comparing with the meteorological-chemical models.

At the beginning of 90' decade these modelling systems are extended and different environmental laboratories in the world start to perform simulations for different episodes on their regional areas, however the environmental modelling systems run on powerful workstations or supercomputers and the maximum interest is having an accurate prognostic of air concentrations of different pollutants when comparing with monitoring data. These modelling systems were running in research laboratories or Universities. During the last 2-3 years a growing interest from the environmental authorities on using these tools for having a better knowledge of the air concentrations in their regions has been observed. The interest of these users is found to be initially on the forecasting capabilities of such a tool. However, the complexity of these modelling systems made very difficult to run the modelling systems quick enough to take advantage of these forecasting capabilities unless a specific design is made.

This contribution deals on the EMMA (San José et al. [8]) tool, which is an operational version of the research model ANA (San José et al. [6]; San José et al. [7]) which was developed previously. EMMA is a European Union DGXIII project is dealing with the development, test and validation of innovative telematics system for monitoring and forecasting of air pollution in urban areas.

2. The operational modelling system EMMA

The ANA (Atmospheric mesoscale Numerical pollution model for regional and urban Areas) model is composed on several different modules. See Figure 1. The chemical module (CHEMA) is a numerical solver for atmospheric chemical reactions based on the SMVGEAR method (Jacobson and Turco [4]). The deposition model DEPO (San José et al. [7]) which incorporates the Wesely [10] dry deposition parameterization, the REMO model which generates off-line the land use types for ANA from the Landsat-V satellite. Chemical module and meteorological modules are connected on-line. Finally the EMIMA model generates the emission data off-line for the ANA simulation. EMIMA is designed for Madrid area although an EMIMA version to be applied over different domains than Madrid is under development.

EMMA is developed under the telematics programme of the European Union DGXIII environmental programme. Telematics components including multimedia products and services are extensively used in the project. The Madrid EMMA application and particularly the EMMA software which has been built to communicate the user and the mathematical models included in this application was built by using Tcl/Tk language which is interpreter and toolkit under the UNIX environment. This application is a friendly-user interface to allow operators in the authorities environmental offices to manage the application.

EMMA was designed to allow forecasting of 24-48 hours by using the initial meteorological and air pollution network information, which is available in the specific site format. The simulation is designed to run during five days -when non-weekend days are involved- and seven days when weekend days are included. The first 24 hours the simulation makes use of meteorological and air pollution network information of the specific site. In case of Madrid, a meteorological vertical sounding every 12 hours a day and surface air concentration from the Madrid air pollution network is available. After the first 24 hours the model continues to run up to 120 hours on pure prognostic mode. See figure 2.

The Tcl/Tk software packages are used to build the interface. Tcl is a simple scripting language for controlling and extending applications; its name stands for "tool command language". Tcl provides generic programming facilities, such as variables and loops and procedures that are used for a variety of applications. Its interpreter is a library of C procedures that can easily incorporated into applications and each application can extend the core Tcl features with additional commands for that application. One of the most useful extensions for Tcl is Tk, which is a toolkit for the X Window System. The combination of Tcl and Tk provides a rapid development, makes it easily for applications to have powerful scripting languages, makes an excellent "glue language" and finally is user convenience. All these characteristics made Tcl/Tk as the selected software package for building the user interface with the mathematical modelling system. Figure 3 shows how the different menus are presented to the operator.

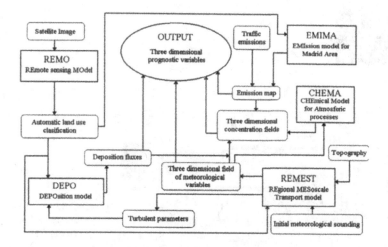

Figure 1. ANA modelling system.

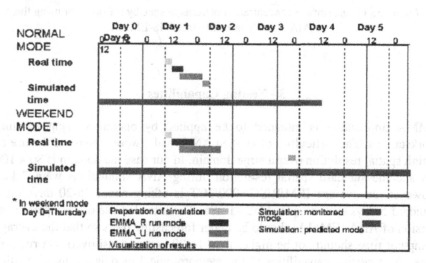

Figure 2. EMMA Operational mode for labour and weekend days.

286

EMMA Regional Domain

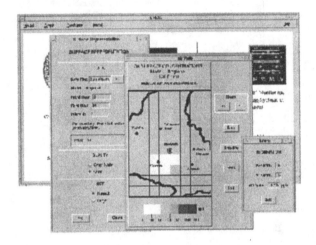

•Regional monitoring
network.

•Detailed 10kmx10km
forecasted 3D
concentrations.

Figure 3. Ozone surface concentration patterns as seen by the operator using the
EMMA friendly user interface VIS-EMMA.

3.- Nesting Capabilities

EMMA application is intended to be applied by operators using medium
workstations. The difficulties of applied ANA model were coming of the use of
a high spatial resolution over a large domain. In our case the domain is 80 x 100
km and the standard spatial resolution during ANA simulations were 2 km.
However, an average IBM/RISC/6000/3CT is taking about 16-20 days on a
standard 5 days episode over Madrid area. The intended use of the operational
version of ANA (EMMA) should be under forecasting basis so that the average
computer time should not be higher than 16-20 hours (quasi-overnight running
time). The nesting capabilities of the meteorological models are used to fulfil
the city domain requirements. Figure 4 shows the Madrid EMMA domain with
the nested domain which includes the Madrid metropolitan area. The mother
domain is having 10-km grid resolution and the city domain is having 1-km grid
spatial resolution. The classical relation of 1:3 between nested and mother

domain is violated but results show that model performs properly although the city version is not yet implemented under operative mode. Figure 5 shows how the VIS-EMMA interface is able to show detailed 1-km surface air concentrations for different times of the simulation.

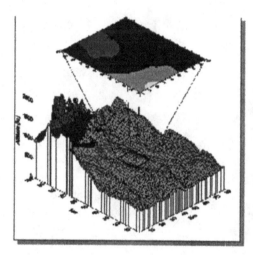

URBAN MODEL (EMMA_U)

Area = 10x12 km

Grid = 10x12 cells

Resolution = 1 km

REGIONAL MODEL (EMMA_R)

Area = 80x100 km

Grid = 8x10 cells

Resolution = 10 km

Figure 4. EMMA nesting solution for operational application over cities with high spatial resolution.

4. Conclusions

A powerful and accurate prognostic urban and regional air concentrations application (EMMA) has been developed based on the ANA research code. EMMA shows that it is possible to develop friendly user interfaces under medium computer platforms to be used by environmental authorities, local environmental administrations and policy makers. The feature has been obtained by applying a nesting submodel hierarchy (EMMAR and EMMAU) which allows a high spatial resolution over the city domain. The EMMA model has been applied successfully over the Madrid Community and it is running in a

288

pre-operational mode during July-November, 1998. A version of EMMA is already installed in the Principado de Asturias Spanish Community (located in the north of Spain) with a nested capability focusing in a smaller Asturias area. Future improvements should be made in order to improve the quality of the predictions by using boundary or vertical meteorological information provided by continental or global meteorological and air chemistry models. A further version of EMMA based on the WorldWide Web possibilities by using JAVA languages is currently under development.

EMMA Urban Domain

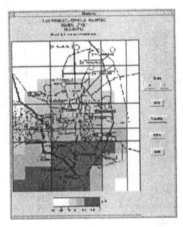

•Urban monitoring network (24 stations)

•Detailed 1km x 1km forecasted 3D concentrations

Figure 5. NO2 surface concentration patterns over Madrid City (nested domain in EMMA application)

References

1. Flassak, T. and Moussiopoulos, N. (1987), An application of an efficient non-hydrostatic mesoscale model, Boundary Layer Meteorology, **41**, pp. 135-147.

2. Gery, M. W., Whitten, G. .Z., Killus, J. P. and Dodge, M. C. (1989), A photochemical kinetics mechanism for urban and regional scale computer modelling, *Journal of Geophysical Research*, **94**, pp. 12925-12956.

3. Grell, G. A., Dudhia, J. and Stauffer, D. R. (1994), A description of the Fifth-Generation Penn State/NCAR Mesoscale Model (MM5). NCAR/TN-398+STR. NCAR Technical Note.

4. Jacobson, M. Z. and Turco, R. P. (1994), SMVGEAR: A sparse-matrix, vectorized gear code for atmospheric models", *Atmospheric Environment*, **28**, pp.273-284.

5. Pielke, R. A. (1984), Mesoscale Meteorological Modelling. Academic Press.

6. San José, R., Rodriguez, L., Moreno, J., Palacios, M., Sanz, M. A. and Delgado, M. (1994), Eulerian and photochemical modelling over Madrid area in a mesoscale context, *Air Pollution II, Vol. 1, Computer Simulation, Computational Mechanics Publications, Ed. Baldasano, Brebbia, Power and Zannetti.*, pp. 209-217.

7. San José, R., Cortés, J., Moreno, J., Prieto, J. F. and González, R. M. (1996), Ozone modelling over a large city by using a mesoscale Eulerian model: Madrid case study, *Development and Application of Computer Techniques to Environmental Studies, Computational Mechanics Publications, Ed. Zannetti and Brebbia*, pp. 309-319.

8. San José, R., Prieto, J. F., Martín, J., Delgado, L., Jimenez, E. and González, R. M. (1997), A Integrated Environmental Monitoring, Forecasting and Warning Systems in Metropolitan Areas (EMMA): Madrid application, *Measurements and Modelling in Environmental Pollution, Ed. San José and Brebbia*, pp. 313-323.

9. San José, R., Prieto, J. F., Castellanos, N. and Arranz, J. M. (1997), Sensititivity study of dry deposition fluxes in ANA air quality model over Madrid mesoscale area, *Measurements and Modelling in Environmental Pollution, Ed. San José and Brebbia*, pp. 119-130.

10. Wesely, M. L. (1989) Parameterization of surface resistances to gaseous dry deposition in regional-scale numerical models. *Atmospheric Environment*, **23**, pp. 1293-1304.

ADAPTIVE ORTHONORMAL SYSTEMS

BL. SENDOV
Bulgarian Academy of Science,
Acad. G. Bonchev str., Bl.25A, 1113 Sofia, Bulgaria
E-mail: sendov@amigo.acad.bg
Fax: +359 707 273

Abstract

A classical linear method for approximation of a function f from a Hilbert space H is the following. Take an orthonormal basis $\{\phi_i\}_{i=0}^{\infty}$ in H. Calculate the Fourier coefficients $c_i(f) = \langle f, \phi_i \rangle;\ i = 0, 1, 2, \ldots, n$. Take for an approximation of f the generalized polynomial

$$P_n(f; x) = \sum_{i=0}^{n} c_i(f)\phi_i(x).$$

The quality of this approximation depend on the function f, on n and on the choice of the orthonormal system $\{\phi_i\}_{i=0}^{\infty}$.

The aim of this paper is to present a method for construction of a orthonormal basis adapted to a given function f, based on a generalization of the Walsh functions, called *Walsh-similar functions*. The advantage of this adapted orthonormal basis is the improvement of the approximation for discontinuous functions and fractal functions. The motivation for this study is the application in the signal and image compression.

1. Introduction

It is natural to choose the instrument for approximation according to the properties of the object to be approximated. Usually, the instrument for approximation is chosen for a given class of functions. The most popular instruments for approximation are the algebraic and trigonometric polynomials, rational functions, splines and wavelets. In this paper we consider an instrument for approximation, which may be adapted to a particular function.

Is it reasonable to choose a particular instrument for approximation of a particular function ? In fact, the methods in the fractal approximation theory [1] are based on the so called *Iterated Function Systems*

Z. Zlatev et al. (eds.), Large-Scale Computations in Air Pollution Modelling, 291–301.
© 1999 *Kluwer Academic Publishers. Printed in the Netherlands.*

(IFS), which are chosen differently for different target functions for approximation. Also, some methods for signal and image compression, based on the so called *wavelet libraries* and *wavelet packets* [9], use in fact instruments for approximation chosen for a particular function.

To construct a particular approximation instrument for any particular function from a functional space, a class of functions, similar to the Walsh functions [12, 4], is used. The scheme is as follows. Consider, for example, the functional space $L^2[0, 1)$ of square integrable real functions defined on $[0, 1)$. We define a subset of functions $S = \bigcup_{n=1}^{\infty} S_n \subset L^2[0, 1)$, $S_n \subset S_{n+1}$, where S_n depend on n parameters. (The constant 1 belongs to S, S is not a linear set, and S is not dense in $L^2[0, 1)$).

Every function $\varphi \in S$ has the property that *starting* with φ one may construct, "fast and easy" as the construction of the Walsh functions [12], an orthonormal basis $\varphi_0(x) = \varphi(x), \varphi_1(x), \varphi_2(x), \varphi_3(x), \ldots$ in $L^2[0, 1)$.

To approximate a particular function $f \in L^2[0, 1)$ two steps are to be made:

1) Solve a "small" non linear approximation problem. Find the function $\varphi \in S_n$, which is the closest to f, in a given sense, i. e., to calculate n parameters defining φ.

2) Solve a "large" linear approximation problem. Find the best L^2 approximation of f in the N-dimensional linear subspace $A_N \subset L^2[0, 1)$, spanned over the orthonormal system $\varphi_0, \varphi_1, \varphi_2, \ldots, \varphi_{N-1}$. That is to calculate the Fourier coefficients $c_i(f) = \int_0^1 f(t)\varphi_i(t)\, dt$; $i = 0, 1, 2, \ldots, N-1$.

We say "small" and "large", because it is natural to have $N = 2^n$.

The first step is the choice of the particular approximation instrument, adapted to a particular function to be approximated. The aim of this step is to find an orthonormal system in respect to which the entropy

$$\mathcal{E}(\varphi; f)_N = -\sum_{i=1}^{N-1} c_i(f)^2 \log_2 c_i(f)^2$$

of the Fourier coefficients of f ($\|f\|_2 = 1$)is smaller than the entropy in respect to a standard, non adapted, orthonormal system. This is important for the compression of the data.

Using this scheme to represent the approximation, one needs $N + n$ numbers, instead of N. But if a smaller entropy of the Fourier coefficients is got, this may pay off. As a matter of fact, we shall show that this increase from N to $N + n$ may be avoided.

To simplify the presentation we limit ourselves to the spaces $L^2[0, 1)$. In general cases, the results are true in $L^p[0, 1)$, $L^p[0, \infty)$; $1 \leq p \leq \infty$ respectively and for functions of 2 and 3 variables.

2. Basic notations

Let $y = y_{-k}2^k + y_{-k+1}2^{k-1} + \cdots + y_0 + y_1 2^{-1} + y_2 2^{-2} + \cdots$ be the dyadic representation of the non negative real number y, where $y_i = 0, 1$ are the dyadic digits of y. If y is dyadic rational, take the finite dyadic representation of y (with finite number of ones). By definition $y_i = 0$ for $i > \log_2(1 + y)$. Pay attention that the binary digits are defined in the same way for real numbers and for natural numbers. The sum of the power of 2 and the corresponding index is always zero.

For the natural numbers, the following unique representation is also used:
$$m = 2^{m(0)} + 2^{m(1)} + 2^{m(2)} + \cdots + 2^{m(p)},$$
where $0 \le m(0) < m(1) < m(2) < \cdots < m(p)$.

The addition $\dotplus$ (addition modulo 2) of two non negative real numbers x, y is defined as follows

$$z = x \dotplus y = z_{-l}2^l + z_{-l+1}2^{l-1} + \cdots + z_0 + z_1 2^{-1} + z_2 2^{-2} + \cdots, \quad (2.1)$$

where $l \le \max\{\log_2(1 + x), \log_2(1 + y)\}$ and

$$z_i = x_i \dotplus y_i, \quad \text{for} \quad i = -l, 1 - l, 2 - l, \ldots \ .$$

If in the representation (2.1) for z, we have $z_m = 0$ and $z_i = 1$ for $i = m + 1, m + 2, m + 3, \ldots$, then we correct the representation (2.1) by the change $z_m = 1$ and $z_i = 0$ for $i = m + 1, m + 2, m + 3, \ldots$.

We shall use the operation $x \dotplus y$ together with the operation $x + y$. The first operation is defined only in $R^+ = [0, \infty)$. Both operations are commutative and associative. Also $x \dotplus x = 0$, or $(x \dotplus h) \dotplus h = x$. As usual, set $y_i \dotplus 1 = \overline{y_i}$.

Define the operation $x \circ y$ for $x, y \ge 0$ as follows

$$x \circ y = \sum_{i=-\infty}^{\infty} x_i y_{1-i}. \quad (2.2)$$

Obviously the sum (2.2) is in fact final and $x \circ y$ is a non negative integer and $x \circ y = y \circ x$. If $[x]$ is the integer part of x, then

$$x \circ y = [x] \circ y + x \circ [y].$$

If m is a non negative integer $m = m_0 2^0 + m_{-1} 2^1 + m_{-2} 2^2 + \cdots + m_{-p} 2^p$, then

$$\check{m} = m_0 2^{-1} + m_{-1} 2^{-2} + m_{-2} 2^{-3} + \cdots + m_{-p} 2^{-p-1} \in [0, 1) \quad (2.3)$$

is a dyadic rational.

From (2.3) it follows that $\check{m}_i = m_{1-i}; \quad i = 1, 2, 3, \dots$.

Let, as usual, $L^2[0, 1)$ and $L^2[0, \infty)$ be the sets of square integrable functions defined on $[0, 1)$ and on $[0, \infty)$ respectively, and $\langle f, g \rangle = \int_0^1 f(x)g(x)\, dx$, $\|f\|_2 = \sqrt{\langle f, g \rangle}$ be the scalar product and the L^2 norm.

N. Fine [4] proved that for every integrable f and every $h \in [0, 1)$, the equality

$$\int_0^1 f(x \dot{+} h)\, dx = \int_0^1 f(x)\, dx. \tag{2.4}$$

holds.

Every function $f(x) \in L_2[0, 1)$ may be considered as the function of the binary digits of $x = x_1 2^{-1} + x_2 2^{-2} + x_3 2^{-3} + \cdots$, or $f(x) = f(x_1, x_2, x_3, \dots)$. Call a function $\varphi \in L^2[0, 1)$ **dyadic factorable** if it have the form

$$\varphi(x) = c \prod_{i=1}^{\infty} \lambda_i(x_i),$$

where c is a real constant and $\lambda_i(\cdot)$; $i = 1, 2, 3, \dots$ are binary functions normalized by the conditions

$$\lambda_i(1)^2 \le 2, \quad \lambda_i(0) = \sqrt{2 - \lambda_i(1)^2}, \quad i = 1, 2, 3, \dots \ . \tag{2.5}$$

A dyadic factorable function (2.5) is called **starting function** if

$$c = 1 \quad \text{and} \quad \lim_{s \to \infty} 2^s \int_0^{2^{-s}} \varphi^{s+}(x)\, dx = 1. \tag{2.6}$$

The set of all starting functions is denoted by $\mathcal{S}$ and the following notations are used:

$$\varphi(x) = \prod_{i=1}^{\infty} \lambda_i(x_i), \quad \varphi(x)^{s-} = \prod_{i=1}^{s} \lambda_i(x_i), \quad \varphi(x)^{s+} = \prod_{i=s+1}^{\infty} \lambda_i(x_i),$$

$$\varphi(x)^{0+} = \varphi(x), \quad \varphi(x) = \varphi(x)^{s-}\varphi(x)^{s+}; \quad i = 1, 2, 3, \dots \ .$$

Thus a starting function $\varphi(\Lambda; x) = \varphi(x) = \prod_{i=1}^{\infty} \lambda_i(x_i)$ is defined by a sequence $\Lambda = \{\lambda_i\}_{i=1}^{\infty}$, where

$$\lambda_i(0) = \sqrt{2 - \lambda_i^2} = \sqrt{2} \sin \alpha_i, \quad \lambda_i(1) = \lambda_i = \sqrt{2} \cos \alpha_i; \quad 0 \le \alpha_i < \pi. \tag{2.7}$$

The most simplest dyadic factorable functions are the Rademacher functions

$$r_0(x) = 1, \quad r_k(x) = (-1)^{x_k}; \quad k = 1, 2, 3, \dots, \ .$$

The Walsh functions are products of Rademacher functions. The Paley ordering [10] of the Walsh functions is

$$w_m(x) = \prod_{i=1}^{p} r_{m(i)}(x) = \prod_{i=1}^{p} (-1)^{x_{m(i)}}, \quad m = 0, 1, 2, \dots .$$

The set S is invariant to the dyadic shifts. For every $h \in [0, 1)$ and for every $\varphi \in S$, the function $\psi(x) = \varphi(x\dot{+}h) = \prod_{i=1}^{\infty} \lambda_i(x_i\dot{+}h_i)$ belongs to S.

Define the **Rademacher** and **Walsh transformations** in $L^2[0, 1)$ as follows

$$(r_s f)(x) = (-1)^{x_s} f(x\dot{+}2^{-s}); \quad s = 0, 1, 2, \dots ,$$

$$(w_m f)(x) = (-1)^{x \circ m} f(x\dot{+}\check{m}); \quad m = 0, 1, 2, \dots .$$

It is obvious that the transformations r_s and w_m produce the Walsh functions from the starting function $w_0(x) = 1$.

From (2.4) there follows that the scalar product is invariant under the transformations of Rademacher and Walsh:

$$\langle r_s f, r_s g \rangle = \langle f, g \rangle, \quad \langle \varphi_m f, \varphi_m g \rangle = \langle f, g \rangle.$$

C. Watary [13] introduced the $L^p; p \in [1, \infty]$ **dyadic module of continuity**

$$\hat{\omega}(f; \delta)_p = \sup_{h \le \delta} \left\{ \int_0^1 (f(x\dot{+}h) - f(x))^p \right\}^{1/p}; \quad 1 \le p \le \infty,$$

for a function $f \in L^p[0, 1)$

Define also the so called **relative** module of continuity.

DEFINITION 2.1. *Let f, g be functions from $L^2[0, 1)$. The L^2 **relative** dyadic module of continuity are defined as follows:*

$$\hat{\omega}_g(f; \delta)_2 = \hat{\omega}_f(g; \delta)_2 = \sup_{h \le \delta} \left\{ \int_0^1 (g(x) f(x\dot{+}h) - g(x\dot{+}h) = f(x))^2 \right\}^{1/2}.$$
$$(2.8)$$

Obviously, for $g(x) = 1$ we have $\hat{\omega}_g(f; \delta)_2 = \hat{\omega}(f; \delta)_2$ and $\hat{\omega}_f(f; \delta)_2 = 0$.

As it will be further shown, the relative module of continuity is useful for measuring the approximation when the approximation instrument is chosen for a particular function.

3. Walsh-similar (WS) functions

Starting with the function $w_0(x) = 1$ and using the Walsh transformations we produce the Walsh orthonormal system. A natural question is: Are there functions, which are not equal to 1 or -1 a. e., and such that the Walsh transformations produced from them orthonormal systems ? The answer is positive. In this section we

show that such functions are the starting functions. These functions play the role of the constant 1 in defining the so called Walsh-similar functions, which depend on a sequence of parameters and have properties similar to these of the Walsh functions.

The L^2 norm of every $\varphi \in S$, according to (2.6) is equal to 1,

$$\|\varphi\|_2^2 = \langle \varphi, \varphi \rangle = 2 \int_0^{1/2} \varphi^{1+}(x) \ dx = \cdots = 2^s \int_0^{2^{-s}} \varphi^{s+}(x) \ dx = 1.$$

A starting function φ may have an arbitrary fractal dimension between 1 and 2 [11], depending on the corresponding sequence Λ.

The following theorem is proved in [11].

THEOREM 3.1. *Let* $\varphi \in S$ *and*

$$\varphi_m(x) = (w_m \varphi)(x) = (-1)^{x \circ m} \varphi(x \dot{+} \breve{m}).$$

Then, the sequence

$$\varphi_0(x) = \varphi(x), \varphi_1(x), \varphi_2(x), \varphi_3(x), \ldots \tag{3.9}$$

is an orthonormal basis in $L^2[0, 1)$.

We call the functions φ_m **Walsh-similar (WS) functions**. Other appropriate names as *Walsh-like functions* and *generalized Walsh functions* are already used [3], [5], [6], [7], [8] for other purposes. For the Kernel

$$K_{2^s}(\Lambda; x, t) = \sum_{m=0}^{2^s-1} \varphi_m(x) \varphi_m(t)$$

we have [11]

$$K_{2^s}(\Lambda; x, t) = \Omega_s(x, t) \varphi^{s+}(x) \varphi^{s+}(t), \tag{3.10}$$

where

$$\Omega_s(x, t) = \begin{cases} 2^s & \text{for } x \dot{+} t \le 2^{-s}, \\ 0 & \text{for } x \dot{+} t > 2^{-s}, \end{cases} \quad \text{and} \quad \int_0^1 \Omega_s(x, t) \ dt = 1. \tag{3.11}$$

THEOREM 3.2. *Let the starting function* $\varphi(\Lambda; \cdot) = \varphi$ *be bounded,* $\sup_{x \in [0,1)} |\varphi(x)| = M$, *and*

$$B_{2^s}(f; x) = \sum_{m=0}^{2^s-1} = c_m(\Lambda, f)\varphi_m(x), \quad c_m(\Lambda, f) = \int_0^1 f(t)\varphi_m(t) \, dt$$

be the 2^s-*th partial Fourier sum of a function* $f \in L^2[0,1]$ *for the orthonormal system* (3.9) *generated from* φ. *Then*

$$\|f - B_{2^s}(f)\|_2 \le M\hat{\omega}_{\varphi^s+}(f; 2^{-s})_2.$$

Proof. From (3.10) and (3.11) there follows that

$$B_{2^s}(f; x) = 2^s \varphi^{s+}(x) \int_{\Delta_s(x)} f(t)\varphi^{s+}(t) \, dt, \tag{3.12}$$

where $\Delta_s(x)$ is the dyadic interval of the form $[p2^s, (p+1)2^s)$ containing x. Then from (3.12) we have

$$\|f - B_{2^s}(f)\|_2^2 = \int_0^1 \left(f(x) - 2^s \varphi^{s+}(x) \int_{\Delta_s(x)} f(t)\varphi^{s+}(t) \, dt \right)^2 \, dx =$$

$$\int_0^1 2^s \left[\int_{\Delta_s(x)} \varphi^{s+}(t) \left(f(x)\varphi^{s+}(t) - f(t)\varphi^{s+}(x) \right) \, dt \right]^2 \, dx \le$$

$$M^2 \int_0^1 \sup_{t \in \Delta_s(x)} [f(x)\varphi^{s+}(t) - f(t)\varphi^{s+}(x)]^2 \, dx = (M\hat{\omega}_{\varphi^s+}(f; 2^{-s})_2)^2,$$

which completes the proof.

4. Adaptation through best L^2 approximation

Consider the starting function $\varphi(\Lambda, \cdot) = \varphi$ as a function of $\alpha_1, \alpha_2, \alpha_3, \ldots$ defined by (2.7). Then, according to (2.7)

$$\frac{\partial}{\partial \alpha_s} \varphi(x) = (-1)^{x_s} \varphi(x \dotplus 2^{-s}) = \varphi_{2^s}(x). \tag{4.13}$$

From (4.13) we obtain that the m-th Walsh-similar function with starting function φ may be represented in the form

$$\varphi_m(x) = \frac{\partial}{\partial \alpha_{m(0)}} \frac{\partial}{\partial \alpha_{m(1)}} \cdots \frac{\partial}{\partial \alpha_{m(p)}} \varphi(x),$$

where $m = 2^{m(0)} + 2^{m(1)} + \cdots + 2^{m(p)}$, $m(0) < m(1) < \cdots < m(p)$.

Let $f \in L^2[0,1)$ and $\varphi \in \mathcal{S}$. Set

$$\Phi(\alpha_1, \alpha_2, \alpha_3, \ldots) = \|f - \varphi\|_2^2. \tag{4.14}$$

The necessary conditions for a minimum of Φ, according to (4.13), are

$$\frac{\partial \Phi}{\partial \alpha_k} = -2 \int_0^1 (f(x) - \varphi(x)) \varphi_{2^k}(x) \, dx = -2 \int_0^1 f(x) \varphi_{2^k}(x) \, dx = 0,$$

or, if the parameters $\lambda_i = 1$ are fixed for $i > s$, one may calculate λ_i for $i = 1, 2, 3, \ldots, s$ from the conditions

$$c_{2^{k-1}}(\Lambda, f) = 0 \quad \text{for} \quad k = 1, 2, 3, \ldots.$$

Call the starting function φ_f a **self function** of f if

$$\|f - \varphi_f\|_2 = \inf \{\|f - \varphi\|_2 : \varphi \in \mathcal{S}\}.$$

If the sequence of the first 2^s Fourier coefficients of the function $f \in L^2[0,1)$, in respect to a self function is

$$c_0, c_1, c_2, \ldots c_{2^s - 1},$$

then $c_{2^{p-1}} = 0$ for $p = 1, 2, 3, \ldots, s$. In the places of these coefficients we put the numbers $\epsilon_p = \lambda_p - 1$, where $\{\lambda_p\}_{p=1}^s$ are the parameters defining the self function ($\lambda_p = 1$ for $p > s$). In this way we preserve the number of coefficients needed for the presentation of the linear approximation of f.

5. Some properties of the Walsh-similar functions

The Walsh-similar functions have properties similar to these of the Walsh functions. It is interesting to study the limits of these similarities.

5.1. RELATIVE DYADIC DERIVATIVE

There are various forms of dyadic differentiation, some of which interact with the Walsh functions similarly as the classical derivative does with the exponential functions, see B. L. Butzer and V. Engels [2] and the references there.

It is possible to define a relative dyadic derivative, which interacts with the Walsh-similar functions as the dyadic derivative does with the Walsh functions.

DEFINITION 5.1. *Let $\varphi(x) = \varphi_0(\Lambda; x) \in \mathcal{S}$ and $f \in L^2[0,1)$. Define the pointwise* **relative dyadic derivative** *as follows*

$$f^{[1,\Lambda]}(x) = f^{[1]}(x) = \sum_{i=1}^{\infty} 2^{i-2} \lambda_i(\overline{x_i}) \left[\lambda_i(\overline{x_i}) f(x) - \lambda_i(x_i) f(x \dotplus 2^{-i}) \right],$$

if the sum is convergent.

THEOREM 5.1. *Let $\varphi(x) = \varphi(\Lambda; x) \in \mathcal{S}$ and $\{\varphi_m\}_{m=0}^{\infty}$ are the WS functions produced from φ. Then*

$$\varphi_m^{[1,\Lambda]}(x) = \varphi_m^{[1]}(x) = m\varphi_m(x).$$

5.2. LEBESGUE CONSTANTS

Now we estimate the Lebesgue's constants

$$L_n(\Lambda) = \sup_{x \in [0,1)} \int_0^1 |K_n(\Lambda; x, t)| \, dt,$$

of the orthonormal system generated by the bounded starting function

$$\varphi(x) = \varphi(\Lambda; x) \in \mathcal{S}, \quad \sup_{x \in [0,1)} |\varphi(x)| = M,$$

generalizing the corresponding result in [4].

THEOREM 5.2. *For every natural n, the inequality $L_n(\Lambda) = M^2 \log_2 n$ holds.*

6. Generalized Walsh-similar functions

Set by definition $\lambda_{1-i} = \lambda_i; \quad i = 1, 2, 3, \dots$.

DEFINITION 6.1. *Let $\varphi = \varphi(\Lambda; \cdot)$ be a starting function with sequence Λ, then the function*

$$\phi_y(\Lambda; x) = (-1)^{x \diamond y} \prod_{i=-\infty}^{\infty} \lambda_i(x_i \dotplus y_{1-i})$$

is defined for every $x, y \geq 0$. Call the function $\phi_y(\Lambda; x)$ **Generalized Walsh-similar** *function, or* **GWS** *function with sequence Λ.*

Every GWS function is represented in the form

$$\phi_y(\Lambda; x) = \varphi_{[y]}(\Lambda; x)\varphi_{[x]}(\Lambda; y), \tag{6.15}$$

where

$$\varphi_{[y]}(\Lambda; x) = (-1)^{x_0[y]} \prod_{i=1}^{\infty} \lambda_i(x_i \dotplus y_{1-i}) = \prod_{i=1}^{\infty} (-1)^{x_i y_{1-i}} \lambda_i(x_i \dotplus y_{1-i}).$$

The set $\{\varphi_k(x)\}_{k=0}^{\infty}$ is a closed orthonormal system in $L^2[0,1]$, see [11] .

Every function $\varphi_k(x)$ is periodic with period 1.

6.1. RIEMANN-LEBESGUE THEOREM

The corresponding Riemann-Lebesgue Theorem is proved following [5].

THEOREM 6.1. *If f is absolutely integrable on $[0, \infty)$, then for a fixed normal sequence Λ, $\phi_y(\Lambda; x) = \phi_y(x)$,*

$$\lim_{y \to \infty} \int_0^{\infty} f(x)\phi_y(x) \, dx = 0.$$

Proof. We may choose n so that for all $y \geq 0$,

$$|J_n| \leq \|\phi(\Lambda)\| \int_n^{\infty} |f(x)| \, dx < \epsilon/2.$$

Now, from (6.15),

$$I_n = \sum_{k=1}^{n-1} \int_k^{k+1} f(x)\varphi_{[y]}(x)\varphi_{[x]}(y) \, dx = \sum_{k=1}^{n-1} \varphi_k(y) \int_k^{k+1} f(x)\varphi_{[y]}(x) \, dx,$$

$$|I_n| \leq \|\phi(\Lambda)\| \sum_{k=1}^{n-1} \left| \int_k^{k+1} f(x)\varphi_{[y]}(x) \, dx \right|. \tag{6.16}$$

On the right side of (6.16) we have a sum of a fixed number of Fourier coefficients of order $[y]$. We may choose y so large that this sum is less than $\epsilon/2$, so that $|I_n| + |J_n| < \epsilon$, which completes the proof.

6.2. FOURIER INTEGRAL THEOREM

THEOREM 6.2. *Let f be absolutely integrable on $[0, \infty)$ and contin-uous in every dyadic irrational $x \in (0, \infty)$. Then, for a fixed normal sequence Λ, $\phi_y(\Lambda; x) = \phi_y(x)$, the spectrum function*

$$F(y) = \int_0^{\infty} f(x)\phi_y(x) \, dx$$

exists and for every dyadic irrational $x \in (0, \infty)$, we have the inversion formula

$$f(x) = \int_0^\infty F(y)\phi_y(x) \ dy.$$

Remark. This theory could be developed for functions of two and three variables. The starting functions in these cases have the form

$$\varphi(x, y) = \prod_{i=1}^\infty \lambda_i(x_i, y_i) \quad \text{and} \quad \varphi(x, y, z_i) = \prod_{i=1}^\infty \lambda_i(x_i, y_i, z_i).$$

References

1. Barnsley, M. F. and Hurd, L. P. (1993), *Fractal Image Compression*, AK Peters, Ltd. Wellesley, Massachusetts.
2. Butzer, P. L. and Engels, V. (1989), *Theory and Applications of Gibbs Derivatives*, Matematički Institut, Beograd.
3. Chrestenson, H. E. (1955), A Class of Generalized Walsh Functions, *Pacific J. Math.*, **5**, 17 - 31.
4. Fine, N. J. (1949), On the Walsh Functions, *Trans. Am. Math. Soc.*, **65**, 373 - 414.
5. Fine, N. J. (1950), The Generalized Walsh Functions, *Trans. Am. Math. Soc.*, **69**, 66 - 77.
6. Lévy, P. (1944), Sur une généralisation des fonctions orthogonales de M. Rademacher, *Comm. Math. Helv.*, **16**, 146 - 152.
7. Larsen, R. D. and Madych, W. R. (1976), Walsh-like Expansions and Hadamard Matrices, *IEEE Trans. Acoust. Speech Signal Processing*, **ASSP-24**, 71 - 75.
8. Madych, W. R. (1978), Generalized Walsh-like Expansions, *IEEE Midwest Symp. Circ. Syst.*, **21**, 378 - 382.
9. Meyer, Y. (1993), *Wavelets*, SIAM, Philadelphia.
10. Paley, R. E. (1932), A Remarkable Series of Orthogonal Functions. *Proc. London Math. Soc.*, **34**, 241 - 279.
11. Sendov, Bl. (1997), Multiresolution analysis of functions defined on the dyadic topological group, *East J. on Approx.*, **3**, 225 - 239.
12. Walsh, J. L. (1923), A Closed Set of Normal Orthogonal Functions, *Amer. J. Math.*, **55**, 5 - 24.
13. Watari, C. (1957), On Generalized Walsh Fourier Series, I, *Proc. Japan Acad.*, **33**, 435 - 438.

ON SOME FLUX-TYPE ADVECTION SCHEMES FOR DISPERSION MODELLING APPLICATION

D. SYRAKOV[1] AND M. GALPERIN[2]

[1]National Institute of Meteorology and Hydrology, Sofia 1784, Bulgaria
[2]Department of Biophysics, Radiation Physics and Ecology, MPhEI

Abstract

The advection scheme TRAP (from TRAPezium) was elaborated for the Bulgarian dispersion model EMAP, a PC-oriented Eulerian multi-layer model. The TRAP scheme is explicit, positively definite and conservative with limited numerical dispersion and good transportivity. Displaying the same properties as Bott's scheme [1], the TRAP scheme turns out to be faster. In the Bott scheme the flux area is calculated by integrating the polynomial fit over the neighbouring grid values. In the TRAP scheme, the flux area is supposed trapezoidal. It is determined as a product of the Courant number and a single value of the approximating polynomial referring the middle of the passed distance. In the TRAP scheme, the same 4th order polynomial is used and Bott's normalisation is also applied.

Some new and faster schemes build on the base of the TRAP concepts are presented and tested here. The performance quality is determined exploiting the rotational test: instantaneous point-shaped and cone-shaped sources are rotated in a 101×101 grid-point field. A set of criteria is used reflecting suitable characteristics of the advection algorithm. Additional demonstration tests are made over one of the schemes found out to be the best one.

1. Introduction

One of the key problems in airborne pollution modelling is the accuracy and the speed of the numerical schemes. The description of the advection processes still keep to be a real challenge for tracer dispersion modelers. Even the best chemical scheme or boundary layer parameterization is useless while the advection scheme produces considerable errors. In the same time, a tendency is observed comprehensive models to be used in long-term integration. This urges

Z. Zlatev et al. (eds.), Large-Scale Computations in Air Pollution Modelling, 303–312.
© 1999 Kluwer Academic Publishers. Printed in the Netherlands.

the modelers to search for compromise between the accuracy and the speed of computations. The same are the requirements of the models designed for performing on small computational platforms. A lot of methods and schemes were proposed in the literature ([4],[5],[12]), but only a small number of them are suitable and are being in practical use. So far, none of the existing schemes possesses all properties of the exact solution of the advection equation.

One of the most widely used schemes is the one, elaborated by Bott [1] and further improved by himself ([2],[3]). In the Bott scheme, the advective fluxes are computed utilising the integrated flux concept of Tremback et al. [11]. The fluxes are normalised and then limited by upper and lower values. The produced scheme is conservative and positively definite with small numerical diffusion. These properties make the Bott scheme very attractive for further improvements and optimisations. In this study, comparative tests of the Bott scheme [1] and some new schemes are provided.

2. Description of Advection Schemes

As splitting is applied in the multi-dimensional case, the one-dimensional advection equation in flux form is considered here:

$$\partial C / \partial t + \partial (uC) / \partial x = 0, \tag{1}$$

$C(x,t)$ being the tracer concentration, $u(x,t)$ - the advection velocity, x - the space co-ordinate and t - the time. Let homogeneous grid is introduced in space and time: $x \to x_i = i\Delta x$, $i=1,N_x$; $t \to t_n = n\Delta t$, $n=0,N_t$., Δx and Δt being the space and time steps. The corresponding grid values of the velocity and the concentration are $u(x_i, t_n) = u_i^n$ and $C(x_i, t_n) = C_i^n$. Keeping the mass balance in cell i, the flux form of Eq.(1) can be discretized as

$$C_i^{n+1} = C_i^n - \frac{\Delta t}{\Delta x}(F_{i+1/2} - F_{i-1/2}) \quad \text{or} \quad C_i^{n+1} = C_i^n - (Fr_i - Fl_i), \tag{2}$$

where $F_{i\pm1/2} = F(u,C)$ are the mass fluxes through the right and the left edges of the cell, Fr and Fl being masses transported through the edges for one time step. Fr and Fl can be positive or negative in dependence on the transport direction and $Fm=\text{sign}(u_m)Am$, $(m=r,l)$, where Am is the so called flux area (the shadowed area in Fig.1). Usually, the problem can be normalised by introducing the Courant number $U_i^n = \Delta t \, u_i^n / \Delta x$ and setting $\Delta x = \Delta t = 1$. The schemes are explicit when the fluxes are calculated for the moment t_n. This is the case for the schemes discussed here, so the upper index n will be omitted further on. A scheme is mass conserving if it is constructed in such a way that always $Fl_i = Fr_{i-1}$.

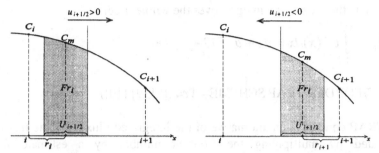

Figure 1. Right edge flux for positive and negative wind velocities.

2.1. BOTT'S SCHEME - **bot**

Bott [1] determines the flux area after the integrated flux concept [11]:

$$Ar_i = \int_{1/2-U_{i+1/2}}^{1/2} C_i^p(x)dx \text{ when } U_{i+1/2}>0, \quad Ar_i = \int_{1/2}^{1/2-U_{i+1/2}} C_{i+1}^p(x)dx \text{ when } U_{i+1/2}<0, \quad (3)$$

where $C_i^p(x)$ is a polynomial of order p referring point i. After testing polynomials of different orders, Bott recommends the 4th order Lagrangean polynomial

$$C^4(x)=a_0+a_1x+a_2x^2+a_3x^3+a_4x^4. \quad (4)$$

If the 5-point pattern <-2,-1,0,1,2> has its origin in point i, the coefficients in Eq.(4) are determined by solving a system of 5 simultaneous algebraic equations. They are ([1]):

$$
\begin{aligned}
a_0 &= C_i, \\
a_1 &= (-C_{i+2} + 8C_{i+1} - 8C_{i-1} + C_{i-2})/12, \\
a_2 &= (-C_{i+2} + 16C_{i+1} - 30C_i + 16C_{i-1} - C_{i-2})/24, \\
a_3 &= (C_{i+2} - 2C_{i+1} + 2C_{i-1} - C_{i-2})/12, \\
a_4 &= (C_{i+2} - 4C_{i+1} + 6C_i - 4C_{i-1} + C_{i-2})/24.
\end{aligned}
\quad (5)
$$

This local approach of polynomial fitting (separate coefficients for every grid point) leads to low numerical diffusion, but in case of steep gradients in concentration field the integrals in Eq.(3) can receive negative or unrealistic high values. Bott introduces limiters for the flux area

$$0<Ar_i<C_i \text{ when } U_{i+1/2}>0 \text{ and } 0<Ar_i<C_{i+1} \text{ when } U_{i+1/2}<0. \quad (6)$$

As to assure the positive definiteness of the scheme, Bott introduces a normalisation of the fluxes, multiplying the flux area by the factor

$$NF_i=C_i/A_i \text{ at } U_{i+1/2}>0 \quad \text{or} \quad NF_i=C_{i+1}/A_{i+1} \text{ at } U_{i+1/2}<0, \quad (7)$$

where A_i is the area of the integral over the whole i-cell

$$A_i = \int_{-1/2}^{1/2} C_i^4(x)dx = a_0 + a_2/12 + a_4/80.$$ (8)

2.2. 4TH ORDER TRAP SCHEME - **Tr4** ([8],[9],[10])

The TRAP concept is an alternative of the Integrated Flux one. The flux area is calculated by multiplying the Courant number by a estimate for the concentration referring the middle of the passed distance. The estimates are obtained on the base of the Bott's polynomial Eq.(4) with coefficients Eq.(5). The arguments in Eq.(4) must be

$$r_i=(1-U_{i+1/2})/2 \quad \text{at} \quad U_{i+1/2}>0 \quad \text{and} \quad r_{i+1}=-(1+U_{i+1/2})/2 \quad \text{at} \quad U_{i+1/2}<0$$ (9)

Bott's limiters Eq.(6) and normalisation Eqs.(7) and (8) are also applied.

2.3. 3RD ORDER BESSEL POLYNOMIAL TRAP SCHEME -**TrB** ([9])

The initial setting of Δt-value for any dispersion model utilising explicit advection scheme must keep the Courant stability condition $u\Delta t/\Delta x<1$ fulfilled in the whole domain during all period of integration. Usually, some value of u greater than the strongest winds is fixed leading to small Δt. During the run, the predominant part of flux area calculations is made at small Courant numbers. This means that very often the flux areas are placed close to the cell edges, around the point $i+1/2$ in the case of i cell right edge flux. High precision of the approximation in this point is necessary for improving the accuracy of the schemes. It is well known that the Lagrange polynomial gives better interpolating quality for $|r|\leq0.25\Delta x$ around the central point i. This is the reason for using different polynomials in the cases of positive and negative velocity. The right edge of the cell is placed at $r=0.5$ and the flux area (respectively, the middle of the flux trapezium) is placed near this point. There is a polynomial approximation giving best results for $0.25\Delta x \leq r\leq0.75\Delta x$. Such is the Bessel interpolation polynomial. As Bessel polynomials are always of an even degree, a third order interpolation is proposed here:

$$C^b(r) = b_0 + b_1r + b_2r^2 + b_3r^3,$$ (10)

where the coefficients $b_k, k = \overline{0,3}$ are determined by (the i cell is referred)

$$b_0 = C_i$$
$$b_1 = (-2C_{i-1} - 3C_i + 6C_{i+1} - C_{i+2}) / 6 \tag{11}$$
$$b_2 = (C_{i-1} - 2C_i + C_{i+1}) / 2$$
$$b_3 = (-C_{i-1} + 3C_i - 3C_{i+1} + C_{i+2}) / 6$$

The flux area is calculated by multiplying the Courant number by the value of Eq.(10) at $r_i=(1-U_{i+1/2})/2$. In this case, one and the same polynomial at positive and negative advection velocities is used when calculating the flux area, but the limiting and the normalisation must be performed taking into account the upstream concept. For the normalization procedure, the cell area is necessary. It is determined by integrating the Bessel polynomial in limits [-0.5,0.5] leading to

$$A_i = (C_{i-1} + 22C_i + C_{i+1}) / 24 \text{ at } U_{i+1/2} > 0. \tag{12}$$

2.4. 3RD ORDER LAGRANGEAN POLYNOMIAL TRAP SCHEME -Tr3.

As the rightmost edge of cell i is just in the middle of the set of 4 neighbouring points (i-1, i, i+1, i+2) it is natural to choose a polynomial of 3rd order as to approximate concentration profile in this region:

$$C^3(x)=c_0+c_1x+c_2x^2+c_3x^3. \tag{13}$$

The 4-point pattern (-3/2,-1/2,1/2,3/2) has its origin at point i+1/2 (shifted pattern) determining the coefficients:

$$c_0 = \left[9(C_{i+1} + C_i) - (C_{i+2} + C_{i-1})\right] / 24$$
$$c_1 = \left[27(C_{i+1} - C_i) - (C_{i+2} - C_{i-1})\right] / 24 \tag{14}$$
$$c_2 = \left[-(C_{i+1} + C_i) + (C_{i+2} + C_{i-1})\right] / 4$$
$$c_3 = \left[-3(C_{i+1} - C_i) + (C_{i+2} + C_{i-1})\right] / 6$$

The flux area is calculated by multiplying the Courant number by the value of Eq.(13) at $r_i=U_{i+1/2}/2$. Bott's limiters Eq.(6) and normalisation Eq.(7) are applied, i cell area (limits [-1,0]) determined as

$$A_i=c_0-c_1/2+c_2/3-c_3/4. \tag{15}$$

2.5. 2ND ORDER LAGRANGEAN POLYNOMIAL TRAP SCHEME - Tr2.

It will be shown later that the decrease of the order of approximation from 4 to 3 leads to acceleration of computations without considerable change for the worse. Further decrease of this order and applying the shifted pattern from Tr3 is worth to be checked:

$$C^2(x)=d_0+d_1x+d_2x^2. \tag{16}$$

The 3-point grid pattern has again its origin in point $i+1/2$, but its orientation will depend on the transport direction. It is natural to use 2 points in upstream direction and one - in downwind one. The polynomial coefficients are:

$$\text{at } U_{i+1/2} > 0 \qquad\qquad \text{at } U_{i+1/2} < 0$$

$$
\begin{aligned}
d_0 &= (-C_{i-1} + 6C_i + 3C_{i+1})/8 \\
d_1 &= -C_i + C_{i+1} \\
d_2 &= (C_{i-1} - 2C_i + C_{i+1})/2
\end{aligned}
\quad \text{and} \quad
\begin{aligned}
d_0 &= (3C_i + 6C_{i+1} - C_{i+2})/8 \\
d_1 &= -C_i + C_{i+1} \\
d_2 &= (C_i - 2C_{i+1} + C_{i+2})/2
\end{aligned}
\qquad (17)
$$

The flux area is calculated by multiplying the Courant number by the value of Eq.(16) at $r_i=-U_{i+1/2}/2$. Bott's limiters Eq.(6) and normalisation Eq.(7) are applied, cell areas (integral limits $[-1,0]$ or $[0,1]$) determined as

$$A_i=d_0-d_1/2+d_2/3 \text{ at } U_{i+1/2}>0, \qquad A_{i+1}=d_0+d_1/2+d_2/3 \text{ at } U_{i+1/2}<0. \qquad (18)$$

3. Numerical Experiments

The schemes from section 2 have passed different tests. Only part of the results, these ones from the two-dimensional rotational test, are presented here. Instantaneous releases with different initial profiles are rotated with constant angular velocity. For each full rotation, the initial condition occurs to be the exact solution of the advection equation. A number of criteria are established for estimation of the schemes properties. Denoting the initial field as C_{ij}^o and the final one - as C_{ij}, a number of estimates are introduced, shown on Table 1.

A grid field of 101×101 points with $\Delta x=\Delta y=1$ is the test domain. The rotational wind field after Smolarkiewicz [6] with constant angular velocity of $\omega\approx0.1$ (628 time steps per 1 rotation) and center in point (51,51) is imposed on this area. Two types of sources - point-shaped one (a limited δ-function) and cone-shaped one - with maximum concentration of $C^o_{max}=3.87\Delta x$ at point (76,51) are supposed. The cone base radius is $15\Delta x$. The quantitative results of all schemes for 1 and 6 rotations are presented on Table 2. The normalisation of the time of integration is made by the time of **TrB**.

It can be noticed that none of the tested schemes describes satisfactorily the advection of a single disturbance in the concentration field (point source). There is no difference scheme, approximating the advection equation, capable to describe adequately this severe discontinuity in the concentration field. This can be seen in **Cmax** values, which fall down to some percents of the initial maximum. The steep gradients are squashed down immediately after the start of movement and after that the new nearly-Gauss-shaped distribution is advected. The cone experiments show much better results.

Table 1. Estimates for simulation quality of advection schemes (rotational test)

$\textbf{Cmax} = \max(\,C_{ij}\,)/\max(\,C_{ij}^{o}\,)$	**Cmax**<1 indicates presence of numerical diffusion
$\textbf{Cmin} = \min(\,C_{ij}\,)/\max(\,C_{ij}^{o}\,)$	**Cmin**<0 indicates absence of positive definitness
$\textbf{SM} = (\sum_{ij} C_{ij}^{o} - \sum_{ij} C_{ij}\,) / \sum_{ij} C_{ij}^{o}$	**SM** - normalized difference of sum of masses. **SM**≠0 indicates absence of conservativity
$\textbf{SM2} = (\sum_{ij} C_{ij}^{o\,2} - \sum_{ij} C_{ij}^{\,2}\,) / \sum_{ij} C_{ij}^{o\,2}$	**SM2** reflects the second moment conservativity. The ideal advection scheme has **SM2**=0.
$\textbf{DXc} = \sum_{ij} iC_{ij} / \sum_{ij} C_{ij} - \sum_{ij} iC_{ij}^{o} / \sum_{ij} C_{ij}^{o}$ $\textbf{DYc} = \sum_{ij} jC_{ij} / \sum_{ij} C_{ij} - \sum_{ij} jC_{ij}^{o} / \sum_{ij} C_{ij}^{o}$	**DXc** and **DYc** estimate the displacement of the mass center due to numerical effects. **DXc** = **DYc** = 0 after a number of full rotations indicate an ideal transportivity of the advection scheme.
$\textbf{DD} = (D - D^{0}) / D^{0},$ where $D = \sum_{ij} C_{ij}\left[(x - \textbf{Xc})^{2} + (y - \textbf{Yc})^{2}\right] / \sum_{ij} C_{ij}$	**D** - mass dispersion arround the mass center **DD** indicates the degree of deconcentration of masses due to the numerical diffusion
$\textbf{RC} = \sum_{ij} \left| C_{ij} - C_{ij}^{o} \right| / \sum_{ij} C_{ij}^{o}$ $\textbf{RC2} = \sum_{ij} (C_{ij}^{2} - C_{ij}^{o\,2}) / \sum_{ij} C_{ij}^{o\,2}$	**RC, RC2** - restoration capabilities (absolute and squared). In case of ideal advection scheme, they are to be equal to zero after a number of full rotations
$\textbf{T} = \Delta T_{calc} / \Delta T^{c}_{calc},$	relative speed of performance

These experiments show that the Bott scheme and the TRAP scheme of 4th order possess practically equal simulation properties, **Tr4** being some times faster. The 3rd orders schemes, built with maximal approximation accuracy between the grid points (Bessel polynomial and Lagrange polynomial with shifted pattern) display almost the same properties as **Bot** and **Tr4**, being faster. The fastest one is the scheme **TrB** and it is used for normalisation of performance times. The 2nd order schemes are more faster but their simulation properties got worse. They still can be used in some applications requiring a high speed of performance.

In Fig.2, a demonstration of the **TrB** capabilities is presented. In the first row of graphs, a cylinder with radius $15\Delta x$ and axes in point (76,51) as well as a centred parallelepiped with base $50\Delta x$ (both with height $3.87\Delta x$) are rotated over a constant background field with height of $1\Delta x$. The one rotation appearance of these figures is quite plausible keeping in mind the discontinuity in the initial distribution. In the second row of graphs, an instantaneous line source and a continuous point source are rotated over the same background field, the first case simulating a front and the second one - a real stack source.

The last case demonstrates the fast decrease of the initial concentration and the transformation of the one-dimensional disturbances in three-dimensional Gauss-shaped objects continuously moving and overlapping one over another.**TrB** is proved also on the Smolarkiewicz's deformational test [6]. In Fig. 3, the numerical solutions after 19, 38, 57 and 75 iterations are shown. These distributions nearly correspond to the analytical solutions, presented in Fig 3a-d of Staniforth et al.[7].

4. Conclusion

The version **TrB** of the TRAP scheme shows almost the same characteristics as the original Bott one, but it is some times faster. Another important advantage of the new low order schemes is the fact that they need only two grid points at the border of the model domain for boundary ones. The 4 order Bott and TRAP schemes need 3 boundary points at every border. The new TRAP scheme is built in the EMAP model ([8],[9]) for further use in air pollution calculations.

Table 2. Estimates [%] of different schemes' performance. Rotating instantaneous point and cone sources.

	Point	Bot	Tr4	TrB	Tr3	Tr2	cone	Bot	Tr4	TrB	Tr3	Tr2
1	Cmax	3.78	3.81	3.53	3.53	2.40	Cmax	91.16	91.19	91.09	91.09	87.75
	Cmin	.00	.00	.00	.00	.00	Cmin	.00	.00	.00	.00	.00
r	CM	.00	.00	.00	.00	.00	CM	.000	.000	.000	.000	.000
o	CM2	-97.99	-97.99	-98.16	-98.16	-98.68	CM2	-.90	-.89	-.89	-.89	-1.95
t	DXc	23.4	22.4	39.3	39.3	21.4	DXc	.0	.0	.0	.0	-.1
a	DYc	-30.8	-26.6	-78.3	-78.3	-25.6	DYc	-.2	.0	-.4	-.4	-.1
t	DD	-	-	-	-	-	DD	1.793	1.819	1.964	1.964	2.952
i	RC	192.4	192.7	193.6	193.6	195.2	RC	3.444	3.443	3.698	3.698	4.582
o	RC2	94.4	94.4	95.4	95.4	96.5	RC2	.107	.107	.114	.114	.182
n	T	766	110	100	109	89	T	659	120	100	108	90
6	Cmax	1.66	1.68	1.56	1.56	.98	Cmax	86.43	86.49	85.93	85.93	81.98
	Cmin	.00	.00	.00	.00	.00	Cmin	.00	.00	.00	.00	.00
r	CM	.00	.00	-.00	-.00	.00	CM	-.001	-.003	-.003	-.003	.000
o	CM2	-99.1	-99.1	-99.2	-99.2	-99.6	CM2	-3.43	-3.40	-3.41	-3.41	-8.19
t	DXc	23.6	21.8	40.4	40.4	19.4	DXc	-.1	-.3	-.3	-.3	-.6
a	DYc	-37.9	-27.6	-101.5	-101.5	-28.2	DYc	-.3	.5	-.1	-.1	.8
t	DD	-	-	-	-	-	DD	6.63	6.65	7.05	7.05	11.61
i	RC	196.7	197.0	197.1	197.1	198.0	RC	8.100	8.126	8.466	8.466	9.425
o	RC2	97.5	97.5	97.9	98.0	98.6	RC2	.415	.420	.441	.441	.643
n	T	515	108	100	108	88	T	500	109	100	109	89

References

1. Bott, A. (1989), A positive definite advection scheme obtained by nonlinear renormalization of the advective fluxes, *Mon.Wea.Rev.* **117**, pp. 1006-1015.
2. Bott, A. (1992), Monotone flux limitation in the area preserving flux form advection algorithm, *Mon.Wea.Rev.* **120**, pp. 2592-2602.

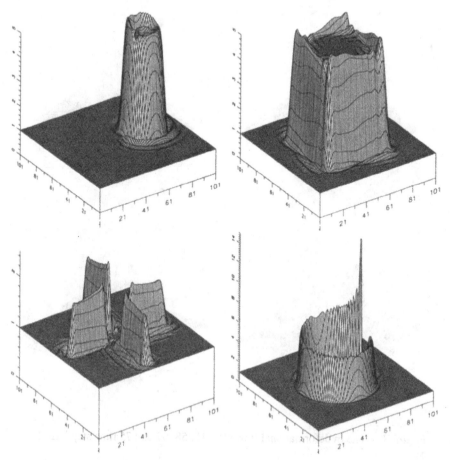

Figure 2. One rotation of instantaneous cylinder, cube and line sources as well as continuous point source over background field

3. Bott, A. (1993), The Monotone Area-preserving Flux-Form Advection Algorithm: Reducing the Time-splitting Error in Two-Dimensional Flow Fields, *Mon.Wea.Rev.* **121**, 2637-2641.
4. Peters, L. K., Berkowitz, C. M., Carmichael, G. R., Easter, R. C., Fairweather, G., Ghan, S. J., Hales, J. M., Leung, L. R., Pennell, W. R., Potra, F. A., Saylor, R. D. and Tsang, T. T. (1995), The current state and future direction of Eulerian models in simulation the tropospheric chemistry and transport of trace species: a review, *Atmos. Environ.*, **29**, pp. 189-222.
5. Rood, R. B. (1987), Numerical advection algorithms and their role in atmospheric transport and chemistry models, *Rev. Geophys.* **25**, pp. 71-100.
6. Smolarkiewiecz, P. K. (1982), The multidimensional Crowley advection scheme, *Mon. Wea.Rev.* **113**, pp. 1109-1130.
7. Staniforth, A., Côté, J. and Pudikiewicz, J. (1978), Comments on "Smolarkiewicz's deformational flow", *Mon.Wea.Rev.* **115**, pp. 894-900.

312

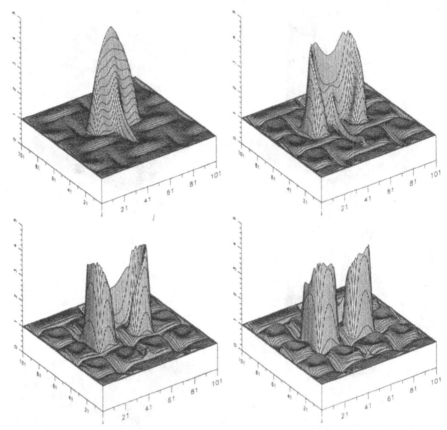

Figure 3. Cone in deformational flow after 19, 58, 57 and 75 time steps (Δ*t*=0.7)

8. Syrakov, D. (1995), On a PC-oriented Eulerian Multi-Level Model for Long-Term Calculations of the Regional Sulphur Deposition, in S. E. Gryning and F. A. Schiermeier (eds.), *Air Pollution Modelling and its Application XI,* NATO • Challenges of Modern Society **21**, Plenum Press, New York and London, pp. 645-646.
9. Syrakov, D. (1996), On the TRAP advection scheme - Description, tests and applications, in G. Geernaert, A. Walløe-Hansen and Z. Zlatev (eds.), *Regional Modelling of Air Pollution in Europe. Proceedings of the first REMAPE Workshop, Copenhagen, Denmark, September 1996,* National Environmental Research Institute, Denmark, pp. 141-152.
10. Syrakov, D. and Galperin, M. (1997) On a new Bott-type advection scheme and its further improvement, in H. Hass and I. J. Ackermann (eds.), *Proc. of the first GLOREAM Workshop, Aachen, Germany, September 1997,* Ford Forschungszentrum Aachen, pp. 103-109.
11. Tremback, C. J., Powell, J., Cotton, W. R. and Pielke, R. A. (1987), The forward-in-time upstream advection scheme: Extension to higher orders, *Mon.Wea.Rev.* **115**, pp. 540-555.
12. WMO-TCSU (1979), Numerical methods used in atmospheric models, Vol. I and II, GARP Publication series, No 17.

A STUDY OF SULFUR DIOXIDE DISTRIBUTION OVER ÝSTANBUL, TURKEY AND PRELIMINARY RESULTS OF NEURAL NETWORK ANALYSIS

METE TAYANÇ[1] AND ARSLAN SARAL[2]

[1]*Marmara University, Department of Environmental Engineering, Göztepe, Ýstanbul, Turkey.*
[2]*Yýldýz Technical University, Department of Environmental Engineering, Beþiktaþ, Ýstanbul, Turkey.*

Abstract

The air quality problems over Ýstanbul are related to the low-quality fossil fuel consumption and atmospheric conditions. Sulfur dioxide concentration levels are investigated over Ýstanbul to assess air pollution during the heating seasons in which the concentration of air pollutants reach high levels due to the consumption of low-quality fossil fuels. Results reveal that the consumption ratio of coal/fuel-oil has increased drastically in 1980s. Optimum interpolation technique, kriging, is used to obtain the spatial distribution of sulfur dioxide over the area. The resultant sulfur dioxide concentration fields showed three critical regions; Haliç basin and Þiþli-Taksim area on the European side and Göztepe on the Asian side. It is found that there is a considerable decrease in air pollution levels over Ýstanbul in the 1995-1996 heating season. Important factors that have been responsible in this decrease of pollutant levels are found to be the favorable weather conditions, switching to natural gas in many buildings and the consumption of pre-treated coal. For the prediction of air pollution, neural networks are used as the modeling tool and the results are found to be very encouraging. A data set of one month period is used which contains meteorological and air pollution parameters. An average error between the actual and the predicted sulfur dioxide concentration levels of 12% shows us the reliability of the neural network modeling.

Z. Zlatev et al. (eds.), Large-Scale Computations in Air Pollution Modelling, 313–324.
© 1999 *Kluwer Academic Publishers. Printed in the Netherlands.*

1. Introduction

In urbanized areas, air pollution adversely affects human health, damages vegetation, deteriorates building materials, reduces visibility, interferes with economic development and devalues the quality of life. Air pollution is created by the emissions of primary pollutants from anthropogenic and natural sources. Subsequently, complex interactions occurs as the pollutants evolve under the influence of meteorological, physical, chemical and biological processes. To develop effective control strategies, it is necessary to understand in detail the physical and chemical processes that govern its formation, dispersion, transportation and consequences.

One of the fastest growing cities of the world is Ýstanbul [1]. The potential for air pollution problems is high because of high emissions and the close geographical proximity of the major industrial and urban centers such as Gebze, Ankara, Athens, Costanza, etc.. Through transportation processes, the pollutants emitted from various countries can also be expected to contribute negatively to the air quality of Ýstanbul. The future air quality of the city is at risk because of Ýstanbul's large and rapidly increasing population, its growing economy, and the associated systems of energy consumption and production.

The behavior and impacts of air pollutants has been widely studied in the past decade, most of these conducted for the industrial and urban areas of the developed world [2]. The inadequate number of studies in Turkey prevents the development of national policies to address the rapidly emerging problems of air pollution in the country.

It is the purpose of this article to gain perspective for the assessment of the spatial distribution of one of the air pollutants, sulfur dioxide, for the Ýstanbul metropolitan region by the use of a spatial prediction scheme, kriging. Kriging methodology has been increasingly used in environmental studies [3,4]. This study focuses on the severity of the level of sulfur dioxide concentrations in Ýstanbul and elaborates on the reasons of this pollution in terms of meteorological factors. Temporal variation of the emissions from the combustion of various fuel types and the relationship of air pollution with ventilation is also investigated.

2. Study Area and Experimental

Ýstanbul (40°N, 29°E), the largest city of Turkey is located at the meeting point of the two continents, Europe and Asia. The total residential area of Ýstanbul in these two continents is about 5712 km^2, and nowadays, the population of the metropolitan area is well over ten million people. Figure 1 shows the study area. Increased immigration from less developed regions of the country since 1950s, especially during 1980s, has been causing a rapid increase in the population that leads to uncontrolled settlement and industrialization, mainly in the suburban parts of the city [1]. Population density at the central residential and business districts has been increased from 3000 people/km^2 to over 6000 people/km^2. In addition to rich trade and commerce, Ýstanbul has been experiencing very high industrial activity. Almost 70 % of Turkey's industry is located in the southern coast of the city.

The climate of Ýstanbul shows a transition between the Mediterranean and temperate climates with cool and wet winters and warm and humid summers. The coldest and the wettest months are January and February while the hottest are July and August. The average annual temperature and total annual precipitation is 13.7°C and 734 mm, respectively.

Continuous daily observations of sulfur dioxide and particulate matter have been carried out in the city by seven stationary stations, Eminönü, Fatih, Þiþli, Zeytinburnu, Kadýköy, Kartal and Ümraniye since 1985, under the control of the Ministry of Health. Measurements before 1985 had also been conducted, but the resultant air pollution data series in that period include many discontinuities and they may be unreliable owing to incorrect calibration of the used instruments. Sulfur dioxide has been measured by the sampling train method which has been a reference method for many years and recommended by the Environmental Protection Agency of US EPA [5]. Hydrogen peroxide solution can be used in the midget impingers to dissolve sulfur dioxide and then the solution can be titrated with barium perchlorate using a thorin indicator to obtain a cumulative concentration of sulfur dioxide.

Particulate matter has been monitored by the popular filtering method. Weighted dry filter paper, adequate to the standards, is used to collect the total suspended particles in the atmosphere on a cumulative basis. Initial weight can be subtracted from the final dried weight of the filter to find the cumulative accumulated weight of particulate matter. To increase the monitoring resolution in the study area, nine more stations were added to the observation network. These are Bakýrköy, Beþiktaþ, Beyoðlu, Baðcýlar, Bayrampaþa, Gaziosmanpaþa, Üsküdar, Maltepe, and Göztepe stations. These places were chosen according to their rapidly increasing population densities and their associated increase in the consumption of lignite as the primary fuel source. The location of these stations can be seen in Figure 1.

Meteorological data such as the wind speed and direction, the mixing height, and the number of cyclones affecting the region in the period of 1985-1994 are obtained from Göztepe and Kandilli meteorological offices. The data of upper air sounding at 00.00 UTC is only supplied by Göztepe synoptic station. State

Institute of Statistics provided the record of the consumed fuel types and amounts for the period of 1980-1991.

Figure 1. Study area and the air pollution monitoring stations.

3. Results

3.1. CONSUMED FOSSIL-FUELS AND SULFUR DIOXIDE EMISSIONS

Air pollution problem in Ýstanbul is a result that links the low quality fuel consumption for industrial and heating purposes of the region with the damage it causes to the environment. Developing an accurate and detailed picture of sulfur dioxide emissions in the city is a crucial component of air pollution assessment and may serve multiple purposes. By studying the use of various fossil fuels in industry, transportation and building heating, it is possible to estimate the shares of different fuels in the releases of sulfur dioxide into the atmosphere.

The contribution of each fuel type used in Ýstanbul for the period of 1980-1991 to sulfur dioxide emissions in million kilograms per year is presented in Figure 2a and b. The impact of gasoline and diesel use on the sulfur dioxide emissions is very low, generally on the order of 1% for gasoline and 3% for diesel, amongst the total contribution of all fossil fuels. Figure 2b depicts the sulfur dioxide emissions released by the gasoline and diesel combustion.

The important fuels having major contribution to sulfur dioxide are coal and fuel-oil. Figure 2a shows the increase in sulfur dioxide emissions caused by coal and fuel-oil use. Although the emissions generated by the fuel-oil consumption increase, their percentage in the total emissions decrease. Main reason for this situation is the much greater increase in the coal consumption. Turkey is rich for lignite reserves, and coal can be made available to customers in a much cheaper way and much shorter time periods than compared with the other fuels. Owing to the above reasons, high sulfur containing lignite has been increasingly used in recent years.

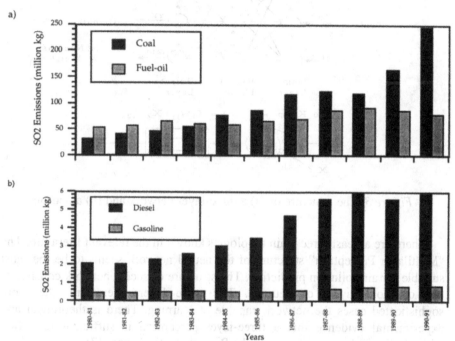

Figure 2. Sulfur dioxide emissions produced by different fuel use in Ýstanbul

3.2. NEURAL NETWORK ANALYSIS

A neural network is a special structure consisting of basic blocks called neurons, organized and interconnected in one or more layers [6]. It imitates the functioning of the human brain. The neural network-based prediction works in the way as follows. A historical set of significant meteorological and air pollution data are used for input to the model, and the outputs are ambient concentrations predicted by the model. First the network is trained with the historical data. By the proper choice of training sets, the network is capable of predicting the ambient concentrations as an output according to the meteorological situation given as an input (after learning process has been completed), and the internal structure of the network established during the learning period.

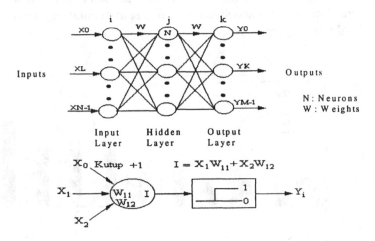

Figure 3. The structure of (a) a three-layer network and (b) a neuron

There are at least three main topologies known in the relevant literature. The "Multilayer Perceptron" structure of the neural network seems to be the most suitable for air pollution prediction. The structure of such a network consists of one or more layers of neurons. The more layers there are, the more sophisticated cases the network capable of learning. There is theoretical and experimental evidence that a three-layer perceptron is sufficient to define arbitrary linear decision regions in the R^n mathematical space [7].

The structure of a three-layer network and that of a neuron are shown in Fig.3a and 3b, respectively.

In our preliminary studies with neural networks for the air pollution prediction in Ýstanbul, we tried to predict the daily average SO_2 concentration with neural networks. The input parameters used for the network are the pressure, dry bulb temperature, relative humidity, cloud level, daily mean wind velocity, first inversion height at 12.00 h, first inversion height at 24.00 h and daily precipitation values. Daily SO_2 concentrations are predicted with these input parameters. The results are very encouraging to forecast air pollution with neural networks. As an example, Figure 4 shows the real values and the network predicted values of sulfur dioxide for November 1994 over Ýstanbul. The Network used is a typical multilayer perceptron with an input layer, a hidden layer with the number of neurons equal to that of input layer and an output layer consisting of only SO$_2$ parameter. The parameters first normalized in their own ranges and then grouped arbitrarily as the training and the testing data sets. The network first rained with the training data set to obtain the optimum values of internal network parameters and then tested for the validation of the optimum network structure. Then whole data set was predicted with the network constructed. As Figure 4 shows, the real and network predicted values are very compatible with each other with an average error of 12%. We used just a month of data set which is considerably small for neural network studies but the results reveal that more data will produce less error and more sophisticated cases will be handled more realistically [8]. These results also show that the input parameters used have good interrelation with the pollution parameter predicted. Neural networks do not give a mathematical relationship between input and output parameters but instead they reveal the interrelation between them with the success of results. That is to say, the smaller the error value, the better the interrelation between input and output parameters. This can be succeeded, in a way, with larger data sets.

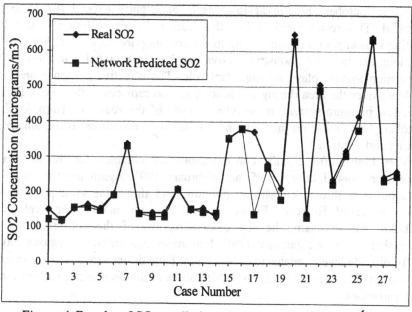

Figure 4. Results of SO$_2$ prediction with neural networks in Ýstanbul.

3.3. DISPERSION FIELDS OF SULFUR DIOXIDE

Monthly sulfur dioxide concentration levels from 15 monitoring stations are used to obtain the spatial patterns of air pollution. The optimum interpolation technique, kriging, is used to obtain spatial distribution fields of sulfur dioxide over Ýstanbul. Figure 5a and b depicts the spatial distribution of the monthly averages of the sulfur dioxide concentration levels over Ýstanbul for December 1993 and January 1994. It is intuitively clear that the sulfur dioxide concentration levels for the heating season months over the Ýstanbul metropolitan area are well over the long-term standard of WHO, 50 µg/m³. Some parts of the region were experiencing concentrations much greater than 300 µg/m³, above which an increase in mortality can be established [9]. Three maximum regions are observed; two in the European side and the other over the Asian side. The peak region over the European side includes Golden Horn valley, Taksim-Þiþli, and Topkapý-Eminönü areas. On the Asian side Göztepe-Bostancý area received major threaten from sulfur dioxide pollution. These areas are generally characterized by very high residential population densities.

However, sulfur dioxide concentration levels in the other months of the year (April to October) remain at low levels or slightly exceed the long-term standard. This result yields a fact that Ýstanbul's air pollution is mainly a heating season phenomenon, owing to the consumption of low quality coal for building heating and industrial activities. The interannual variation in the circulation system plays an important role. The effective cyclones and anti-cyclones over the area during the heating season replaces with a weak Persian Gulf low pressure system in the other months of the year, i.e., from April till October. This condition prevents the formation of inversion layers, enhancing mixing and dilution.

Very high sulfur dioxide concentrations are obtained for the months of December 1994, January 1995, and February 1995. Again most parts of the Ýstanbul were threatened with air pollution and almost the same peak regions can be observed. However, Figure 6a and b presents much lower levels of air pollution for the same heating season months of the following season; December 1995 and January 1996. Maximum concentration regions having sulfur dioxide levels greater than 300 µg/m³ have disappeared in these months. This leads us to investigate the possible causes of the decrease in sulfur dioxide concentrations.

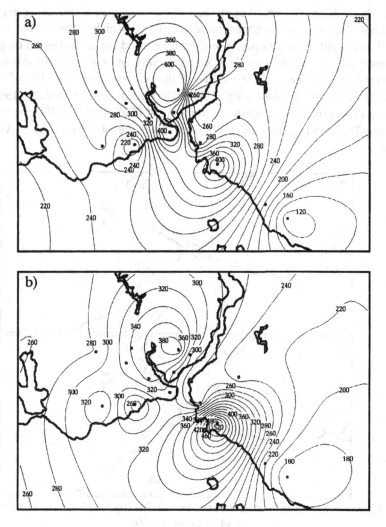

Figure 5. Sulfur dioxide concentration fields (µg/m³) for a) December 1993
and b) January 1994.

3.4. REASONS OF THE DECREASE IN SULFUR DIOXIDE LEVELS

The impact of the atmospheric circulation and stability conditions on the
dilution of air pollution has long been known. Here we attempt to isolate the
consequences of changes in meteorological parameters on sulfur dioxide
concentrations. The State of Colorado Department of Health in Denver uses an
air pollution dispersal index (ventilation index) for the goals of forecasting

[10]. The technique is based upon the individual concepts of mixing depth and wind speed as well as the atmospheric ventilation which is the product of wind speed and night mixing depth. The categories used for air pollution dispersion are *poor*, *fair*, *good*, and *excellent*. The daily mixing heights used for producing the ventilation indices for the months of the periods December 1993-February 1994, December 1994-January 1995, December 1995-February 1996 are obtained from the data of upper air sounding at 00.00 UTC of Göztepe synoptic station. The wind speeds are all derived from the observed data of Göztepe station.

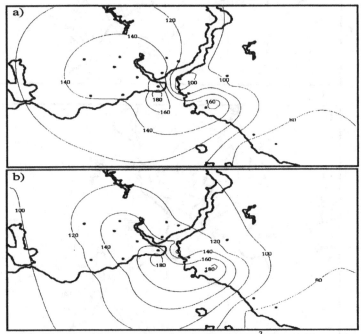

Figure 6. Sulfur dioxide concentration fields ($\mu g/m^3$) for a) December 1995 and b) January 1996.

The number of days belonging to each category is determined for the above considered periods and the results are shown in Figure 7. The figure depicts that the air pollution dispersion index is worse in the 1993-1994 and 1994-1995 periods than the 1995-1996 period. Although the frequency of days having *fair* dispersion conditions is higher in the 1995-1996 period than the 1993-1994 and 1994-1995 periods, the low ventilation coefficients are considered to be dominant over the higher ones in the terms of air pollution dispersion. Thus, an increase in the ventilation during the last period explains the decrease in sulfur dioxide levels to a considerable extend. However, switching from the use of coal to natural gas for space heating in many buildings has also an impact in the decrease of air pollution levels. A natural gas distribution pipeline network has been constructed to cover the regions having severe air pollution problems and it has been operated since 1995.

4. Summary and Conclusions

In this study the relationships between meteorology, pollutant distribution and air quality over Ýstanbul are investigated. Spatial and temporal distribution of sulfur dioxide on 1993-1994, 1994-1995 and 1995-1996 heating periods have been analyzed in detail. The study has demonstrated that air pollution levels over Ýstanbul has increased considerably in the 1985-1991 heating seasons owing to switch to the use of low quality fossil fuels and uncontrolled immigration to the city and increase in the population densities.

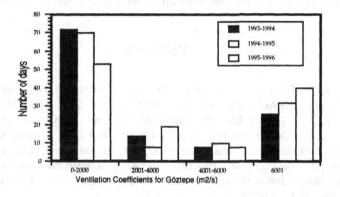

Figure 7. Ventilation coefficients (m^2/s) for the last three heating seasons

Spatial analysis of sulfur dioxide concentration levels reveal that sulfur dioxide problem is specific to the heating periods of 1993-1994 and 1994-1995, and the other months of the year (April-October) are generally characterized by clean air. High consumption of lignite for heating purposes in the heating season period and the dominant Persian Gulf low pressure system in the summer months are the main reasons of the occurrence of this situation. This study also indicates that the long-term WHO standards of particulate matter and sulfur dioxide have been severely violated in Ýstanbul. Generally, the observed maximum concentration regions are associated with the highest population densities.

In the 1995-1996 heating season period, however, the sulfur dioxide concentration levels over Ýstanbul has decreased considerably possibly owing to the decrease in the *poor* and increase in the *fair* and *excellent* ventilation conditions. Another important reason in this decrease of air pollution levels is the switching to the natural gas as a home and business heating fuel in many buildings instead of the low quality lignite.

Air pollution modeling and prediction with neural networks is the other part of this study. In this part, the concentration of SO_2 is related with some meteorological parameters such as pressure, temperature, relative humidity, inversion levels and the amount of precipitation. The simulated values are found to be very similar to the real ones; an average error of 12% is estimated for the data having a period as small as a month. This amount of error can be reduced with larger data sets enabling the greater optimization of the network. Also this amount of error calculated with small amount of data, shows that the output parameter is well dependent on input parameters used. Furthermore, the weights of the input layer indicate the relative dependency of the output parameter on the input parameters. The bigger the weight of the input parameter, the more dependent the output parameter to the considered input.

References

1. Karaca, M., Tayanç, M. and Toros H. (1995), The effects of urbanization on climate of Ýstanbul and Ankara: a first study, *Atmos. Environ.*, **29B**, 3411-3421.
2. Dennis, R. L. (1996), Multi-pollutant modelling and policy: dressing the emperor and avoiding false paradigms, in S. E. Gryning and F. A. Schiermeier (eds.), *Air Pollution Modeling and Its Application XI*, , ed., Plenum Press, New York.
3. Schaug, J., Iversen, T. and Pedersen U. (1993), Comparison of measurements and model results for airborne sulfur and nitrogen components with kriging, *Atmos. Environ.*, **27A**, 831-844.
4. Tayanç, M., Karaca, M. and Yenigün, O. (1997), Annual and seasonal air temperature trend patterns of climate change and urbanization effects in relation with air pollutants in Turkey, *J. Geophys. Res.*, **102**, 1909-1919.
5. De Nevers, E. (1995), *Air Pollution Control Engineering*, McGraw-Hill, New York.
6. Rumelhart, D. E. and McClelland, J. L., (1986), *Parallel Distributed Processing 1, 2.*, MIT Press, Cambridge, MA.
7. Niemann, H. (1989), *Pattern Analysis and Understanding*, 2nd Edn., pp.199-205, Springer, New York.
8. Boznar, M., Lesjak, M. and Mlakar, P. (1993), A Neural Network-Based Method For Short-Term Predictions of Ambient SO2 Concentrations in Highly Polluted Industrial Areas of Complex Terrain, *Atmos. Environ.*, **27B**, 221-230.
9. WHO (1987), *Air quality guidelines for Europe*, WHO Regional Publications, European Series No. 23, Copenhagen.
10. Eagleman, J. R. (1991), *Air Pollution Meteorology*, Trimedia Publishing Company, Lenexa, Kansas.

KRYLOV SUBSPACE METHODS FOR THE SOLUTION OF LARGE SYSTEMS OF ODE'S

P. G. THOMSEN and N. H. BJURSTRØM
Inst. for Mathematical Modelling, DTU, Building 305, DK2800 Lyngby

Abstract

When solving large systems of ODE's like those appearing in air-pollution problems some kind of implicitness in the solver is necessary. The implicit equations that have to be solved are large and often it is preferred to use parallel computers for the solution. It is necessary to use linear systems solvers that parallellize well. Among such methods are different kinds of Krylov subspace methods. The presentation will show several examples of preconditioned conjugate gradient type methods and examples from air pollution problems will be used to make comparisons on the efficiency of such methods. The platform that has been used for carrying out the tests is a general DAE-solving platform developed for such purposes. GODESS is a Generic ODE Solving System implementing an Object Oriented platform for solving systems of ODE's and DAE's. The choice of method is made by the user and several methods have been tried out. A comparison between methods with different linear solvers are presented.

1. Introduction.

Large systems of ODE's arise in many technical applications and the numerical solution of such systems is done by using implicit methods since they are nearly always stiff. When the solution by such methods is carried out a large proportion of the computational work is done in solving linear systems in internal iterations. These linear systems may be solved using direct methods and sparse matrix techniques, a possibility that works well for traditional computer architechtures. Another approach is to apply iterative methods like Krylov subspace methods for the linear systems. This will work well when the number of iteratins in each system is small and when the dimension of the krylov subspace is very small compared to the number of equations.

Z. Zlatev et al. (eds.), Large-Scale Computations in Air Pollution Modelling, 325–338.
© *1999 Kluwer Academic Publishers. Printed in the Netherlands.*

2. Solution methods.

We consider a system of non-linear differential equations

$$y' = f(t, y), \ y(0) = y_o, \quad y, y_o \in \mathbf{R}^n, \quad f : \mathbf{R}^{n+1} \to \mathbf{R}^n.$$

The Jacobian of the system will be needed in the derivations

$$\mathbf{J} = \frac{\partial f}{\partial y}$$

we may safely consider at least one of the eigenvalues of the Jacobian to have a large negative real part.

2.1. ONE STEP METHODS.

The methods that are used, see (2.1), are most often from the group of one step methods are from the family of semi-implicit Runge-Kutta methods. We write the method for the autonomous case

$$y_{n+1} = y_n + h \sum_{i=1}^{r} b_i f(Y_i)$$

$$Y_i = y_n + h \sum_{j=1}^{i-1} a_{ij} f(Y_i) + h\gamma_i f(Y_i) \qquad (2.1)$$

Several possibilities are available for the choice of the method coefficients. We may characterize these methods using the Butcher tableau:

Butcher scheme.

c_1	γ_1	0	.	.	
c_2	$a_{2,1}$	γ_2	.	.	
c_i	$a_{i,1}$	$a_{i,2}$	. . .	γ_i	.
$\underline{b}$	b_1	b_2	. .	b_r	
$\underline{d}$	d_1	d_2	. .	d_r	

Each stage 2.1 is a system of nonlinear equations that has to be solved iteratively. The iterative method used will be a modified Newton iteration.

$$Y^{s+1} = Y^s + M^{-1}R(Y^s)$$

Where the residual $R(Y)$ is defined by

$$R(Y^s) = Y^s - h\gamma f(Y^s) - \phi$$

In the SDIRK-type methods preferred by many authors all the γ_i are equal which means that all the nonlinear systems share the same iteration matrix, and the iteration matrix M is found from the most recent Jacobian matrix using a past y in the derivation of

$$M = I - h\gamma J(y)$$

An iteration step is computed using the following sequence

$$M\delta^s = -R(Y^s), \quad Y^{s+1} = Y^s + \delta^s$$

In the first equation the matrix M is factorised only when the stepsize is changed or when the matrix is updated if the convergence is bad. The iteration is stopped when a residual test is satisfied $\|R(Y^s)\| \leq \epsilon$. The norm used will depend on the application and implementation.

3. Linear systems.

In the methods considered we have to solve systems of linear equations in each Newton iteration step. The structure of the matrices involved in the linear system corresponds to the structure of the Jacobian matrix for the differential system. Traditionally these systems have been solved using direct methods taking advantage of block structure or sparseness. It is however also interesting to study the use of iterative solvers for the systems and we have been interested in the preconditioned Krylov subspace methods. Assume we want to solve the system

$$\mathbf{Ax} = \mathbf{b}$$

If we premultiply the system by a matrix P^{-1} we get

$$\mathbf{P^{-1}Ax} = \mathbf{P^{-1}b}$$

This system has the same solution as the original and may be easier to solve. We use preconditioning to reduce the dimension of the subspace in which the searches and projections are carried out.

4. Krylov subspace methods.

Projection for the solution of linear systems may be very useful when implementing solution methods for large systems to be used in connection with vector or parallel computer architechtures. They are iterative and attempt to find the solution in a sequence of spaces called Krylov subspaces. We will look at different ways of creating these subspaces.

4.1. ARNOLDI'S METHOD.

The Arnoldi algorithm construct the Krylov sequence from an orthonormal basis found by a Gram-Schmidt orthogonalization process.

Algorithm 1 (Arnoldi)

$$choose \ a \ vector \ \mathbf{v}_1 \ with \ norm \ 1$$

$$For j = 1, 2, \ldots, m$$

$$h_{ij} = (\mathbf{A}\mathbf{v}_j, \mathbf{v}_i) \ for i = 1, 2, \ldots, j$$

$$\mathbf{w}_j = \mathbf{A}\mathbf{v}_j - \sum_{i=1}^{j} h_{ij}\mathbf{v}_i$$

$$h_{j+1,j} = \|\mathbf{w}_j\|_2$$

$$if \ h_{j+1,j} = 0 \ then \ stop$$

$$\mathbf{v}_{j+1} = \frac{\mathbf{w}_j}{h_{j+1,j}}$$

$$End$$

4.2. INCOMPLETE ORTHOGONALIZATION METHOD

This variant of the Arnoldi idea is using the following

Algorithm 2 (IOM)

$$calculate \ the \ residuum \ \mathbf{r}_0 = \mathbf{b} - \mathbf{A}\mathbf{x}_0$$

$$\beta = \|\mathbf{r}_0\|_2 \ , put \mathbf{v}_1 = \frac{\mathbf{r}_0}{\beta}$$

$$put \ \rho = \beta \ , \ l = 1$$

$$while \ \rho > \delta$$

$$h_{il} = (\mathbf{A}\mathbf{v}_l, \mathbf{v}_l)$$

$$\mathbf{w}_{l+1} = \mathbf{A}\mathbf{v}_l - \sum_{i=1}^{l} h_{il}\mathbf{v}_i$$

$$h_{l+1,l} = \|\mathbf{w}_l\|_2$$

$$update \ the \ ILU - factorization \ of \ \mathbf{H}$$

$$compute \ \rho = \|\mathbf{b} - \mathbf{A}\mathbf{x}_l\|_2$$

$$l = l + 1$$

$$End$$

If the inner loop fails to converge in say l_{max} steps we make a restart to save work. The effect of only doing an incomplete LU-factorization is to save time and accepting orthogonality to a limited set of basis vectors.

4.3. GMRES.

The Generalize Minimum Residual method is another Krylov subspace method that uses the Arnoldi method to generate an orthonormal basis in the search subspace.

Algorithm 3 (GMRES)

$$calculate\ the\ residuum\ \mathbf{r}_0 = \mathbf{b} - \mathbf{A}\mathbf{x}_0$$
$$\beta = \|\mathbf{r}_0\|_2 \ , put\mathbf{v}_1 = \frac{\mathbf{r}_0}{\beta}$$
$$put\ \rho = \beta \ , \ l = 1$$
$$while\ \rho > \delta$$
$$h_{il} = (\mathbf{A}\mathbf{v}_l, \mathbf{v}_l)$$

$$\mathbf{w}_{l+1} = \mathbf{A}\mathbf{v}_l - \sum_{i=1}^{l} h_{il}\mathbf{v}_i$$

$$h_{l+1,l} = \|\mathbf{w}_{l+1}\|_2$$
$$\mathbf{v}_{l+1} = \frac{\mathbf{w}_{l+1}}{h_{l+1,l}}$$
$$update\ the\ QR-factorization\ of\ \mathbf{H}$$
$$compute\ \rho = |g_l(l+1)|$$
$$l = l + 1$$
$$End$$

For the QR-factorizations we have used plane Givens rotations . A special variant using truncation has been implemented also.

4.4. DESCENT METHODS

Two types of descent methods Orthomin(k) and Young(k) have been implemented They use the left subspace generated by $\mathcal{L} = \mathcal{A}\mathcal{K}_m$

Algorithm 4 (Descent)

$$choose\ \mathbf{x}_0\ and\ put\ \mathbf{r}_0 = \mathbf{b} - \mathbf{A}\mathbf{x}_0\ and\mathbf{p}_0 = \mathbf{r}_0$$
$$For\ i = 0\ until\ convergence$$
$$a_i = \frac{(\mathbf{r}_i, \mathbf{A}\mathbf{p}_i)}{(\mathbf{A}\mathbf{p}_i, \mathbf{A}\mathbf{p}_i}$$
$$\mathbf{x}_{i+1} = \mathbf{x}_i + a_i\mathbf{p}_i$$
$$\mathbf{r}_{i+1} = \mathbf{r}_i - a_i\mathbf{A}\mathbf{p}_i$$
$$compute\ \mathbf{p}_{i+1}\ new\ search\ direction$$
$$End$$

The new search directions can be computed in two different ways using a minimum residual as in ORTHOMIN or a more general Young type of minimization.

5. GODESS.

The Generic ODE Solving System called GODESS is an Object Oriented software package for the solution of Initial Value problems. It was developed as a platform for testing ODE/DAE methods for performance and robustness under ceteris paribus conditions. From being a benchmarking tool it has developed into a platform for testing new ideas and algorithms as well as an Object Oriented platform for development of simulation software. The different methods of which there are many to choose between can be specified at the time of execution, in the present version any Runge-Kutta or multistep method can be specified. The methods may be tested in identical environments on the same problems whereby direct comparisons between methods is possible.

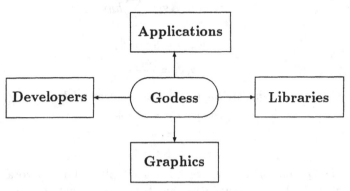

Godess - structure diagram

The diagram represents the way that Godess should be understood and put to work. Godess is a kernel of objects that can be used to solve problems from many applications which may have seperate user interfaces. It delivers results to a graphical system and makes use of one or more libraries of objects and databases with methods and data. The complex process of developing software from a mathematically specified method entails constructing control structures and objectives, selecting termination criteria for iterative methods, choosing norms and many more decisions. Most software constructors have taken a heuristic approach to these design choices, and as a consequence two different implementations of the same method may show significant differences in performance

The software development process has traditionally been carried out by specialists on the numerical analysis involved and often in academic environments. In the GODESS case this is also true resulting in certain limitations on the accessibility and user friendliness of the package. It is the job for specialists to develop a simulation tool for a particular field of application based on GODESS. It has however been used to develop a general Component Based Modelling System (CBMS) and a Dynamic Analysis package called GODYNA. It lies in our plans for the near future to develop a distilled product that will be easier to use for general development purposes.

The linear algebra libraries available in Godess are based on blas/lapack and has facilities for full, banded and sparse structured matrices. A special library adds the Krylov subspace methods to the selection.

6. ESDIRK-3,4 method

In order for a numerical method to be robust in connection with non-linear systems for ODE's and DAE's a set of requirements must be fulfilled. Among the family of one-step methods it has been found that the socalled ESDIRK methods with stage order higher than 1 are candidates and we are introducing one with four stages of order three and stage order two for this purpose. The method has the Butcher tableau shown below:

Table I. Coefficients for the 4-stage ES-DIRK method of order 3 (Thomsen3,4).

0	0			
$\frac{5}{6}$	$\frac{5}{12}$	$\frac{5}{12}$		
$\frac{10}{21}$	$\frac{95}{588}$	$-\frac{5}{49}$	$\frac{5}{12}$	
1	$\frac{59}{600}$	$-\frac{31}{75}$	$\frac{539}{600}$	$\frac{5}{12}$
y_{n+1}	$\frac{59}{600}$	$-\frac{31}{75}$	$\frac{539}{600}$	$\frac{5}{12}$
$\tilde{y}_{n+1}$	$\frac{96}{600}$	$\frac{6}{75}$	$\frac{4116}{6000}$	$\frac{18}{132}$

The method has been developed with the purpose of deriving a low order robust method for general use when the accuracy requirements are not too severe and has shown in tests to be competitive for this type of application.

7. Examples and results.

A number of tests have been carried out to evaluate the methods that have been implemented.

7.1. OZONE PRODUCTION IN THE STRATOSOSPHERE

The problem is a two-dimensional model of Ozone production consisting of two coupled partial differential equations with chemical reactions.

$$\frac{\partial c^i}{\partial t} = K_h \frac{\partial^2 c^i}{\partial x^2} + \frac{\partial}{\partial z}\left(K_v(z)\frac{\partial c^i}{\partial z}\right)$$

$$+V\frac{\partial c^i}{\partial x} + R^i(c^1, c^2, t) \qquad (i = 1, 2) \qquad (1)$$

$$K_h = 4 * 10^{-6}, \quad K_v(z) = 10^{-8}e^{z/5}, \quad V = 0.01$$
$$R^1(c^1, c^2, t) = -k_1 c^1 - k_2 c^1 c^2 + k_3(t) * 7.4 * 10^{16} + k_4(t)c^2$$
$$R^2(c^1, c^2, t) = k_1 c^2 - K_2 c^1 c^2 - k_4(t)c^2$$
$$k_1 = 6.031, \quad k_2 = 4.66 * 10^{-16}$$
$$k_3 = \begin{cases} exp(-22.62/sin(\pi t/43200)) &, \quad t < 43200 \\ 0 &, \quad otherwise \end{cases}$$
$$k_4 = \begin{cases} exp(-7.601/sin(\pi t/43200)) &, \quad t < 43200 \\ 0 &, \quad otherwise \end{cases}$$

In this model c^1 is the concentration of O_1 while c^2 is the concentration of O_3. The model expresses reaction-transport equations with hirizontal diffusion and advection combined with non-uniform vertical diffusion.

7.2. LOTKA-VOLTERRA IN 2D

The second testproblem is a Predator-Prey model with diffusion effects in two dimensions described by two PDE's.

$$\frac{\partial c^i}{\partial t} = d_i \left(\frac{\partial^2 c^i}{\partial x^2} + \frac{\partial^2 c^i}{\partial y^2}\right) + f^i(c^1, c^2), \quad i = (1, 2) \qquad (2)$$
$$f^1(c^1, c^2) = c^1(b_1 - a_{12}c^2), \quad f^2(c^1, c^2) = c^2(b_2 - a_{21}c^1)$$
$$d_1 = 0.05, \quad d_2 = 1.0, \quad b_2 = -1000, \quad a_{12} = 0.1, \quad a_{21} = -100$$

The two species appear as concentrations c^1 and c^2. The space discretization is by central differences controlled by the number N of boxes in each direction. For testing we have used N= 20.

7.3. RESULTS

The tests have been quite extensive , we will include some of the summarized tables and graphs from using the method shown in the previous section (Thomsen3,4). Key to the tables are the following:

AFE The number of function evaluations
AJE The number of Jacobian evaluations
ALL The number of linear solves
GKD The average dimension of the Krylov subspace
AMV The number of vector matrix operations
ALA The number of operations in the linear algebra in millions of OP's
GNI the average number of Newton iterations per nonlinear system

7.4. STOPPING CRITERION

The first series of tests have been concerned with the effects of varying the stopping criterion on the iterations in the linear solver.

Table II. Testresults from solving testproblem 1 using Thomsen3,4 with varying stopping criteria in the unscaled IOM method.

Tolerance	Steps	AFE	AJE	ALL	GKD	AMV	ALA
10^{-14}	518	3460	93	3974	15.6	62037	1023
10^{-10}	511	3425	102	3932	12.1	47605	631
10^{-8}	518	3464	98	3978	10.1	40430	455
10^{-6}	649	4429	205	5025	7.6	37953	345

Table III. Testresults from solving testproblem 1 using Thomsen3,4 with varying stopping criteria in the scaled IOM method.

Tolerance	Steps	AFE	AJE	ALL	GKD	AMV	ALA
0.001	518	3446	104	3960	4.72	18721	115
0.01	507	3388	112	3891	4.14	16129	89.7
0.1	517	3434	110	3952	3.69	14333	69.3
1.0	515	3433	105	3944	3.16	12466	51.6
10.0	502	3363	104	3161	2.70	10420	37.9
100.0	660	4246	157	4902	2.14	10519	30.9

7.5. RESTARTING

Table IV. Effects of restarting when solving testproblem 1, using Thomsens method. Unscaled IOM using abs. tolerance 10^{-8}. The first column indicates how many iterations that were used before the restart.

Restart	Steps	AFE	AJE	ALL	GKD	AMV	ALA
∞	527	3527	109	4049	9.8	39874	439
8	552	3687	118	4262	11.8	49852	270
5	529	3539	104	4064	12.7	51723	189

7.6. TRUNCATION

The effect of truncating the Krylov series before full dimension is reached has been tested.

Table V. Effect of truncating when solving testproblem 1 using ESDIRK45 and scaled IOM with tolerance 0.1. The first column indicates how many orthogonal vectors that were used.

Ort. Vec.	Steps	AFE	AJE	ALL	GKD	AMV	ALA
∞	250	3407	80	3404	3.03	10312	41.7
3	236	3414	82	3411	3.06	10427	33.9
2	278	3932	104	3924	3.29	12928	45.1

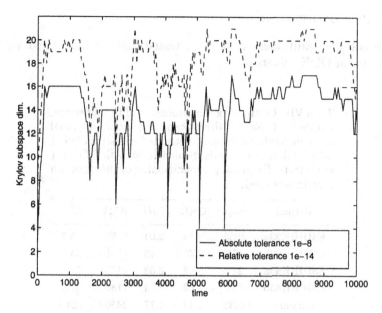

Figure 1. Stepsize for two different tolerances in IOM on integrating testproblem 1 using Thomsens method. TOL is here stopping criterium δ.

7.7. DIFFERENT KRYLOV SUBSPACE METHODS

A comparison of the different Krylov methods is interesting. For the Ozone production problem we have found the following results.

Table VI. Comparison of scaled Krylov subspace methods using diagonal prekondisjoning solving testproblem 1 by ESDIRK45b.

Method	Steps	AFE	AJE	ALL	GKD	AMV	ALA
IOM	237	3245	83	3242	3.62	11731	57.0
GMRES	235	3213	77	3210	3.57	11445	55.6
Orthomin	244	3336	74	3333	3.56	15182	54.2
Young	235	3213	77	3210	3.57	22896	47.5

7.8. PRECONDITIONING

Different preconditioners have been tested and we have tried with a selection of ODE-solvers.

Table VII. Comparing the dimension of the subspace, number of Newton-iterations and computational work in Krylov subspace methods for testproblem 1 using different methods. Scaled IOM with TOL 0.1 in all tests. Diagonal prekonditioning of the iteration matrix was used.

Method	Steps	GKD	GNI	AMV	ALA
ESDIRK23a	1052	2.56	2.04	16505	55.7
ESDIRK23b	865	2.75	2.10	15015	54.5
ESDIRK45a	257	3.62	2.53	11731	57.0
ESDIRK45b	239	3.62	2.74	11800	57.7
Hairwann	1430	2.49	2.77	34500	123.2
sd34var	414	3.02	2.66	14800	57.7
Thomsen	515	3.16	2.55	12466	51.1
BDF	610	2.61	2.18	3476	10.5

The average dimension of the search subspace varies with the preconditioner .

Table VIII. Average dimension of the search space with and without diagonal preconditioning.

Testproblem	GKD no prekond.	GKD med diagonal prekond.
1	3.03	2.50
2	5.95	5.80
3	6.67	3.78

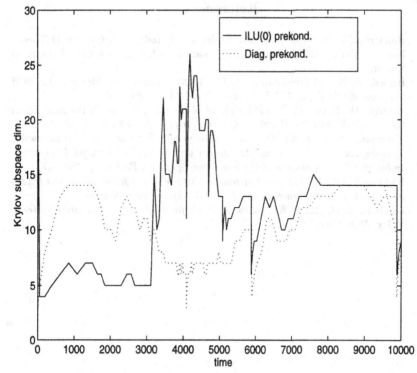

Figure 2. Subspace dimension for truncated Krylov series as function of timetwo types of preconditioners for problem 1.

8. Conclusion.

In conclusion it has been shown that Krylov subspace methods are a relevant alternative to traditionally applied direct solvers in connection with the solution of large systems of ODE's like those appearing in air pollution modelling.

9. Acknowledgements

Nils Henrik Bjurstrøm was on leave from the Deprtment of Mathematics (NTNU, Trondheim, Norway) when this research has been carried out.

References

1. Bjurstrøm, N. H. (1997), *Krylov Underrum Metoder i Godess*, Ph D Thesis, Inst. for Mathematical Modelling, Denmark's Technical University, Lyngby, in Danish.
2. Brown, P. N and Hindmarsh, A. C. (1986), Matrix-Free Methods for Stiff Systems of ODE's, *SIAM J. Numer. Anal.*, **23**, pp. 610-638.
3. Arnoldi, W. E. (1951), The Principle of Minimize Iterations in the Solution of the Matrix Eigenvalue Problem, *Quart.Journ.Appl.Math.*, **9**, pp. 17-29.
4. Thomsen, P. G. (1996), Dynamic Systems and Software, in J. Wasniewski, J. Dongarra, K. Madsen and D. Olesen (eds.), *Applied Parallel Computing. Industrial Computation and Optimization*, Springer, Berlin, pp. 661-667.
5. Kværnø A. (1992), *The order of Runge Kutta Methods for semi-explicit DAE's of index 1*, Report Numerics no. 2, NTNU, Trondheim, Norway.
6. Lambert, J. D. (1991) *Numerical Methods for Ordinary Differential Systems*, Wiley, New York.

THE USE OF 3-D ADAPTIVE UNSTRUCTURED MESHES IN AIR POLLUTION MODELLING

A. S. TOMLIN [1], S. GHORAI [1], G. HART [1] and M. BERZINS [2]
[1] Dept. of Fuel and Energy and [2] School of Computer Studies,
Univ. Leeds, Leeds, LS2 9JT

Abstract

High resolution models of air pollution transport and transformation are necessary in order to test possible abatement strategies based on pollution control and to forecast high pollution episodes. Models are especially relevant for secondary pollutants like ozone and nitrogen dioxide which are formed in the atmosphere through nonlinear chemical reactions involving primary pollutant species often far from their sources. Often we are trying to resolve the interactions between plumes from point sources such as power stations and regional pollution tides of ozone formed in other European countries. One method of tackling this problem of different scales is to use different grid sizes, using highly resolved grids in regions where the structure is very fine. Telescopic gridding is currently used in high emission areas or around sensitive receptor points. However, since meteorological conditions vary, this method cannot resolve a priori highly structured regions away from sources, e.g. along plumes. Such refinement can be achieved using adaptive methods which increase resolution in regions of steep spatial gradients. This paper describes the use of 3-D adaptive gridding models for pollution transport and reaction using both a layered and a fully adaptive 3-D tetrahedral approach. Examples which show the effect of grid resolution on secondary pollutant formation will be shown.

1. Introduction

One of the greatest numerical challenges in air pollution modelling is to achieve a high resolution solution without over-stretching current computational resources. This is a difficult task when often we are considering large numbers of species and many different scales of source types. An obvious way to tackle this problem is to concentrate the computational grid in regions where we gain accuracy from doing so, and to use a coarse mesh elsewhere, thus reducing the total number of solution nodes. Adaptive gridding techniques have been developed as an attempt to automate the grid refinement process so that a priori decisions need not be made about where to place extra mesh elements.

Z. Zlatev et al. (eds.), Large-Scale Computations in Air Pollution Modelling, 339–348.
© 1999 Kluwer Academic Publishers. Printed in the Netherlands.

The effect of grid resolution on solution accuracy has been considered for advection schemes previously. In the context of air pollution modelling we are interested in two main areas - local concentrations and regional species budgets of reactive pollutants. This paper will aim to address a number of questions related to these two areas in the context of the effect of grid resolution in models describing the transport of chemically reacting species.

2. Model Equations and Solution Strategy

The atmospheric diffusion equation in three space dimensions is given by:

$$\frac{\partial c_s}{\partial t} = -\frac{\partial(uc_s)}{\partial x} - \frac{\partial(vc_s)}{\partial y} - \frac{\partial(wc_s)}{\partial z}$$

$$+\frac{\partial}{\partial x}\left(K_x\frac{\partial c_s}{\partial x}\right) + \frac{\partial}{\partial y}\left(K_y\frac{\partial c_s}{\partial y}\right) + \frac{\partial}{\partial z}\left(K_z\frac{\partial c_s}{\partial z}\right) \tag{1}$$

$$+R_s(c_1, c_2, ..., c_q) + E_s - (\kappa_{1s} + \kappa_{2s})c_s,$$

where c_s is the concentration of the s'th compound, u,v,w are wind velocities, K_x, K_y and K_z turbulent diffusivity coefficients and κ_{1s} and κ_{2s} dry and wet deposition velocities respectively. E_s describes the distribution of emission sources for the s'th compound and R_s is the chemical reaction term which may contain nonlinear terms in c_s. For n chemical species an n-dimensional set of partial differential equations (p.d.e.s) is formed describing the rates of change of species concentration over time and space, where each is coupled through the nonlinear chemical reaction terms.

Two approaches have been adopted in the solution of the equation system. The first is to restrict the solution of equation (1) to 2 dimensions and describe the vertical transport using a parametrised approach similar to that used in the LOTOS model and described by van Loon[1]. This is essentially a 2-D approach using a triangular mesh but with 4 vertical layers describing the surface, mixing, reservoir and upper layers of the troposphere. The equations are discretised on a triangular unstructured mesh using the finite volume method of SPRINT-2D described in detail in (Berzins et al, 1992, 1995) [2, 3] and Tomlin et al. [4, 5]. Although the mixing layer height is diurnally varying the number of vertical layers remains at 4 and grid refinement is only possible in the horizontal direction (the toblerone approach). Operator splitting here is achieved at the level of the solution of the nonlinear equation system resulting from the method of lines, thus reducing splitting errors (Tomlin at al 1997). The second approach uses a fully 3-D unstructured mesh based on tetrahedral elements. A cell vertex finite volume scheme has been chosen so that the number of mesh elements and therefore flux calculations can be reduced when compared to a cell centred scheme. The dual mesh is constructed by dividing each tetrahedron into four hexahedra of equal volumes, by connecting the mid-edge points, face centroids and the centroid of the tetrahedron. The flux evaluations are cast into a edge based operation. The advective flux is discretized using a 2nd order upwind scheme with limiter proposed by Barth and Jesperson[6]. This scheme first performs a linear reconstruction to interpolate data to the control volume faces. Monotonicity principles are enforced to ensure that the reconstructed values are bounded by the values of a cell and its neighbours. To this end, multi-dimensional limiter functions are used. The diffusive term is discretized using a central differencing scheme. Operator splitting is here done using a standard approach so that the chemistry and transport steps are treated separately.

3. Mesh Generation and Adaption

3.1. THE MESHES

The choice of an unstructured mesh over a regular Cartesian mesh has been made so that resolution of small scale structures such as those due to a point source can be achieved even in a large domain. Early tests using regular meshes revealed that the same levels of refinement cannot be achieved as with the unstructured approach without refining large areas of the total domain. There is however an overhead resulting from the more complex data structures required for an unstructured mesh, and from the complexity of the description of the numerical scheme.

The initial unstructured triangular meshes used in SPRINT-2D are created from a geometry description using the Geompack (Joe and Simpson, 1991) [7] mesh generator. The initial tetrahedral meshes are generated by dividing the whole region into cuboids and then subdividing a cuboid into 6 tetrahedral elements. Extra mesh points are placed in the lower regions of the boundary layer.

Local refinement of the mesh is then achieved by subdividing the triangles or tetrahedra using data structures to enable efficient mesh adaptation. For triangles regular subdivision results in 4 sub triangles but this may leave some nodes which are unconnected on the edges of the refinement region. These nodes are removed by "Green" refinement which divides triangles into 2. Figure 1 below demonstrates regular and "Green" refinement for the tetrahedral mesh. Green refinement is used on the edges of the adapted region so that hanging nodes are removed and each node is connected in the mesh.

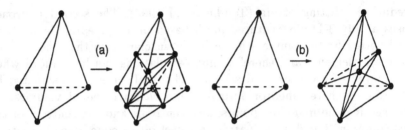

Figure 1. (a) Regular refinement based on the subdivision of tetrahedra by dissection of interior diagonal (1:8) and (b) "Green" refinement by addition of an interior node (1:6)

3.2. ADAPTION CRITERIA

Two methods of adaption criteria have been implemented. The first calculates a first and second order solution and bases the current spatial error on the difference between the two. The spatial error for the next time-step is then predicted using interpolation and compared to a user defined tolerance level. In regions where the spatial error is predicted to exceed the tolerance level the mesh element is flagged for refinement. The details of this procedure are outlined in Tomlin et al (1997). The second is somewhat simpler and is based on the calculation of the solution gradient between neighbouring nodes. Both techniques show similar refinement behaviour and adapt in regions of steep spatial gradients. For reactive problems we are interested in a number of species and a user decision has to be made about which species or group of species to refine around. Tolerance values are then set by the user for a single species or a group of species. The results shown here use gradients in total NOx (i.e. $NO + NO_2$) as the criteria and this seems to give good resolution of structures such as plume characteristics.

4. Test cases

The following test cases have been designed to investigate the effects of mesh resolution on the solution for situations representative of those found in regional scale air pollution models.

4.1. EFFECTS OF HORIZONTAL REFINEMENT FOR THE LAYERED MESH

As emissions inventories improve it will become common to represent data as point, line and area sources on a regional basis. It is therefore important to investigate the effects of mesh resolution on the representation of transport and reaction from various source types. The following results show the effect of resolution for the transport and

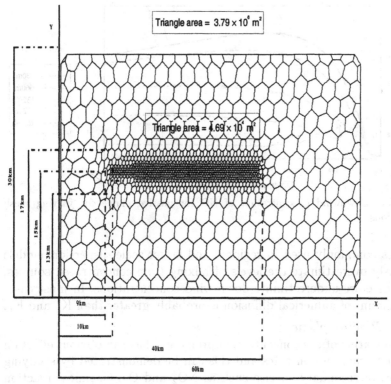

Figure 2. Dual mesh for passive dispersion from a single NOx source showing refinement along plume centre

reaction from a single NOx source through a background of VOC's and ozone.

4.1.1. *Passive dispersion*

If we neglect chemical reaction terms we first see the effect of the mesh on turbulent transport. Figure 2 shows the dual mesh for the refined simulation.

Figure 3 shows the down-wind concentrations of NO for different levels of mesh refinement ranging from 300m up to 10km in edge length. The difference in NO concentrations between the high and low resolution simulations is around a factor of 4 at a distance of 30 km from the source.

The reason for this high level of error can be attributed to high levels of numerical diffusion from the coarse mesh. A simple calculation based on the determination of plume width for different meshes and a Gaussian approach demonstrates that the numerical diffusion for the coarsest mesh is in fact 8 times that represented by K_x and K_y. The conclusion from this simple test case must be that the types of

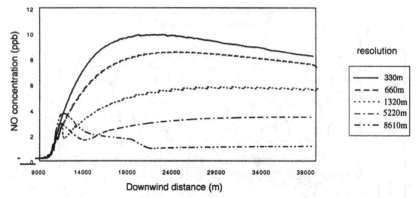

Figure 3. Downwind concentration of NO for passive plume for varying mesh resolutions

meshes commonly used in regional and larger scale dispersion models seriously overestimate the levels of mixing from concentrated sources, and therefore underestimate down-wind concentrations. In most cases the amount of numerical diffusion is probably greater than K_x and K_y.

4.1.2. *Reactive plume*
This smearing of the concentration profile will have an obvious effect on the chemical reaction rates and this can be demonstrated by studying profiles of secondary species such as NO_2 and O_3. Chemical reaction terms in the model are represented by both the GRS [8] and a carbon bond type scheme [9] showing similar local concentrations for O_3. The local concentrations of ozone are clearly affected by grid resolution as shown in Figure 4 as are the total species budgets for ozone across the domain although space does not permit a demonstration of this effect.

4.1.3. *Regional scale model*
The results from the layered model clearly demonstrate the effects of mesh resolution on secondary species predictions. Regional scale comparisons have also been made using the same approach and multi-scale emissions data - EMEP emissions data on a European scale and point and area emissions data (10x10km) from NETCEN for the UK region. Dry deposition and vertical dispersion are included following the approach of van Loon [1]. The meshing approach here has been to use a large scale grid over Europe in order to provide boundary conditions for the regional scale model. A nested region with some refinement has been defined over the UK and transient refinement to finer meshes is allowed in this region according to the spatial error of the solution. A maximum refinement level has to be provided by the user since point sources will otherwise reach high gradients leading to further refinement requirements. Usually there are about 5-6 levels of

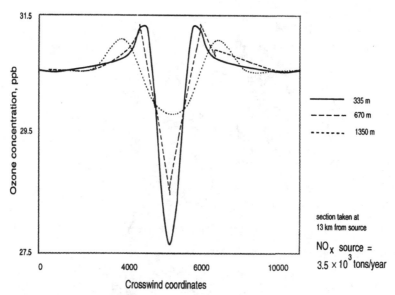

Figure 4. Cross plume ozone profile

refinement from the European to the local scale with mesh elements ranging from 150km to 2km in terms of edge length. Space does not permit a detailed discussion of the effects of mesh on for the regional scale model. In essence the results are similar to the simple test case. Refinement generally takes place in regions with high spatial gradients such as along the edges of power station and urban plumes. Figure 5 shows a typical mesh for a regional scale calculation.

Local concentrations are significantly affected by the mesh resolution so that any comparison with measured data will be better achieved using better resolved meshes. As in the simple test example there is some effect on the total ozone concentration integrated over the entire refinement region especially when extremely coarse meshes are used. The effect is not as significant however as for local concentrations, suggesting that coarse resolution models are a reasonable way to estimate general ozone trends over large time-periods but will not be suitable for local ozone forecasts and for comparison with monitored data, particularly in areas where many sources of different scales are present.

5. Vertical Refinement using tetrahedral meshes

The layered model uses highly parametrised description of mixing in the vertical direction. This will have a large effect on near source concentrations as enhanced mixing caused by coarse layering will change the vertical profile of concentrations. A tetrahedral model has therefore

Figure 5. Typical mesh used in the refinement region for the UK.

been developed which can adapt in the vertical plane. The advection scheme has first been tested using a standard rotation problem.

5.1. MOLENKAMPF TEST

To verify the advective scheme using a limiter, we used the linear advection equation of Molenkampf[10]. The dimensions of the base grid in the horizontal, lateral and vertical directions are 60 km, 15 km and 3 km respectively. The initial conditions describe a sphere with centre at $(x_0, y_0, z_0) = (30, 3, 0.3)$ km,

$$c(x, y, z) = 10^{11} \exp(-\kappa[(x - x_0)^2 + (y - y_0)^2 + 100(z - z_0)^2]),$$

where $\kappa = 7.5 \times 10^{-11}$ and c is in molecules cm^{-3}. The test equation is

$$\frac{\partial c}{\partial t} + (y - y_1)\omega\frac{\partial c}{\partial x} + (x_1 - x)\omega\frac{\partial c}{\partial y} = 0,$$

where $\omega = 10^{-4}$ and $(x_1, y_1) = (30, 7.5)$ km. The period of the rotation is approximately 17.44 hours. The initial grid consisted of 4,961 tetrahedra and the final grid after approximately one rotation contains

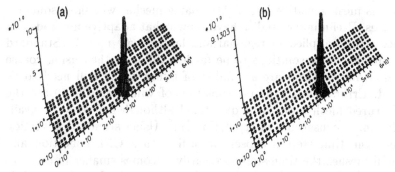

Figure 6. (a) Initial concentration of peak and (b) Peak concentration after one rotation

11,638 tetrahedra. The grid is refined by 3 levels if the concentration gradient and concentration values exceed some tolerance parameters.

Figure 6 shows the peak concentrations at t= 0 and after one rotation of the sphere. The final peak is about 91% of the original showing that little numerical diffusion is taking place. The Molenkampf problem demonstrates the importance of using a safety layer around the high resolution region. This means that as well as adapting mesh elements of high spatial error their neighbours are also refined. The advantage of this is that as the region of steep gradients advances due to advection, the cells ahead of the front are already of high resolution and the plume front is not diluted into large cells. Without the safety layer the mesh remains coarse since as the steep gradient is advected into a large grid cell, the solution is smeared and the gradients become low. Further refinement will then be prevented. This might imply that the best strategy to start a model with multiple source points and a complex initial concentration field would be to use a refined mesh and then coarsen the mesh as the solution proceeds where the gradients are low. This way it will be certain that initial steep gradients are maintained throughout the simulation.

A small amount of vertical numerical diffusion resulting from a horizontal advection scheme is a side effect of using fully unstructured meshes. As Cartesian meshes are usually aligned to the surface we would not expect them to demonstrate this problem for modelling dispersion over flat terrain. However, the results show that the total overall diffusion can be kept small if sensible mesh refinement is used.

6. Discussion and Conclusions

The test cases studied clearly demonstrate the need for finer meshes when predicting local concentrations. For ozone, total species budgets are not as mesh sensitive but highly coarse meshes will show some inaccuracies. Preliminary studies have shown that adaptive methods can be successfully applied in regional and larger scale models. Standard test problems show solutions to be far less diffusive then using coarse meshes. However there are a number of issues which still need to be resolved. Errors due to the interpolation of wind field data onto the unstructured mesh need to be quantified although methods are available for mass conserving interpolation [11]. Using standard advection schemes the time-step is chosen according to a CFL condition and for small meshes the time-step necessarily becomes smaller in order to preserve stability. There are limits to the amount of refinement possible therefore if long-time runs are to be carried out. These are solvable problems however and adaptive methods would seem to be well suited for air pollution problems.

References

1. van Loon, M. (1996), *Numerical Methods in Smog Prediction*, Ph D thesis, CWI, Amsterdam.
2. Berzins, M., Lawson, J. and Ware, J. (1992), Spatial and Temporal Error Control in the Adaptive Solution of Systems of Conversation Laws, In R. Vichnevetsky, D. Knight, and G. Richter (eds.), *Advances in Computer Methods for Partial Differential Equations VII*, IMACS, pp. 60-66.
3. Berzins, M. and Ware, J. M. (1995), Positive cell-centred finite volume discretization methods for hyperbolic equations on irregular meshes. *Appl. Numer. Math.*, **16**, pp. 417-438.
4. Tomlin, A. S., Berzins, M., Ware, J., Smith, J. and Pilling, M. J. (1997), On the use of adaptive gridding methods for modelling chemical transport from multi-scale sources, *Atmos. Environ.*, **31**, pp. 2945-2959.
5. Hart, G., Tomlin, A. S., Smith, J. and Berzins, M. (1999), Multi-scale atmospheric dispersion modelling by use of adaptive gridding techniques. *Environmental Monitoring and Assessment*, in press.
6. Barth, T. J. and Jesperson, D. C. (1989), The Design and Application of Upwind Schemes on Unstructured Meshes, AIAA-89-0366, Jan. 9-12.
7. Joe, B. and Simpson, R. B. (1991), Triangular meshes for regions of complicated shape, *Int. J. Numer. Meth. Eng.*, **23**, pp. 987-997.
8. Venkatram, A., Karamchandani, P., Pai, P. and Goldstein, R. (1994), The development and application of a simplified ozone modelling system (SOMS), *Atmos. Environ.* **27B**, pp. 3665-3678.
9. Heard, A. C., Pilling, M. J. and Tomlin, A. S. (1998), Mechanism reduction techniques applied to tropospheric chemistry, *Atmos. Environ.*, **32**, pp. 1059-1073.
10. Molenkampf, C. R. (1968) Accuracy of finite-difference methods applied to the advection equation, *J. Appl. Meteor.*, **7**, pp. 160-167.
11. Mathur, R. and Peters, L. K. (1989), Adjustment of wind fields for application in air pollution modelling, *Atmos. Environ.*, **24A**, pp. 1095-1106.

ATMOSPHERIC ENVIRONMENTAL MANAGEMENT EXPERT SYSTEM FOR AN OIL-FIRED POWER PLANT

E. A. UNZALU

Environmental Laboratory
LABEIN. Technological Research Centre
Bilbao, Basque Country, Spain

Abstract

The Atmospheric Environmental Management Expert System, actually in operation in a 1,000 MW Power Plant, reckons in real time the atmospheric impact caused by the Plant in the affected surrounding area, an urban and industrial zone of complex terrain, located in an estuarine valley by the Atlantic Ocean, in the Bay of Biscay's nook.

The expert system is based on a real time reception of meteorological and SO_2 emission data, which are used to automatically select the most appropiate meteorological and dispersion models from the set implemented in the system, and to execute them in order to estimate the atmospheric impact caused by SO_2. It includes the automatic calculation of dispersion parameters such as atmospheric stability and mixing height , by means of advanced methods implemented in the software.

Along the project, the implemented meteorological and dispersion models have been calibrated and validated through several multidisciplinary experimental campaigns.

Keywords: Atmospheric Environmental Management, Expert System, Oil-fired Power Plant.

An earlier version of this article has been published in the 3rd International Conference on Air Pollution 95, held in Greece, "Atmospheric environmental expert system for the operation of a powerplant" Picazo M. Air Pollution III, Volume 2, pp. 501-508. Editors: H. Power, N. Moussiopoulos, C. A. Brebbia. Computational Mechanics Publication.

Z. Zlatev et al. (eds.), Large-Scale Computations in Air Pollution Modelling, 349–358.
© 1999 *Kluwer Academic Publishers. Printed in the Netherlands.*

1. Introduction

Generally, potential atmospheric environmental impact is attributed to the electricity generation and mainly to the related oil-fired power plants. Therefore the efforts, devoted to achieve any improvement in the design and implementation of the systems dedicated to the environmental management systems for these type of plants, are well worthy. This has been the main scope of this Expert System design and implementation.

Furthermore, it is said that the achievement of this Expert System has joined under a common target activities of complementary disciplines such as Meteorology, Modelling and Information Technologies, that cover in this case and among other ones, fundamental functions as data acquisition and intermodular and overall communications.

1.1. GENERAL CHARACTERISTICS

This Expert System has been set up in one specific Oil-fired Power Plant, consisting of two groups of 541 MW and 377 MW of installed power. The Plant is located on the Bay of Biscay seashore, in a central location of the Bilbao city connurbation. In fact the Expert System provides one continous surveying of the atmospheric impact caused by the SO_2 emitted from the Power Plant, in its surronding area of 1,225 sq.km.

The peculiar location of the Plant, a coastal zone with strong land-sea interactions and complex land morphology , as well as the existence of multiple pollutant sources in the area, due to its great industrial and commercial activity of the area, gives an outstanding importance to the correct performance of the system.

Furthermore, also the existence of one extensive SO_2 Monitoring Network in the area has contributed in a good way to develop the whole project very satisfactorily.

2. Project Development

The complete project has been developed in practically 6 years(1991- 1997), and its structural scheme was configurated in three main sections, all of them very crucial for the reliability of the whole system:

1. Analysis and Characterization of the Historical Data, Meteorological Parameterization and Selection of Dispersion Models.

2. Evaluation of the Models, covering Validation and Sensitivity.
3. Expert System : Design of the Architecture, Development, Test and Setting-up.

2.1. METEOROLOGICAL FEATURES

On the basis of the historical meteorological data of the area, covering a 15 year period , as registered by the existing Meteorological Network of the Metropolitan Bilbao Area, it has been possible the parameterization of the atmospheric situations, based on the wind velocity and direction, permitting in this manner the selection of a set of patterns. Listed in accordance with their occurrence frequency the selected patterns are : Northwest, Calm, Drainage 1, Drainage 2, Breeze, West, South 1, South 2, East 1, East 2 and Northeast.

The most representative situation, Northwest, reaches a frecuency of 30 %. When that occurs the stack plume is directed to the most populated zone of the conurbation ,that is, central Bilbao .

2.2. EXPERIMENTAL CAMPAIGNS

In order to obtain experimental data from the lower atmosphere and the real SO_2 impact caused by the Power Plant, a set of four field cases or experimental campaigns have been carried out in the area. These campaigns have served as well to calibrate the models , to study their sensitivity and the last one, practically developed under dominant winds, that is, the meteorological situation corresponding to Northwest pattern has been used to accomplish the validation for this prevalent situation in the area.

The equipments used in the field cases have been the following:
• Meteorological Towers
• SO_2 Monitors
• Tethered balloons.
• Free ballon stations and theodolites
• Sodars
• Instrumented Vehicle for SO_2 ground level and overhead concentrations

3. Systems Specifications and Components

As mentioned above, the Expert system was conceived first of all, to give practically in real time the numerical and graphical reliable calculation of the Power Plant atmospheric impact, meaning at the same time that required calculation time on the basis of real time meteorological and emission data acquisition was to be reasonably low. For the fulfilment of these purposes, and taking also into account the array of equipments and models offered by the market the selected ones were as follows:

3.1. HARDWARE & SOFTWARE

3.1.1. Hardware features:
- Intergraph 6880314 Model Workstation
- Processor and Memory
        - RISC Microprocessor
        - RAM 64 Mb Memory
        - 2 Gb H.D.
        - 85 Mips
        - 16.3 Mflops
- EDGEII Graphical Sistem
- CD-ROM Unit
        - 600 Mb
- 19" Single Monitor (1 Mpixel   resolution / 76 Mhz)
- 2.3 Gb. Cartridge Tape
- HP Paintjet XL300 color Printer

3.1.2. Software features:
- UNIX O.S.
- Intergraph MGE Geographical Information System, with the following modules:
        - Microstation 32
        - Relational Interface System   (RIS)
        - Modular System Nucleus   (MGE/SX).
        - MGE Terrain Modeler
        - MGE Grid Analyst
        - MGE DTM Development   Platform (MDDP)
        - Microstation 32 Customer   Support Library (MCSL)
        - MDL (Microstation   Developopement Language)
        - ORACLE Data Base

- C and Fortran 77 Compilers

3.2. LIBRARY OF MODELS

Models have been selected as explained below:

Meteorological Models   **Dispersion Models**

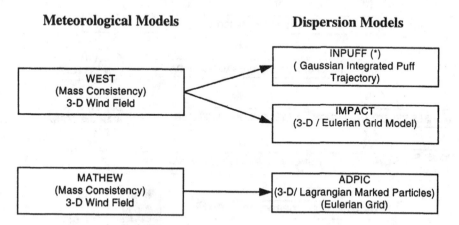

Models resolution corresponds to x: 0.5 Km, y: 0.5 Km and z: 0.5 Km.
(*) Modified, including topographical adjustment.

3.3. SYSTEM OPERATIONAL SCHEME

The modular structure of the Expert System makes feasible the updating of its components, that is, algorithms of calculation, meteorological and dispersion models, data bases etc, favourizing its application for whichever location.

The multiprocess and continous execution of this modular system, covers 5 modules: Data Acquisition and Depuration, Meteorological Classification and Models Selection, Models Execution, Graphical Post-Process and User's Interface.

The system reckons automatically two of the most important parameters for the plume dispersion, such as the atmospheric stability and the mixing height, on the basis of the available meteorological data in real time, for the concerned hour, using the implemented routines and programmes . Operational ranking preference for these calculations is given to the Similarity Theory, covering both thermal and mechanical turbulences, followed by Sodar dedicated own programmes and Bulk Richardson number method. Furthermore, the system is provided with a set of default values based on local historical data.

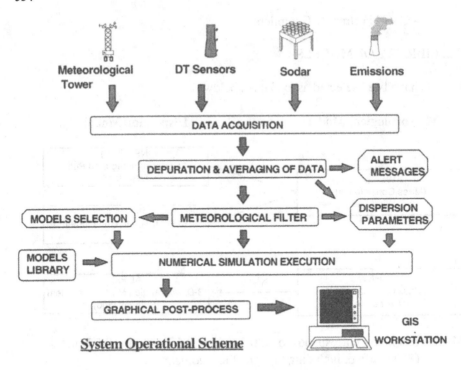

<u>System Operational Scheme</u>

4. System Validation

As said before, the set up cases have served , mainly the last one, to validate the global system and everyone of its components, including modular communications, software programmes and obviously the meteorological and dispersion models. Due to the ocurring atmospheric conditions through the field campaigns, the validation of two meteorological situations corresponding to Northwest winds, that is, the fourth quadrant dominant situation and the calm period have been carried out.

The results obtained by the validation present a good correlation of estimated and measured values for the set of the Network central stations, with a global coefficient of r = 0.55, even more for some stations the correlation coefficient reaches up to r = 0.76, denoting a very acceptable system efficiency.

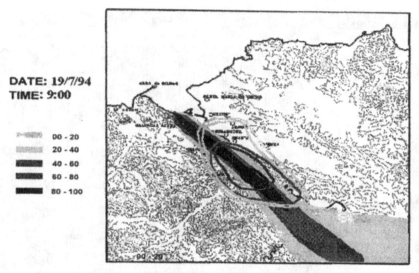

DATE: 19/7/94
TIME: 9:00

00 - 20
20 - 40
40 - 60
60 - 80
80 - 100

Comparison of estimated & measured SO$_2$ concentrations

5. Monitor Graphs & Pictures

As represented below, the System monitor presents hourly the picture of the ground level SO$_2$ concentrations and wind field at the plume transport height. Furthermore, as complementary information is available a complete set of options , such as: Grill Cell Examination, Zooms, SO$_2$ concentration value for each cell, etc.

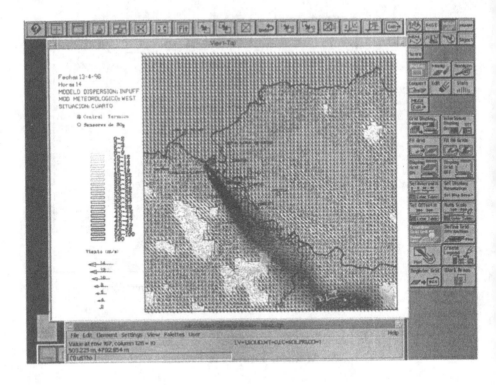

6. User's Interface

In order to give complementary information of the upmost interest for the analysis of any historical situation, the Expert System has been endowed with the module named "User's Interface". By means of this a set of additional actions, as listed below, can optionally be accomplished:

- Visualization of any previous simulation.
- Four cross sections of the plume SO_2 concentrations.
- Visualization of wind fields and SO_2 concentrations for levels other than the automatic hourly picture and their respective zooms.
- 3D plume plotting and zooms.
- Data treatment for historical simulations (back-up, restore and restored erased).
- Treatment of historical depurated data (back-up, restore and restored erased).

- Selection of the Power Plant Operational Unit, including Stop option.
- Management Expert System switch-off.

7. Concluding Remarks

As a reliable tool, this Expert System obviously permits the Power Plant Management to act in the correct way to accomplish any temporary emission limits, that occasionaly could occur under peculiar atmospheric conditions hampering good pollutant dispersion, as was the case of several European and Latin - American cities in the years 1997/ 1998, normally with occurrence under strong high pressure and calm atmospheric conditions.

It makes possible as well, that in the future, the system duely provided with the suitable forecast meteorological models, can obviously be used for the purpose of the pollution impact forecast in the area.

As a summary of the actual and further system application can be listed the following:

- Integration of photochemical modeling , mainly NO_x and VOC_s , into the actual System.
- Development of Expert Systems for the Automatic Operation of Power Plants on the basis of predicted atmospheric impacts and applicable legislation.
- Know-how transferable to whichever utilities or industries, previously adapted to their features and locations.
- Integration of the Expert System into the air quality surveillance official networks.

Acknowledgements

OCIDE and Iberdrola, owner of the Power plant, are greatly aknowledged for financial support of the project where the Expert System has been set up and presently is in operation.

Very special gratitude is extended to all project participants from LABEIN, IBERDROLA, ASINEL and NILU, and for the kind disposition received from the Basque Autonomous Government, providing free access to Historic Data Bases and the Lower Nervion/ Ibaizabal Monitoring Network.

358

References

1. Sozzi, R. (1994), *Il Planetary Boundary Layer : Metodi Per la Stima Della Turbolenza*, Informe Técnico SA-RI-06/94, Servizi Territorio s.c.r.l.
2. Foster, C. (1992), *Users Guide to the Mathew/Adpic Models*, University of California Press, U.S. Department of Energy document, UCRL-MA-103581.
3. Thykier-Nielsen, S. and Mikkelsen, T. (1992), *Fitting of Pre-calculated Wind Fields*, Risø National Laboratory, Roskilde, Denmark.
4. Stull, R. (1989), *An Introduction to Boundary Layer Meteorology*, Kluwer Academic Publishers, Dordrecht, The Netherlands.
5. Petersen, W. (1986), *Inpuff, A Multiple Source Puff Model*, Trinity Consultants Inc. Dallas.
6. Panovsky, H. and Dutton, J. A. (1983), *Atmospheric Turbulence : Models and Methods for Engineering Aplications*, Wiley, New York.
7. Hanna S. R., Briggs G. A. and Hosker R. P., Jr. (1982), *Handbook on Atmospheric Diffusio*, Technical Information Center, U.S.Department of Energy, Washington.
8. Fabrick, A. and Haas, P. (1980), *Users Guide to Impact*, Radian Corporation, DCN 80-241-403-01.

A COLLABORATIVE FRAMEWORK
FOR AIR POLLUTION SIMULATIONS

E. VAVALIS
Purdue University, Computer Science Department
W. Lafayette, IN 47907, U.S.A.

1. Introduction

Simulation, now well-accepted as the third paradigm of science in support of theory and experimentation, holds great promise as a powerful tool for modeling complex systems. Simulating the pollution in the atmosphere is a big, difficult and scientifically interesting problem with substantial impact to both the national economy and the public health. The air pollution process is very complex and several important aspects of it are understood vaguely, at best. As such its modeling requires substantial computational and experimental facilities to support theoretical and experimental researchers.

The main objective of our study is to propose a general approach to developing a computational methodology and a software environment that offers efficient, portable support for scalable collaborative Long Range Air Transport Pollution (LRTAP) [14] simulations on heterogeneous network platforms. We are interested in collaborative operations of two kinds:

- Cooperative operations between distinct but neighboring LRTAP models (simulation engines).

- Collaboration between users and/or modelers who interact with each other during model development, experimentation, validation and "production mode" operation.

The rest of this note is organized as follows. In section 2 we present our motivation and we argue the necessity of collaborative air pollution models. In section 3 we discuss ways to apply an agent–based approach to LRTAP modeling and in section 4 we describe the agent and runtime support needed for the efficient implementation. Section 5 contains our conclusions.

Z. Zlatev et al. (eds.), Large-Scale Computations in Air Pollution Modelling, 359–367.
© *1999 Kluwer Academic Publishers. Printed in the Netherlands.*

2. Motivation

We start by stating the main objective in LRTAP simulation area

> *Build a physically detailed and computational efficient air pollution model*

which in many cases is unfortunately not feasible to accomplish since: The critical factors leading to the formation of many pollutants are not fully understood, the pollutant dynamics are complicated, the mathematical formulation is difficult, the code implementation/maintenance is very costly and the computational complexity is high. It is worth to point out that the computational complexity of a LRTAP third generation model [4] mainly depends on:

- The length of the period of the simulating episode.

- The finesh of the 3–dimensional discretization of the air region.

- The number of spices and the number of chemical reactions considered.

It is safe to predict that the CPU requirements for a typical air pollution "production run" will very soon be in the range of teraFLOPS.

The above led scientists to adopt either of two suboptimal approaches by using: a simple global model patched by source terms and ad hoc post-processing interpretations, or a complex global model with limited resolution.

Additional difficulties come from the facts that: We often lack a central database that accommodates the various meteorological, physical and geographical data that are extremely crucial in simulating air pollution. The necessary coordination between local authorities/scientists is usually very time consuming in particular when the LRTAP simulation involves several different countries. The above two difficulties are well capable of proving any model/code useless in the case when immediate action is needed (e.g. simulating or predicting huge environmental catastrophes, like the Chernobyl accident).

The above described pessimistic picture can be drastically changed if our initial objective is revised into

> *Develop a collaborative framework that is based on the cooperation of different air pollution models and/or different research teams in a natural, effective and timely fashion on heterogeneous networked computer platforms*

to take proper advantages of the following observations [11]:

- Individual group of physical or chemical processes in LRTAP are fully understood.

- Groups of processes can be easily uncoupled over certain areas.

- Not-required processes can be easily identified over particular geographic regions.

- Legacy codes exist or can be easily build for any simple LRTAP model.

- Very detailed, specialized local models are widely available.

- The air region under investigation can be partitioned into subregions in an easy and/or natural way.

- TeraFLOP computing is available through networks of computers.

In the rest of this node we briefly describe our efforts -still in their early stages– to build an air pollution simulation engine out of existing legacy models and software for effective and accurate simulations over an extensive network of heterogeneous computer machines.

3. Collaborative air pollution agents

Currently, the simulation of LRTAP entails development of task-specific, monolithic software systems that are expensive to build, cumbersome to maintain, and difficult (or even impossible) to map onto distributed environments. The latter difficulty, coupled with the inherent ability of interacting systems to cooperatively reach common ground at their boundaries, suggests that complex models may be represented in a natural way: a mathematical network of interacting nodes, (see Fig. 3(b)) where each node contains a relatively simple model of the physics and chemistry of the particular component it represents. The idea is to first partition a physical problem domain into disjoint subdomains and impose appropriate boundary conditions on interfaces between subdomains. Then, given an initial guess, simulation proceeds, imitating the physics of the system by iterating the the following steps until convergence.

Step 1 Solve the sub–model in each subdomain and obtain a local solution.

Step 2 Use the solution values to evaluate how well the interface conditions are satisfied. Use an *interface relaxation* formula to compute new values of the boundary conditions.

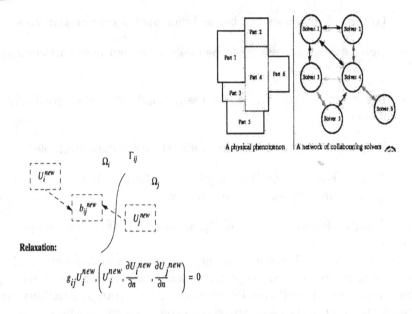

Figure 1. (a) Interface Relaxation Mechanism (b) A Network of Cooperating Solvers

Note that iteration occurs at the continues level and there is no numerical approximation at this level.

Interface relaxation –a step beyond domain decomposition– is one of the key ideas of our proposed framework [7]. There are several known methods for interface relaxation [3, 8], and the general mechanism is illustrated in Fig. 3(a). Here, the generic relaxation formula $g_{i,j}$ (based on the current solutions U_i^{new} and U_j^{new} of the two neighboring subdomains Ω_i and Ω_j) yields successive approximations $b_{i,j}^{new}$ to solution values along their common interface $\Gamma_{i,j}$. The project builds on our expertise obtained in building two prototypes [1, 13] of SCIAGENTS , an agent based system for differential equation problems. They accurately and effectively couple ordinary and partial differential equation models that have two-way interaction. These SCIAGENTS systems and related literature can be found in the web pages http://www.cecs.missouri.edu/~joshi/sciag/ and http://www.cs.purdue.edu/homes/mav/projects/dom_dec.html.

Before considering the implementation of our simulation engine we state and briefly comment on the following interesting practical questions and mathematical problems:

How we partition the physical problem domain? There are several natural ways to split the 3-dimensional LRTAP domain. Vertical 3-

dimensional partitions can be implicitly defined by associated partitions of the terrain based on various political, geographical,ecological, atmospheric and meteorological characteristics like country/authority borders, coast lines, urban area limits, dominant wind directions and strong variations in land morphology and land use. Vertical (advection dominated) partitions can be coupled with 3-dimensional horizontal (diffusion dominated) partitions that slit the the atmosphere in several layers according to their air pollution mechanisms.

What type of interface relaxation mechanisms should be used? The relaxers might range from just interchanging values among subproblems to simple averaging schemes [8] that impose the continuity of the concentration and its flux of each spices under consideration to sophisticated relaxers like the ones considered for fluid flow problem [6].It is our believe that the above will be sufficient for fast convergence. It remains to be seen if the convergence of the interface relaxation iterations can be further accelerated by developing new relaxation schemes unique to air pollution simulations like applying "naive" Lagrangian models on the interfaces [11].

How the relaxers can be tuned up? Most of the known relaxation methods involve parameters [8, 10]. Roughly speaking they play the role of preconditioning or domain overlapping. Careful selection of good values for these parameters can significantly speed up the convergence. These optimum values depend on virtually all the important characteristics of the physical problem and its partition. and therefore their theoretical determination is a hard problem even for simple model cases [10]. Nevertheless, one can couple the limited theoretical results available with practical adaptive algorithms using automatic differentiation with great success [9].

How we match unstructured grids on the interfaces? Since one can apply different space discretization methods on each subproblems there should be powerful interpolation schemes that match values on interfaces. There is a significant amount of research done in this direction in both general and interface relaxation frameworks [7] which can be readily used.

4. The Agent and run–time support systems

The above core methodology will be insufficient unless coupled with powerful collaboration, data communication, load balancing, and other runtime support systems. Borrowing from our experience with our two prototypes of SCIAGENTS [7, 1, 13] we envision the proposed simulation engine as a network of Agents of various types.

An **initiator Agent** will be responsible, besides initiating the network of agents, on global actions like dynamically create or delete other agents, determining the global convergence of the simulation and visualizing the global result. Several other **control Agents**, after linking themselves together, will supervise the execution of the *contract* by the rest of the agents. The *contract* is generated during the definition of the physical problem and its partition usualy via a GUI. It should contain information about computing resources, connectivity, control tolerances ....

A **scheduling Agent** contacts the hosts on which the processes will be executed and starts up the corresponding **execution Agents**. These execution agents are of various types. For each sub-problem we might have a **solver Agent** that corresponds to the local LRTAP solver which actually solves the mathematical model, a **data-mining Agent** responsible for accessing the various local geographical, meteorological and satellite data and a **visualization Agent** for viewing local results, and monitoring the global execution. **Mediator Agents** will also be execution agents and they are realizations of the interface relaxers.

Almost all execution agents will be build by properly wrapping existing legacy codes. These wrappers act as an interface layer generating agents out of monolithic codes. They are relatively inexpensive to build. Mediator agents need to be created from scratch but their complexity is minimal comparing to the overall problem. They are usually build on top of existing interpolation libraries. It is worth to mention here that the whole SCIAGENTS system which involves more than 1.5 million lines of mainly C and Fortran code contains less than 2 thousand lines of "wrapping" and "mediator" code. We plan to use the BOND agent system[1] to implement our simulation machine. It provides automation on building the various agents and effective agent support for message passing, threads and data migration, object replication, monitoring and directory services that hide the low level details of the distributed system.

Heterogeneous systems are usually poor performers. We therefore need powerful runtime support to achieve realistic simulation times.

Deterministic performance modeling and bit–reproducible results may need to be sacrificed to performance and optimal arithmetic complexity may need to be sacrificed to optimal overall (wall-clock) solution time. Realize that a solver agent might very well have the capability of "quiting" itself in the case he observes (based on given tolerances contained in the *contract*) local convergence or it might decide to proceeding past an outstanding communication (containing interface data)

---

[1] see http://bond.cs.purdue.edu

from a mediator agent. Both situations have been already observed in our prototype implementations and both might require check-pointing and restoration mechanisms transparent to the user.

Parallel libraries now need multi-threaded extensions. Consider, as an example, a solver agent. It can be implemented as three fat threads: (1) an "input" thread that a user can interact with for dynamic parameter modification, (2) an "output" thread which offers dynamic display of simulation results, and (3) a "solver" thread that iterates over its subdomain, given boundary values. Depending on the legacy code, more application-level threads can be defined but even with as few as three threads important separate actions (execution, output visualization and runtime modification of parameters) can be performed concurrently.

Success in promoting locality on the global memory system, which in our case is distributed over a large area network with a high degree of hierarchy, is also essential. A new notion of data and temporal locality is need and process/thread migration is required. Our runtime support is based on the PaCS system[2] for multi-threaded, multi-way protocols, speculative, migrant-threads based distributed simulation.

Existing dynamic load balancing schemes [2] will serve as a basis and will be coupled with interdomain information that will increase their effectiveness. For example cloud formation or strong winds in selective subregions can be very easily determined by consulting the monitoring agent early enough to allow local adaption and process reallocation. Note that besides the parallelism due to the different agents there is fortunately a lot of inherent parallelism inside agents. For example such parallelism has been exploited for solver agents [12, 5, 2]. Each such agent can be mapped onto a parallel machine too. The load balancing problem is expected to be easier due to the this multi level parallelism.

5. Conclusion

We propose a methodology for building a powerful LRTAP simulation engine by properly combining existing models and software components. Our approach enjoys several very desirable properties:

Problem simplification. It dramatically simplifies the complexity of the physical problem by (1) considering subproblems that involve simpler local physical rules acting on simpler geometries, and (2) providing a convenient abstraction of the modeling and solution processes while simultaneously providing a modeling practice that yields a closer representation of the physical world.

---

[2] see http://www.cs.purdue.edu/research/PaCS/PaCS.html

Reduction in software development time. It drastically reduces the time to develop a simulation engine by permitting the heavy reuse of legacy scientific software. Note that the basic (software) building blocks of a system are typically already existing problem solving environments.

Parallelism and scalability. It allows heterogeneous distributed resources to be harnessed by using a naturally parallel and highly scalable approach. The network of collaborative solvers is easily mapped onto a wide variety of distributed high performance computing architectures. Challenging parallelization issues like data partitioning, assignment and load balancing are easier handled on the level of physics rather than on the level of computational abstractions.

Cooperability and adaptivity. Due to natural problem partitioning, collaboration of groups of modelers is naturally promoted and also simplifiable at each stage of the modeling process. Each modeler can select the particular component of the physical artifact that fits his expertise, build his own local model and adapt any of the local parameters involved, in a dynamic manner, at any stage of the computation. The only information he needs to give his neighbors, for collaboration, is the information on relaxation at the interfaces.

Numerical efficiency. It increases the efficiency of the overall numerical scheme by allowing one to use the most appropriate and most efficient numerical method for each particular subproblem.

We should mention that the abstraction inherent to our agent approach allows one to further expand our proposed methodology by using collaborative techniques to split the differential equations involved in the chemical sub–model and/or build hybrid models that combine both Lagrangian and Eulerian features [11].

It is important to make clear that the proposed simulation engine is not to replace existing models and/or software but to provide the general framework and the practical tools to build on top of them by properly integrating them into a common distributed system and to effectively orchestrate them in a simple and effective way.

Acknowledgments. This work was supported in part by the National Science Foundation through grants CCR–9202536 and CDA–9123502 and in part by the Secretariat of Research and Development Greece though PENED grants 95–107 and 95–602. Author's permanent address: Institute of Applied and Computational Mathematics, FORTH, 711 10 Heraklion, GREECE and University of Crete, Mathematics Department, 714 09 Heraklion, GREECE.

References

1. Drashansky, T. T. (1986), *An Agent-based approach to building multidisciplinary problem solving environments*, PhD thesis, Computer Science Department, Purdue University, 1996.
2. Elbern, H. (1997), Load balancing of a comprehensive air quality model, in G. Hoffmann and N. Kreitz (eds.), *Making its mark – The use of parallel processors in meteorology*. World Scientific.
3. Mu, M. and Rice, J. R. (1995), Modeling with collaborating PDE solvers - theory and practice. *Computing Systems in Engineering*, **6**, pp. 87–95.
4. Peters, L. K., Berkowitz, C. M., Carmichael G. R., Easter, R. C., Fairweather, G., Ghan, S. J., Hales, J. M., Leung, L. R., Pennell, W. R., Potra, F. A., Saylor, R. D. and Tsang, T.T. (1995), The current state and future direction of Eulerian models in simulating the tropospheric chemistry and transport of trace species: A review, *Atmos. Environ.*, **29**, pp. 189–222.
5. Pai, P. and Tsang, T. H. (1993), On parallelization of time–dependent three-dimensional transport equations in air pollution modeling. *Atmos. Environ.*, **27A**, pp. 2009–2015.
6. Quarteroni, A., (1999), *Domain decomposition methods for flow problems*, Oxford Univeristy Press, Oxford, UK, in print.
7. Rice, J. R. (1998), An Agent–based architecture for solving partial differential equations, *SIAM News*, **31**, 1998.
8. Rice, J. R., Tsompanopoulou, P. and Vavalis, E. (1997), Review and performance interface relaxation methods for elliptic differential equations, Technical Report CSD-TR-97-004, Purdue University, West Lafayette, IN, USA.
9. Rice, J. R., Tsompanopoulou, P. and Vavalis, E. (1998), Automated estimation of relaxation parameters for interface relaxation, Technical Report CSD-TR-98-018, Purdue University, West Lafayette, IN, USA.
10. Rice, J. R., Tsompanopoulou, P. and Vavalis, E. (1998), Fine tuning interface relaxation methods for elliptic differential equations, Technical Report CSD-TR-98-017, Purdue University, West Lafayette, IN, USA.
11. Rice, J. R. and Vavalis, E. (1998), Collaborative agents for modelling air pollution, *Systems Analysis Modelling Simulation*, **32**, pp. 93–101.
12. Shin, W. and Carmichael, G. R. (1992), Comprehensive air pollution modeling on a multiprocessor system computer, *Comp. Chem. Engng.*, **16**, pp. 805–815.
13. Tsompanopoulou, P. (1999), *Collaborative PDEs: Theory and Practice*, PhD thesis, Math. Dept., Univ. Crete, Greece, in preparation..
14. Zlatev, Z. (1995), *Computer treatment of large air pollution models*, Kluwer Academic Publishers, Dordrecht.

INCLUDING OF SURFACE SOURCE IN A SURFACE LAYER PARAMETERIZATION

D. YORDANOV[1] AND D. SYRAKOV[2]
[1]Geophysical Institute, Acad.G.Bonchev Street, Bl.24, Sofia 1113, Bulgaria
[2]National Institute of Meteorology and Hydrology

Abstract

A PC-oriented Eulerian multi-layer model EMAP [5,6] was developed. The vertical diffusion block of the model uses a 2nd order implicit scheme realised on a non-homogeneous staggered grid which includes the dry deposition flux as a bottom boundary condition. Experiments with EMAP show that, if the concentration at the first computational level is used for calculation of the dry deposition flux, the deposited quantity changes when the height of the level is changed. Roughness level is necessary to calculate properly the dry deposition. It is impossible to have such a model level, since roughness changes from one grid point to another. On the other hand, because of the steep gradients in the surface layer (SL), many levels must be introduced near the ground for adequate description of pollution profiles. This would enormously increase memory and time requirements without any practical need. The first computational level is usually placed high enough above the surface and parameterization of all transport processes in the layer bellow is applied. The problem becomes much more complex if sources at the surface have to be treated. Such are the processes of evaporation and re-emission of the tracer under consideration. An effective parameterization of SL diffusion processes, based on similarity theory, was developed and tested in [8,9]. It allows to have the first computational level at the top of this layer. Here, an upgrade of this parameterization is presented, taking into account the presence of a continuous surface source. This parameterization is built in in EMAP and applied to different pollution problems. In addition, a simple re-emission scheme is built in to make use of this SL parameterization.

Z. Zlatev et al. (eds.), Large-Scale Computations in Air Pollution Modelling, 369–380.
© 1999 Kluwer Academic Publishers. Printed in the Netherlands.

1. Introduction

The dry deposition flux of gases and particles from the atmosphere to a receptor surface depends on their concentration in the air, on the turbulent transport processes in SL, on the physical and chemical nature of depositing species and, finally, on the absorbing and capturing abilities of the surface. Presently, the resistance approach for describing these processes is largely used in air pollution numerical modelling. A complete scheme can be found in [3]. This scheme is realised in the EMEP/MSC-W models. Following this approach the surface layer flux F can be presented as

$$F = V_d(z)C(z) \tag{1}$$

where $C(z)$ is the air concentration at height z, $V_d(z)$ - the dry deposition velocity and z - some reference height. Usually, z is placed at the top of SL and coincides with the lowest computational level of the dispersion model in use. Eq.(1) is used both as boundary condition to the vertical diffusion equation and dry deposited quantity estimator.

In the resistance model only the most important pathways through which compounds are transported and deposited are taken into account. The parameter $V_d(z)$ is presented as the inverse of the sum of three resistances:

$$V_d(z) = [\rho_a(z) + \rho_b + \rho_s]^{-1}, \tag{2}$$

where ρ_a is the SL aerodynamic resistance (it represents the resistance against the turbulent transport in the surface boundary layer), ρ_b is the viscous resistance (it accounts for the resistance against the transport in the viscous laminar sub-layer adjacent to the surface where molecular diffusion dominates, i.e. the roughness layer r_0) and ρ_s is the surface resistance accounting for the uptake or destruction of the pollutant by the surface. It is obvious that $V_d(z)$ in Eq.(2) depends on z only through ρ_a.

The aim of this study is to include the presence of continuous source acting at the ground surface in the SL parameterization. For the purpose, z in Eq.(2) is put to tend to r_0. As a result, the roughness level dry deposition velocity can be determined as

$$V_d(r_0) = [\rho_b + \rho_s]^{-1} \tag{3}$$

and can be calculated following the procedures described in [3]. Further, $V_d(r_0)$ will be denoted simply as V_d.

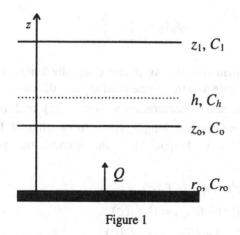

Figure 1

2. Parameterization of Surface Layer Diffusion Processes with Surface Source

Let us have a vertical grid, hanged over the ground surface with roughness r_0, the boundary level (zero level) being at height z_0 and the first inner level - at height z_1, as shown in Fig. 1. Here, the roughness height is denoted as r_0 instead of the common z_0, because z_0 will be used to denote the bottom boundary level of the vertical grid. C_{r_0}, C_0 and C_1 denote the concentrations at heights r_0, z_0 and z_1, respectively. For numerical solving of the diffusion equation, some bottom boundary conditions have to be defined, reflecting the relation between C_0 and C_1 after discretization. The aim of the parameterization is to find this relation, together with a method for determining of C_{r_0}, having in mind that the dry deposition flux (at $z = r_0$) is proportional to C_{r_0}.

The interaction of air masses with ground surface can be assessed using a SL model and supposing that the concentration distribution is similar to the temperature one. For the fluxes in SL, the following can be written [1]

$$\kappa z \frac{d|u|}{dz} = u_* \varphi_u(\varsigma), \tag{4a}$$

$$z \frac{d\theta}{dz} = \theta_* \varphi_\theta(\varsigma), \tag{4b}$$

$$z\frac{dC}{dz} = C_*\varphi_\theta(\varsigma), \quad \varsigma = \frac{z}{L}, \tag{4c}$$

where $\kappa=0.4$ is Karman constant; $\mathbf{u}$, θ and C are the horizontal wind velocity, the potential temperature and the concentration; u_*, θ_* and C_* are the friction velocity, temperature and concentration scales, $\varphi_u(\varsigma)$ and $\varphi_\theta(\varsigma)$ - universal functions, L - Monin-Obukhov length. After integration of Eqs.(4) from the roughness surface r_0 to height z, one obtains the profiles of these characteristics

$$\kappa|\mathbf{u}(z)| = u_* f_u(\varsigma), \tag{5a}$$

$$\theta(z) - \theta(r_0) = \theta_* f_\theta(\varsigma), \tag{5b}$$

$$C(z) - C(r_0) = C_* f_\theta(\varsigma), \tag{5c}$$

where the dependence on r_0/L is not expressed explicitly. The vertical exchange coefficients can be written as

$$k_u(z) = u_*\kappa z / \varphi_u(\varsigma), \tag{6a}$$

$$k_\theta(z) = u_*\kappa z / \varphi_\theta(\varsigma), \tag{6b}$$

$$k_c(z) = u_*\kappa z / \varphi_\theta(\varsigma) = k_\theta(z). \tag{6c}$$

The mass flux at some height h between z_0 and z_1 is defined as bottom boundary condition to the diffusion equation, realised on this grid. To find it, Eq.(5c) is written for height h and the scale C_* is determined as

$$C_* = (C_h - C_{ro}) / f_\theta(\varsigma_h), \quad \varsigma_h = h/L. \tag{7}$$

This quantity is replaced in Eq.(4c), written for height h

$$z\frac{dC}{dz}\bigg|_{z=h} = \frac{\varphi_\theta(\varsigma_h)}{f_\theta(\varsigma_h)}(C_h - C_{ro}). \tag{8}$$

The quantity $\varphi_\theta(\varsigma_h)$ can be obtained from Eq.(6c), written for height h. Its replacement in Eq.(8) leads to

$$k_c\frac{dC}{dz}\bigg|_{z=h} = \alpha(C_h - C_{ro}), \quad \alpha = \kappa u_* / f_\theta(\varsigma_h). \tag{9}$$

According to the main SL characteristics - the constancy of flux F from Eq.(1) with height - Eq.(9) can be made equal to the roughness level dry deposition flux $V_d C_{ro}$. When a surface source with strength Q is available, the equation is

$$k_c \frac{dC}{dz}\bigg|_{z=r_0} = V_d C_{ro} - Q = \alpha(C_h - C_{ro}) = k_c \frac{dC}{dz}\bigg|_{z=h}. \tag{10}$$

Eq.(10) expresses the relation between the roughness level concentration and the concentration at height h

$$C_{ro} = \gamma C_h + Q / (\alpha + V_d), \qquad \gamma = \alpha / (\alpha + V_d). \tag{11}$$

Replacing Eq.(11) in the second half of Eq.(10), the mass flux at height h can be found:

$$k_c \frac{dC}{dz}\bigg|_{z=h} = \gamma V_d C_h - \gamma Q. \tag{12}$$

The discretization of the left side of this boundary condition leads to

$$K(C_1 - C_o) = \gamma V_d C_h - \gamma Q, \qquad K = k_c(h) / (z_1 - z_o). \tag{13}$$

To make Eq.(13) suitable for a boundary condition, C_h must be expressed as a function of C_o and C_1. This can be done, assuming that level z_1 also belongs to the surface layer. Having Eq.(5c) written for levels z_0 and z_1 (ς_0 and ς_1), C_{ro} and C_* can be determined as functions of C_o and C_1. Replacing these quantities in Eq.(5c), at $z=h$,

$$C_h = aC_o + bC_1, \quad a = \frac{f_\theta(\varsigma_1) - f_\theta(\varsigma_h)}{f_\theta(\varsigma_1) - f_\theta(\varsigma_0)}, \quad b = \frac{f_\theta(\varsigma_h) - f_\theta(\varsigma_0)}{f_\theta(\varsigma_1) - f_\theta(\varsigma_0)} \tag{14}$$

The replacement of Eq.(14) in Eq.(13) provides the bottom boundary condition

$$C_o = B_1 C_1 + B_0 Q, \quad \begin{cases} B_1 = (K - \gamma V_d b) / c \\ B_0 = \gamma / c \end{cases}, \quad c = K + \gamma V_d a. \tag{15}$$

At each moment, the roughness level concentration can be calculated as

$$C_{ro} = G_1 C_1 + G_0 Q, \quad \begin{cases} G_1 = \gamma(aB_1 + b) \\ G_0 = (\alpha a B_0 + 1) / (\alpha + V_d) \end{cases}, \tag{16}$$

obtained by replacing Eq.(15) in Eq.(11). At $Q=0$, the coefficients B_1 and G_1 from Eqs.(15) and (16) coincide with the coefficients G_{bot} and G_{grn} from [8,9], so the presented parameterization of SL diffusion processes appears to be a generalisation of the parameterization, presented there.

3. One-Dimensional Numerical Experiments

Eqs.(15) and (16) are the base of the proposed SL parameterization. In order to prove this parameterization, a proper shell is constructed. It consists of:

- vertical diffusion block DIFFZ;
- meteorological block MET;
- source block SOU and
- input-output main routine.

DIFFZ subroutine (the vertical diffusion block of EMAP) performs one time of the vertical diffusion equation, discretized over a non-homogeneous grid. The simplest implicit scheme is realised and treated by means of Thomas algorithm. The dry deposition flux is set as a bottom boundary condition.

The proposed parameterization includes the friction velocity u_* and the universal temperature profile f_θ as surface layer characteristics. In order to supply these parameters, a proper model, producing consistent u_* and f_θ, has to be conjugated to the diffusion block. Here we use the MET-block of EMAP. A simple, but very PBL model [10] was realised in it. The model, based on the similarity theory, is one-dimensional, steady-state and barotropic. The U- and V-components of the geostrophic wind (the 850 hPa wind) and the potential temperature difference between 850 hPa level and the ground level are inputs to the model as well as the roughness height r_0 and Coriolis parameter f. The vertical profiles of the U-, V- and W-components and the vertical profile of the diffusion coefficient k_c are the outputs. In addition, the internal PBL parameters - the drag coefficient $c_g = u_* / |u_g|$, the friction deviation α of the surface wind from the geostrophic one and Kazanski-Monin stability parameter $\mu = Lf / \kappa u_*$ - are determined. The surface layer height and temperature profile are also calculated. Calculation of the coefficients B_0, B_1, G_0 and G_1 is added to this block. The universal SL temperature profile is given by

$$f_\theta(Z, R_0, \mu) = \left\{ \begin{array}{ll} \ln(Z / R_0) + \beta_1 \mu(Z - R_0), & \mu > 0 \\ \ln(Z / R_0), & R_0 < Z < H_t \\ \ln(H_t / R_0) + 3(1 - H_t / Z)^{1/3}, & H_t < Z \end{array} \right\} \mu \le 0 ,$$

$$(17)$$

where $Z = z/H$ and $R_0 = r_0/H$ are non-dimensional heights; $H = \kappa u_*/f$ is a height scale; $H_t = \beta_2/\mu$ is the dynamic sub-layer height; $\beta_1 = 10$, $\beta_2 = -0.07$ are universal constants.

In the source block different types of sources in the air are modelled. The source profile can be chosen interactively by the user and its parameters must be specified. The main routine links these blocks and provides input and output procedures. Special conservativity check is built in it. The structure of the vertical grid is also determined in this module. A log-linear vertical grid is constructed, providing for insertion of new levels close to the ground.

The proposed parameterization in the case of $Q=0$ is tested in [8,9]. An eight level grid is used. The level h, where the flux is calculated, is set at roughness level, so the boundary condition flux is equal to the dry deposition flux. The solutions, obtained on this grid are considered to be "exact" ones. A process of removing the lowest levels one by one is started. Each time the solutions are compared to the "exact" ones. Different source types, stability and top boundary conditions are combined. All the other parameters (geostrophic wind, dry deposition velocity etc.) are fixed. Two kinds of initial source profiles are tested - a point one and a Gaussian one, both with maximal concentration at height 200 m and 1000 units of mass released. Instantaneous and continuous versions of these sources are considered. Open and closed top boundary conditions are used.

All the experiments show similar behaviour. Concentration profiles and dry deposed mass are described adequately by a grid, which practically has no levels in the surface layer (the 4-level grid starting from 50 m). The parameterization is effective for a great variety of situations. The only exception is in the beginning of integration, but soon the errors in concentration profiles as well as in deposited quantities decrease to negligible values.

In Fig.2, examples of the calculated profiles evolution are presented, demonstrating the ability of the proposed parameterization to account for the action of surface source. A 4-level case (computational level heights 50,200,650,1450m, boundary levels at 20 and 2500 m) is shown, the lowest inner level placed at the top of SL. Eq.(15) is used as a bottom boundary condition. The surface concentrations (roughness height of 0.1 m assumed) are calculated after Eq.(16). The continuous source strength is 1 unit/s in the case when single source acts (cases **a.**: high source; case **b.**: surface source) and 0.5 unit/s in the combined case of both sources acting together (case **c.**). The high source is places at 200 m, the time step is set to 0.5 h. The roughness level dry deposition velocity is set to 0.01 m/s. The presented profiles are rather realistic but for the coarse vertical resolution. As it can be expected, the slope of the profile curve in SL is positive when only high source is available (case **a.**), and negative in the opposite case (case **b.**). In the combined case this slope depends on the ratio between the fluxes created by the sources. It can be noticed, that after 300 time steps, stationarity is reached in all cases.

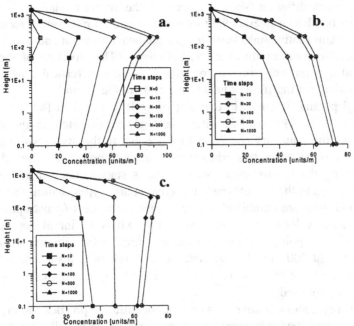

Figure 2. Concentration profiles for the cases of high source (a), surface source(b) and high+surface sources (c).

4. Description of a Re-Emission Scheme

Recent results, and first of all those in [12] and [2] indicate the key role of exchange processes between natural compounds and in particular the role of re-emission in airborne transport. The sequence "emission - airborne transport - re-emission - airborne transport etc." can be repeatedly realised, causing a dramatic change in pollution pattern in air, sea and land. The sensitivity study in [4] shows that even slight re-emission from land causes severe redistribution of the pollutant.

The main aim of this study is to check how well the surface layer parameterization scheme can handle a potential re-emission. The complex processes and fluxes in the boundaries "soil-atmosphere" or "sea-atmosphere" determining the re-emission rate are not reflected here at all. In this sense, this study have to be considered as a kind of sensitivity analysis.

The scheme from [4] for description of re-emission and degradation is adopted. It can be presented by the following system of equations:

$$\partial M_{av} / \partial t = D + W - \alpha_{reemis} M_{av} - \alpha_{degrad} M_{av},$$
$$\partial M_{fx} / \partial t = \alpha_{degrad} M_{av}, \tag{18}$$
$$Q = \alpha_{reemis} M_{av},$$

where $M_{av}(t)$ is that part of the deposited substance that is capable of re-emitting; $D(t)$ is the dry deposition flux; $W(t)$ - the wet deposition flux; α_{reemis} and α_{degrad} - re-emission and degradation factors, correspondingly, depending on temperature T and on the underlying surface type and state R_S (sea, soil, vegetation, glacier etc.). $M_{fx}(t)$ is the substance quantity, degraded and fixed in the soil (note, that the introduced M_{fx} characterises not only the degraded part of the substance but also its accumulation in biota); $Q(t)$ - the re-emission flux.

5. Influence of Re-Emission on Pollution Distribution - One-Dimensional Test

The shell, described previously, is used for testing of the scheme. The re-emission block, containing the discretized Eqs.(18), is built in the main routine, so as to be run just after the diffusion block at each time step, and the calculated re-emission flux is input in the MET-subroutine in the next time step as a surface source.

The vertical grid has 6 levels placed at 2, 10, 50, 200, 640 and 1430 m; the roughness length is set to 0.1 m, the dry deposition velocity - to 0.01 m/s. The meteorological conditions are: neutral stratification and geostrophic wind of 10 m/s. The time step is 1800 s (0.5 hour). The top boundary is open. The source model is a high point source with strength of 0.55555 units/s. This value ensures a release of 1000 mass units per one time step. The height of the source is set to 200 m. Instantaneous and continuous action of the source is modelled. The re-emission and degradation factors are assigned values from 10^{-4} to 10^{-3} h^{-1} for : α_{reemis} and values of 0 and 10^{-3} h^{-1} for α_{degrad}.

The case of instantaneous point source and with no degradation is investigated first.

In Fig.3, the concentration profiles created by an instantaneous point source for some periods of time at various values of α_{reemis} and $\alpha_{degrad}=0$ are selected and presented together. In case of no re-emission, the concentrations in air drop rapidly to zero. The re-emission plays the role of a not very intensive continuous source with changeable intensity. Slowly, all the mass deposited on the ground is emitted back to the atmosphere. At small time intervals, the profiles do not differ considerably with α_{reemis}, because the flux from the high source predominates. When the process goes on, the intensity of re-emission, reflected in the value of α_{reemis}, affects concentration profiles in two directions: first, the mass in the air is much more than in the case of no re-emission, the concentrations differing in orders of magnitude; secondly, the slope of the profiles in SL, which is positive in the beginning because of the high source domination, become greater and greater tending to negative values, which is typical for the case of a surface source (see, for example, Fig.2).

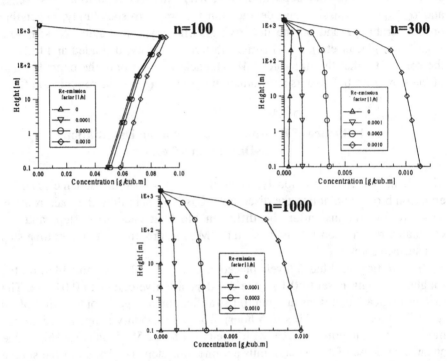

Figure 3. Concentration profiles at different α_{reemis} after 100, 300 and 1000 time steps. Instantaneous point source, $\alpha_{degrad}=0$.

In Fig.4, the influence of the magnitude of the re-emission factor on some integral characteristics of the process of "diffusion - deposition - re-emission" is demonstrated. These characteristics are: integral mass in the grid domain, the quantity of mass going out through the top boundary of the domain and accumulating in the free atmosphere (up-going mass), and the deposited mass. Further on, the values of these characteristics, normalised by the emitted mass, are called **totals**. The sum of totals has to be equal to 1 at each moment so as to have concervativity ensured. It can be noticed from the first graph of Fig.4 that the process is practically finished after 300 time steps when there is no re-emission. The mass injected in the initial moment is already partly deposited, partly gone out of the grid. There is no tracer in the air and the two other characteristics remain unchanged after that moment. Re-emission breaks down this stationarity. It injects continuously small amounts of mass in the air. Diffusion transfers this mass up through the whole boundary layer and delivers it to the free atmosphere. This leads to decrease of the deposited mass and increase of the up-going mass; the bigger the re-emission factor, the better the demonstrated effect. The behaviour of the system in cases with continuous source and when degradation is included are similar to the just commented one. More details can be found in [6]

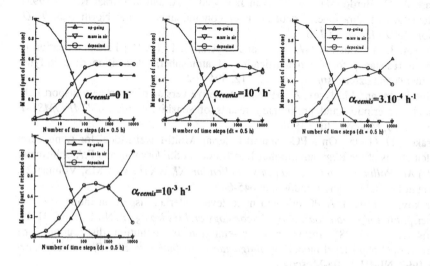

Figure 4. Totals' evolution at different α_{reemis}. Instantaneous source, $\alpha_{degrad}=0$.

6. Conclusion

The tests of the proposed SL parameterization verify its good quality. Concentration profiles and dry deposited mass are described adequately by a grid, which practically has no levels in the surface layer. The parameterization is effective for a great variety of situations, for airborne, surface and combined sources. The parameterization is built in the 3D Eulerian dispersion model EMAP for estimating regional loads of harmful pollutants.

The simple re-emission and degradation scheme Eq.(18) can be combined easily with the SL parameterization. The sensitivity tests performed here show a behaviour of the examined characteristics which is very reasonable from the physical point of view. At the same time they point out that even slight re-emission can cause severe redistribution of the pollutant. The scheme can be used for long term integration of persistent organic pollutants as well as some heavy metals on regional and continental scale.

References

1. Aloyan, A. E., Yordanov, D. L. and Penenko, V. V. (1981), A numerical model of pollutant transport in the atmospheric boundary layer, *Meteorology and Hydrology*, No 8, Gidrometeoizdat, Moscow.

2. Baart, A. C., Berdowski J. J. M., van Jaarsveld J. A. and Wulffraat K. J., (1995), Calculation of atmospheric deposition of contaminants on the North Sea, *TNO-report TNO-MEP-R 95/138*.

3. Erisman, J. W., van Pul, W. A. J., and Wyers, G. P. (1994), Parameterization of surface resistance for the quantification of atmospheric deposition of acidifying pollutants and ozone, *Atmos. Environ.*, **12**, 394-362.

4. Galperin, M. and Maslyaev, A. (1996), A tentative version of airborne transport POP model; Description and preliminary results of sensitivity studies, *EMEP/MSC-E report 3/96*, Moscow, June 1996.

5. Syrakov, D. (1995), On a PC-oriented Eulerian Multi-Level Model for Long-Term Calculations of the Regional Sulphur Deposition, in S.E.Gryning and F.A.Schirmeier (eds) *Air Pollution Modelling and its Application XI*, NATO • CCMS, Volume 21, Plenum Press, N.Y. and London, pp. 645-646.

6. Syrakov, D. (1997), A PC-oriented multi-level Eulerian dispersion model - model description, *Bulgarian Journal of Meteorology and Hydrology* **8**, No 1-2, pp. 41-49.

7. Syrakov, D. (1998), Influence of re-emission on pollution distribution: one dimensional multi-level model, in: *Bulgarian contribution to EMEP, Annual report for 1997*, NIMH, Sofia-Moscow.

8. Syrakov, D. and Yordanov, D. (1996), On the surface layer parameterization in an Eulerian multi-level dispersion model, in: *Bulgarian contribution to EMEP, Annual report for 1995*, NIMH, Sofia-Moscow.

9. Syrakov, D. and Yordanov, D. (1996), On the surface layer parameterization in an Eulerian multi-level model, *Proceedings of the 4th Workshop on Harmonisation within Atmospheric Dispersion Modelling for Regulatory Purposes*, v.1, 6-9 May 1996, Oostende, Belgium.

10. Yordanov, D., Syrakov, D. and Djolov, D. (1983), A barotropic planetary boundary layer, *Boundary Layer Meteorology* **25**, 363-373.

11. Wesely, M. L. (1989), Parameterization of surface resistances to gaseous dry deposition in regional-scale numerical models, *Atmos. Environ.*, **23**, 1293-1304.

12. Wania, F. and Mackay, D. (1995), A global distribution model for persistent organic chemical, *Sci. Total Environ.* **160/161**, 211-232.

CONCLUSIONS OF THE NATO ARW ON LARGE SCALE COMPUTATIONS IN AIR POLLUTION MODELLING

Z. ZLATEV[1], J. BRANDT[1], P. J. H. BUILTJES[2], G. CARMICHAEL[3],
I. DIMOV[4], J. DONGARRA[5], H. VAN DOP[6], K. GEORGIEV[4],
H. HASS[7] AND R. SAN JOSE[8]

[1] National Environmental Research Institute, Roskilde, Denmark
[2] TNO Institute for Environmental Sciences, Apeldoorn, The Netherlands
[3] University of Iowa, Iowa City, Iowa, USA
[4] Bulgarian Academy of Sciences, Sofia, Bulgaria
[5] University of Tennessee, Knoxville, Tennessee, USA
[6] Utrecht University, Utrecht, The Netherlands
[7] Ford Forschungzentrum, Aachen, Germany
[8] Technical University of Madrid, Madrid, Spain

Every day was finished by a common formal discussion. The major problems discussed are listed below.

1. Discussion on Monday, July 6. The main topic of the discussion in the first day was the possibility for running comprehensive environmental models on long time-periods. An attempt to answer the following questions was made.

A. Shall we try (a) to improve the performance of the code on the available computer, (b) to select faster numerical algorithms, (c) to simplify the chemical and/or physical processes or (d) to achieve improvements in more than one of these directions?

B. If we choose some of these directions, (a), (b), (c) and/or (d), is it clear how the improvements can be achieved?

C. The perfect situation is to describe in a sufficiently accurate way all involved physical and chemical processes. Can such a model be run over a long time period?

D. Is the use of general solvers (general-purpose algorithms and software), possibly with a few modifications, a good strategy to develop application codes (high-level software) for Air Quality simulations?

Z. Zlatev et al. (eds.), Large-Scale Computations in Air Pollution Modelling, 381–384.
© 1999 *Kluwer Academic Publishers. Printed in the Netherlands.*

2. Discussion on Tuesday, July 7. The topics of organizing parallel and vector computations and numerical methods for solving big systems of partial differential equations, which can be used in air pollution modelling were discussed at the end of the second day. The following major questions which were asked and discussed.

A. The new supercomputers are rather complicated architectures. How can a large air pollution code be designed for a given architecture: (a) start from a scratch, (b) try to identify the most time consuming parts and optimize only them, (c) try to replace some standard operations with subroutines from well-known libraries or (d) there is no sense to do anything (when you are ready the architecture will be changed)?

B. Where to put the major efforts: (a) vector processors, (b) parallel machines with shared memory, (c) parallel machines with distributed memory, (d) heterogeneous computers, (e) something else?

C. New approaches, such as Problem Solving Environments, based on low- and medium-level software for parallelism, solution of numerical kernels, visualization, I/O, etc., seem to offer the   possibility of building, running and analyzing applications on a variety of parallel and sequential computers, without dealing with details related to numerical analysis and computer science. Can this approach be a "solution" for Air Quality applications?

3. Discussion on Wednesday, July 8. The major topic of the discussion during the third day was: "Solving big optimization problems in connection with data assimilation and/or the distribution of new economical objects (especially for economical programs in developing countries). Coupling existing comprehensive models with economical models". In particular, the following questions were asked and discussed.

A. Is it possible to combine a large scale air pollution code with some economical models? How? When will this become possible?

B. Is it necessary to introduce more optimization methods in the area of air pollution modelling? What kind of answers can we achieved by using such methods?

4. Discussion on Thursday, July 9. The discussion in the end of the fourth day was about the possibilities for running big air pollution models in real time (important for preparing predictions in case of huge environmental catastrophes, like the accident in Chernobyl). Mainly two important questions were discussed.

A. How complex should a model that runs in real time be?

B. When will realistic air pollution forecasts be available? Are such forecasts needed and why?

5. Discussion on Friday, July 10. This was the last day of the workshop. However, most of the participants were present at the discussion, where the topic was: *"Sensitivity analysis of large air pollution models"*. An attempt to answer the following questions was made.

A. What do we want from a sensitivity analysis procedure?

B. Is it necessary (or at least advisable) to try to develop a standard sensitivity analysis procedure for checking the response of the models to variations of different variables?

C. How many runs have to be performed with a perfect sensitivity procedure?

It is impossible to present details about the discussions of the five working days. The duration of every discussion was about one hour (the duration of some of the discussions was even longer). The most important conclusions are summarized below.

Conclusion 1. If all important physical and chemical processes are adequately described in the models, then huge computational tasks are to be solved. If we do not want to simplify the problems by inserting non-physical assumptions only in order to make it possible to handle the models on the computer available, then we have to use parallel and/or vector computers.

Conclusion 2. The combination of new modern computers and fast numerical methods do allow us to solve more problems and bigger problems. Thus, long simulations, which could not be carried out a few years ago, can be performed now. However, it is normally very difficult to adjust the existing models to the new complicated computer architectures and/or to insert in them new faster numerical algorithms. On the other hand, the performance can be improved very considerably if this is properly done. In fact, some kind of long series of runs with different scenarios can only be carried out by using efficiently the great potential power of the new computers and fast numerical algorithms.

Conclusion 3. Optimal solutions are needed; in some cases such solutions are urgently needed. It is not enough to know that certain critical levels are exceeded and where they are exceeded. It is necessary to find out how to reduce in an optimal way the pollution to acceptable levels and to keep them there. Optimization methods will give an answer to such questions. However, it is very difficult to incorporate optimization techniques directly in large air pollution models. Much more efforts are needed in this direction.

Conclusion 4. Running models in real time can be of essential and even live saving importance in the case of accidental, hazardous releases (as, for example, the accident in Chernobyl in 1986). Sometimes air pollution forecasts are needed (in a typical case, in order to predict high, and harmful, ozone concentrations). In both cases it is desirable to carry out the computations with sufficiently large models in which the underlying physical and chemical processes are well described. This leads to very big and very challenging computational tasks. Additionally, some networking seems to be necessary in order to generate input data for the air pollution models; both meteorological data from weather forecasts and measurements of some air pollutants which are needed to get, by using some data assimilation techniques, reliable initial values of the most important concentration fields.

Conclusion 5. Sensitivity analysis is a powerful means for checking the response of the model to variation of different important parameters. A carefully performed sensitivity analysis may give valuable information about the reliability of the model results (as well as for the importance of some physical and chemical mechanisms). The standard sensitivity analysis, which is based on the application of different ideas from perturbation theory and on some hypotheses about the distribution of the studied quantities, is not always sufficient when air pollution problems are handled. If this is the case, then more precise, and more time consuming, techniques for performing sensitivity analysis have to be applied. This is why the sensitivity analysis is often a very demanding computational procedure (when properly designed), which requires both big and fast computers and, of course, codes which run efficiently on the computers used.

Conclusion 6. The tasks which have to be solved (many of them urgently solved) need co-operative efforts of specialists from several scientific fields: air pollution modellers, meteorologists, chemists, computer scientists, numerical analysts, statisticians and economists. The participants in the workshop agreed to continue the efforts towards the establishment of such a co-operation. They shall exchange results in the future. An attempt to organize a second workshop on *"Large Scale Computations in Air Pollution Modelling"* in year 2000 will be made. If such a workshop takes place, then reporting the results achieved in the efforts to solve the problems stated in the conclusions of the first workshop will have highest priority there.

LIST OF THE PARTICIPANTS

Jose Libano Alonso
IASC (Inst. Adv. Sci. Comput.), Univ. Liverpool,
Liverpool, **United Kingdom**
E-mail jlibano@liv.ac.uk

Artash E. Aloyan
Inst. Numer. Math.,
Russian Academy of Sciences,
Gubkin str., 8, Moscow, 117333, **Russia**
Phone: +7 095 938 1823,
Fax:: +7 095 938 1821,
E-mail: aloyan@inm.ras.ru

Ekaterina Batchvarova
National Institute
of Meteorology and Hydrology,
66 Tzarigradsko chaussee, Sofia, **Bulgaria**
Phone: +359 2 975 3986,
Fax: +359 2 880 380,
E-mail: Ekaterina.Batchvarova@meteo.bg
Web-site: http://www.meteo.bg

Claus Bendtsen
Danish Comput. Centre for Res. and Education,
DTU, Buildg. 304, DK-2800 Lungby, **Denmark**
Phone: +45 3587 8969, Fax: +45 3587 8990,
E-mail: Claus.Bendtsen@uni-c.dk
Web-site: HTTP://unidhp.uni-c.dk/~nicbe

Mariana Bistran
Artificial Intelligence Applications Laboratory,
IPA SA Institute, Bucharest, **Romania**
Phone: +401 230 05 91, Fax: +401 230 78 70,
E-mail: sand@automation.ipa.ro
Web-site: http: //www.ipa.ro

Ramaz D . Bochorishvili
Vekua Inst. Appl. Math., Tbilisi State University,
University Str. 2, Tbilissi, **Georgia**
Phone: +99 532 304 081,
E-mail: rdboch@viam.hepi.edu.ge

Lambros Boukas
Dept.Informatics, Univ. Athens,
Panepistimioplis, 157 10 Athens, **Greece**,
Phone: +(301) 7275 100-2,
Fax: +(301) 7231 1569,
E-mail: lbouk@di.uoa.gr

Jørgen Brandt
National Environmental Research Institute
Frederiksborgvej 399, P. O. Box 358,
DK-4000 Roskilde, **Denmark**
Phone: +45 4630 1157, Fax: +45 4630 1214,
E-mail: lujbr@sun4.dmu.dk

Peter Builtjes
TNO-MEP, Dept. Environmental Quality,
P.O.Box 342, Apeldoorn, **The Netherlands**
Pnone: +31 555493038, Fax: +31 555493282,
E-mail: builtjes@mep.tno.nl

Gregory R. Carmichael
Center for Global and Regional Environmental
Research, Iowa University,
Iowa City, Iowa 52242-1000, **USA**
Phone: 319/335-1399, Fax: 319/335-1415,
E-mail: gcarmich@icaen.uiowa.edu

Pasqua D'Ambra
CPS (Center for Research on Parallel Computing
and Supercomputers) - CNR
Via Cintia, Monte S.Angelo, 80126 Naples, **Italy**
Phone: +39 81 675 624, Fax: +39 81 766 2106,
E-mail: pasqua@matna2.dma.unina.it
Web-site: http://pixel.dma.unina.it

Vladimir F. Demyanov
Appl. Math. Dept., St.Petersburg State Univ.,
198904 Staryi Peterhof, St.Petersburg, **Russia**
Phone: +7 812 3442186, Fax:: +7 812 4287179,
E-mail: demyanov@dem.apmath.spb.su

Ivan Dimov
Central Laboratory for Parallel Processing,
Bulgarian Academy of Sciences
Acad. G. Bonchev Str. 25 A, Sofia, **Bulgaria**
Phone: +359 2 70 84 94, Fax: +359 2 70 72 73,
E-mail: dimov@amigo.acad.bg, Web-site:
http://www.acad.bg/BulRTD/math/dimov2.html

Jack Dongarra
Univ. Tennessee and ORNL,
104 Ayres Hall, Knoxville TN, 37996, **USA**
Phone: 423-974-8295, Fax: 423-974-8296,
E-mail: dongarra@cs.utk.edu, Web-site:
http://www.netlib.org/utk/people/JackDongarra/

Han van Dop
Institute for Marine and Atmospheric Research
Utrecht, P.O. Box 80.005 3508 TA Utrecht
Buijs Ballot Laboratorium, Princetonplein 5,
Utrecht, De Uithof, **The Netherlands**
Phone: +31 30 2533154, Fax: +31 30 2543163,
E-mail: dop@fys.ruu.nl

Hendrik Elbern
Inst. Geoph. Meteor., EURAD, Univ. Cologne,
Albertus Magnus Plat, 50923 Koeln, **Germany**
Phone: +49 221 400 2220, Fax: +49 221 400
2320, E-mail: he@eurad.uni-koeln.de

Michael Galperin
Dept. Biophys., Radiation Phys. and Ecology,
Moscow Physical - Engineering Institute,
Studencheskaya street, 31-9,
Moscow, 121165, **Russia**
Fax: +7 095 251 64 38,
E-mail: kastrel@glas.apc.org

Maidarjav Ganzorig
Informatics and Remote Sensing Institute,
Mongolian Academy of Sciences
Erkhtaivan Avenue 54B, Ulaanbaatar-51,
210351 **Mongolia**
Phone: 976 1 453660, Fax: 976 1 458090,
E-mail: ulz@csg.mn

Tevfik Gemci
Department of Environmental Engineering,
Engineering Faculty, Sakarya University
Esentepe Campus, TR-54040 Adapazari, **Turkey**
Phone: +90 (264) 343 16 02 Ext. 291,
Fax: +90 (264) 343 14 50

Krassimir Georgiev
Central Laboratory for Parallel Processing,
Bulgarian Academy of Sciences
Acad. G. Bonchev Str. 25 A, Sofia, **Bulgaria**
Phone: +359 2 713 6612, Fax: +359 2 707 273,
E-mail: georgiev@parallel.acad.bg

Sven Hammarling
NAG Ltd., Wilkinson House, Jordan Hill Road,
Oxford, OX2 8DR, **United Kingdom**
Phone: +44 1865 511 245,
Fax: +44 1865 311 205,
E-mail: sven@nag.co.uk

Heinz Hass
Ford Forschungszentrum Aachen,
Dennewartstr. 25, 52068 Aachen, **Germany**
Phone: +49 241 9421 203,
Fax: +49 241 9421 301,
E-mail: hhass@ford.com

Jan Eiof Jonson
The Norwegian Meteorological Institute, P.O.
Box 43 Blindern, N-0313 Oslo, **Norway**
Phone: +47 22 96 33 24,
Fax: +47 22 69 63 55,
E-mail: jan.eiof.jonson@dnmi.no

George Kallos
Dept. Appl. Phys., Meteor. Lab.,
Univer. Athens,
Panepistimioupolis, Athens 15784, **Greece**
Phone: +30 1 728 4923, Fax: +30 1 963 1967,
E-mail: kallos@etesian.meteolab.ariadne-t.gr,
Web-site: http://forecast.uoa.gr

Kazimir A. Karimov
Inst. Phys., National Academy of Sciences,
Chui Prosp. 265-A, Bishkek, **Kyrgyz Republic**
Phone: (996-3312) - 243661,
Fax: (996-3312) - 243607,
E-mail: eco@kyrnet.kg

Petros Katsafados
Dept . Appl. Phys., Meteor. Lab.,
Univer. Athens,
Panepistimioupolis, Athens 15784, **Greece**
Phone: +30 1 728 4832, Fax: +30 1 729 5281,
E-mail: pkatsaf@etesian.meteolab.ariadne-t.gr
Web-site: http://forecast.uoa.gr

Sergey Konkov
Ministry of Health,
Miasnikova Str. 39, Minsk, **Belarus**
Phone: +375 172 22 26547,
Fax: +375 172 22 62 97

Vladimir K. Kouznetsov
Institute of Atmosphere Protection,
7, Karbisheva str., St Petersburg, **Russia**
E-mail: sriatm@main.mgo.rssi.ru

Ayse Gulcin Kucukkaya
Dept. Restoration, Faculty of Engineering and
Architecture, Trakya Univ., Edirne, **Turkey**
hone: 90 284 213 8513, Fax: 90 284 212 6067,
E.mail: tummf@superonline.com

Thelma Mavridou
Technical University of Crete,
Heracleon, **Greece**

Andrzej Mazur
Meteor. Centre, Inst. of Meteorology and Water
Management, Podlesna 61, Warszawa, **Poland**
Phone: 022 34 1651 ext. 205, Fax: 022 35 2813,
E-mail: Andrzej_Mazur@imgw.pl, Web-site:
http://www.enviroware.com/enviweb/source/
people.htm

Niki Mimikou
Department of Informatics, University of Athens,
Panepistimioplis, 157 10 Athens, **Greece**,
Phone: +301 7275 100-9, Fax: +301 7231 1569,
E-mail: nmis@di.uoa.gr

Nikolaos Missirlis
Department of Informatics, University of Athens,
Panepistimioplis, 157 10 Athens, **Greece**,
Phone: +301 7275 100-3, Fax: +301 7231 1569,
E-mail: nmis@di.uoa.gr

Jonas Mockus
Dept. Optimization, Inst. Math. and Informatics,
Akademijos 4, Vilnius 2600, **Lithuania**
Phone: 3702 729616,
Fax: 3702 729209,
E-mail: jonas@optimum.mii.lt

Stanislav K. Myshkov
Appl. Math. Dept., St.Petersburg State Univer.,
198904 Staryi Peterhof, St. Petersburg, **Russia**
Phone: +7 812 428 5042,
Fax: +7 812 428 7179,
E-mail: demyanov@dem.apmath.spb.su

Francisco Jose de Magalhaes Neves
Departemento de Matematica,
Universade de Evora,
Largo dos Colegias 2, 700 Evora, **Portugal**
Fax: +351 066 744 546,
E-mail: francisco_neves@hotmail.com

Magdalena Palacios
Environ. Inst., CIEMAT (Centre for Research on
Energy, Environment and Technology)
Avda. Complutense, 22, 28040 Madrid, **Spain**
Phone: +34 1 346 61 74, Fax: +34 1 346 61 21,
E-mail: magda ciemat.es

Maria Prodanova
National Institute
of Meteorology and Hydrology,
66 Tzarigradsko chaussee, Sofia , **Bulgaria**
Phone: +359 2 975 3433, Fax: +359 2 880 380,
E-mail: Maria.Prodanova@meteo.bg
Web-site: http://www.meteo.bg

Olga Puentedura
INTA, 28850 Madrid, **Spain**
Phone: +34 1 520 2009, Fax: +34 1 520 1317,
E-mail: puentero@inta.es

Tamas Rapcsak
Laboratory of Operations Research and Decision
Systems, Hungarian Academy of Sciences
H-1518 Budapest, P.O. Box 63, **Hungary**
Phone: 361 209 5266, Fax: 361 209 5267,
E-mail: RAPCSAK@oplab.sztaki.hu

Angelo Riccio
Department of Chemistry, University of Napoli,
Via Mazzocannone 4, 80134 Naples, **Italy**
Phone: +39 81 547 650, Fax: +39 81 552 7771,
E-mail: riccio@risc55.dichi.unina.it

Gheorghe M. Sandulescu
Artificial Intelligence Applications Laboratory ,
IPA SA Institute, Bucharest, **Romania**
Phone: +401 230 05 91, Fax: +401 230 78 70,
E-mail: sand@automation.ipa.ro
Web-site: http: //www.ipa.ro

Roberto San Jose
Environmental Software and Modelling Group,
Computer Science School, Techn. Univ. Madrid
Campus de Montegancedo
Boadilla del Monte-28660, Madrid, **Spain**
Phone: +34 91 336 7465, Fax: +34 91 336 7412, E-mail: roberto@fi.upm.es
Web-site: http://artico.lma.fi.upm.es

Arslan Saral
Yidiz Teknik Universitesi, Insaat Fak.,
Cevre Muh. Bol., Yildiz
80750 Besiktas Istanbul, **Turkey**
Phone: 90 212 259 7070, Ext. 2839,
Fax: 90 212 259 6762,
E.mail: saral@yildiz.edu.tr

Blagovest Sendov
Central Laboratory for Parallel Processing,
Bulgarian Academy of Sciences
Acad. G. Bonchev Str. 25 A, Sofia, **Bulgaria**
Phone: +359 2 70 84 94, Fax: +359 2 70 72 73,
E-mail: sendov@amigo.acad.bg
Web-site:
http://www.acad.bg/BulRTD/math/sendov2.html

Irena Stoycheva
Institute of Economics, Bulgarian Academy of
Sciences, Aksakov 3, 1040 Sofia, **Bulgaria**
Phone: +359 2 84 121 831, Fax: +359 2 882
108, E-mail: ineco@iki.bas.bg

Dimiter Syrakov
National Institute
of Meteorology and Hydrology,
66 Tzarigradsko chaussee, Sofia , **Bulgaria**
Phone: +359 2 975 3986, Fax:: +359 2 88 0380,
E-mail: Dimiter.Syrakov@meteo.bg

Mete Tayanc
Dept. Envir. Engng., Marmara Univer.,
Kuyubasi, 81040, Kadikoy, Istanbul, **Turkey**
Phone: 90 216 349 2711,
Fax: 90 216 348 0293,
E-mail: mtayanc@marun.edu.tr

Per Grove Thomsen
Inst. Math. Modelling, Techn. Univer. Denmark,
Build. 305, DK-2800 Lyngby, **Denmark**
Phone: +45 4525 3073, Fax: +45 4593 2373,
E-mail: pgt@imm.dtu.dk

Alison Tomlin
Dept of Fuel and Energy, University of Leeds,
Leeds LS2 9JT, **United Kingdom**
Phone: 01132 332500, Fax: 01132 440572,
E-mail: alisont@chemistry.leeds.ac.uk

Ilia Tzvetanov
Institute of Economics, Bulgarian Academy of
Sciences, Aksakov 3, 1040 Sofia, **Bulgaria**
Phone: +359 2 84 121831, Fax: +359 2 882108,
E-mail: ineco@iki.bas.bg

Eloy A. Unzalu
Environmental Laboratory, LABEIN,
Technological Research Centre, Bilbao, **Spain**
Phone: +34 4 489 24 26 or +34 4 489 24 00,
Fax: +34 4 489 24 60, E-mail: eloy@labein.es

Nedialko Valkov
National Institute
of Meteorology and Hydrology,
66 Tzarigradsko chaussee, Sofia, **Bulgaria**
Phone: +359 2 975 3427, Fax:: +359 2 880 380,
E-mail: Nedialko.Valkov@meteo.bg
Web-site: http://www.meteo.bg

Emanuel Vavalis
Purdue University, Computer Science
Department, West Lafayette, IN 47 907 USA
Phone: 1-765-494-0739, Fax: 1-765-494-0739,
E-mail: mav@cs.purdue.edu
Web-site: http://www.cs.purdue.edu/homes/mav/

Lex Wolters
Computer and Software Systems Division, Dept.
of Computer Science, Leiden University
P.O. Box 9512, Leiden, **The Netherlands**
Phone: +31 71 527 7054,
Fax:: +31 71 527 6985, E-mail:
llexx@cs.leidenuniv.nl
Web-site: http://www.wi.leidenuniv.nl/~llexx

Dimiter Yordanov
Geophysical Institute
Bulagarian Academy of Sciences,
Acad. G.Bonchev str., bl. 3, Sofia, **Bulgaria**
Phone: +359 2 971 3023, Fax: +359 2 700 226,
E-mail: emilia@geophys.acad.bg (za Yordanov)

Zahari Zlatev
National Environmental Research Institute
Frederiksborgvej 399, P. O. Box 358,
DK-4000 Roskilde, **Denmark**
Phone: +45 4630 1149, Fax: +45 4630 1214,
E-mail: luzz@sun2.dmu.dk
Web-site: http://www.dmu.dk/
atmosphericenvironment/staff/zlatev.htm

SUBJECT INDEX

AUTHOR INDEX